Xylanase from Microorganisms

Abdul Sattar Jatoi
Editor

Xylanase from Microorganisms

Trends and Future

Springer

Editor
Abdul Sattar Jatoi
Department of Chemical Engineering
Dawood University of Engineering
and Technology
Karachi, Pakistan

ISBN 978-981-95-0823-5 ISBN 978-981-95-0824-2 (eBook)
https://doi.org/10.1007/978-981-95-0824-2

This Springer imprint is published by the registered company Springer Nature Singapore Pte Ltd.
The registered company address is: 152 Beach Road, #21-01/04 Gateway East, Singapore 189721, Singapore

In the name of Allah, the Most Gracious and the Most Merciful. First of all, I would like to raise my heartfelt gratitude and appreciation to Allah SWT for the permission, guidance, wisdom and blessing for all these years till now, when I have reached this important destination of my journey in life to accomplish my goal. Finally, I would likc to present my most heartfelt and warmest appreciation to the great parents and parents-in-law (may ALLAH SWT bless and reward them).

Dr. Abdul Sattar Jatoi

Preface

Xylanases, enzymes that catalyze the hydrolysis of xylan into xylose, have emerged as indispensable tools in diverse industrial sectors including paper and pulp, textiles, animal feed, biofuel production, and food processing. The demand for eco-friendly and sustainable alternatives to harsh chemical treatments has driven significant interest in microbial xylanases due to their specificity, efficiency, and adaptability under various operational conditions.

This book, *Xylanase from Microorganisms*, is a comprehensive effort to consolidate the current knowledge and emerging advancements in the microbial production and application of xylanases. Microorganisms—ranging from bacteria and fungi to genetically modified strains—serve as prolific producers of xylanase, offering high yield, stability, and activity under industrial settings. Understanding the genetics, physiology, and biochemistry of these microbes is crucial for optimizing enzyme production and tailoring it for specific applications.

The chapters in this volume explore fundamental aspects of microbial xylanase biology, including gene expression, enzyme structure, substrate specificity, and catalytic mechanisms. Advanced bioprocessing strategies, fermentation optimization, and the role of bioinformatics and protein engineering in enhancing xylanase functionality are also discussed. Furthermore, the book delves into real-world industrial applications, highlighting both the achievements and challenges faced in commercial deployment.

This work is intended for researchers, students, and professionals in microbiology, biotechnology, enzymology, and industrial processing. It aims to bridge the gap between laboratory-scale research and large-scale applications, providing a valuable reference for academic and industrial stakeholders seeking to harness the potential of microbial xylanases.

We extend our gratitude to the contributors whose expertise and dedication have made this book possible. It is our hope that this compilation will inspire further innovation and collaborative efforts in the field of microbial enzyme technology.

Karachi, Pakistan Dr. Abdul Sattar Jatoi

Acknowledgments I thank Vice chancellor and higher management of Dawood University of engineering and technology and colleagues in the department of chemical engineering for the support and encouragement.

Contents

About the Editor

Dr. Abdul Sattar Jatoi is an assistant professor in the chemical engineering department at Dawood University of Engineering and Technology, Karachi, Pakistan. He has over 13 years of experience in Academics and Research. Dr. Jatoi has expertise in food bioprocess and their application in various food industries. Microbial electrochemical systems for environmental and energy applications. He has published over 140 papers, including 30 book chapters and four books. He is a peer-reviewed member of more than 50 scientific journals. Seventy works were published in various international conference proceedings.

His research interests include food and bioprocessing industries, microbial electrochemical system, emerging pollutant treatment via different approaches, sustainable nanomaterials development, wastewater treatment, green-water treatment technology, separation and purification techniques, enviro-analytical techniques.

email: abdul.sattar@duet.edu.pk

Chapter 1
Xylanase from Microorganisms: Trends and Future

Tahseen Fatima Miano

Abstract An essential enzyme in the breakdown of xylan, a crucial hemicellulosic component of plant walls, xylanase has become a mainstay in a number of commercial applications, such as the pulp and paper, food processing, and biofuels sectors. Numerous xylanase-producing species with distinct enzymatic characteristics can be found in microbial sources, including bacteria, fungus, and actinomycetes. The biochemical traits of microbial xylanases that make them suitable for a variety of industrial processes are examined in this chapter, including their activity profiles, substrate selectivity, and ideal working conditions. The stability, efficiency, and specificity of these enzymes have been further enhanced by advancements in protein and genome engineering, expanding the range of uses for them. New methods like metagenomics and omics-driven approaches have made it easier to find new xylanases with better characteristics, but immobilization immobilization strategies have enhanced enzyme reusability and cost-efficiency. The integration of microbial xylanases into biorefinery systems underscores their role in promoting sustainable industrial practices and a circular bioeconomy. This chapter highlights current trends and provides a roadmap for future innovations in microbial xylanase research and applications, emphasizing their potential to transform industrial processes and address global sustainability challenges.

Keywords Xylanase · Microbes · Production · Industrial processing · Enzymes

1 Introduction

Xylan found in algae and plant cell wall it is made up of long chain polysaccharide molecule a compact structure. Xylan hydrolyzed to form xylanase enzyme that used in metabolic reactions taken place in plant cell, further the xylan molecules

T. F. Miano (✉)
Institute of Food Sciences and Technology, Sindh Agriculture University, Tandojam, Sindh, Pakistan
e-mail: tahseenfm@yahoo.com

A. Sattar Jatoi (ed.), *Xylanase from Microorganisms*,
https://doi.org/10.1007/978-981-95-0824-2_1

Xylanases, also known as endo-β-1,4-xylanases, enzymes that catalyze the hydrolysis of xylan, a major hemicellulosic component of plant cell walls [1]. These enzymes have garnered significant interest due to their diverse industrial applications, including biofuel manufacture, paper bleaching, textile processing, and food and feed industries. The bacterial use for fermentation and hydrolysis use bacteria, fungi, and actinomycetes are the primary sources of xylanase production [2]. This chapter delves into the trends in xylanase research, explores the molecular and biotechnological advancements, and discusses future perspectives for this enzyme [3]. The depolymerization of xylan into its constituent components facilitates the production of various bioproducts [4]. However, the heterogeneous and structurally complex nature of xylan poses significant challenges for its complete hydrolysis, necessitating the involvement of a diverse array of enzymes [5]. Given its intricate chemical composition, the efficient degradation of xylan involves the concerted act multiple conversion as hydrolytic enzymes [5] the xylases composition is given in Fig. 1.

Among these, endoxylanases also β-xylosidases are collectively known for example xylanases as primary xylanolytic reacting enzymes responsible for xylan hydrolysis [7]. Endoxylanases cleave in addition to homopolymeric compound present to create backbone of 1,4-linked β-D-xylopyranose, generating xylo-oligomers, while β-xylosidases further hydrolyze these oligomers to release xylose [8]. Notably, endoxylanases play a crucial role in breaking glycosidic bonds, leading to the liberation of short-chain xylo-oligosaccharides [9].

Biocatalysts, particularly xylanases, have garnered increasing attention over chemical reagents consists of high specificity, operational efficiency under mild pH and temperature conditions, reduced energy requirements, and environmentally friendly nature [10].

The depolymerization of xylan into its constituent components facilitates the production of various bioproducts [11]. However, the heterogeneous and structurally

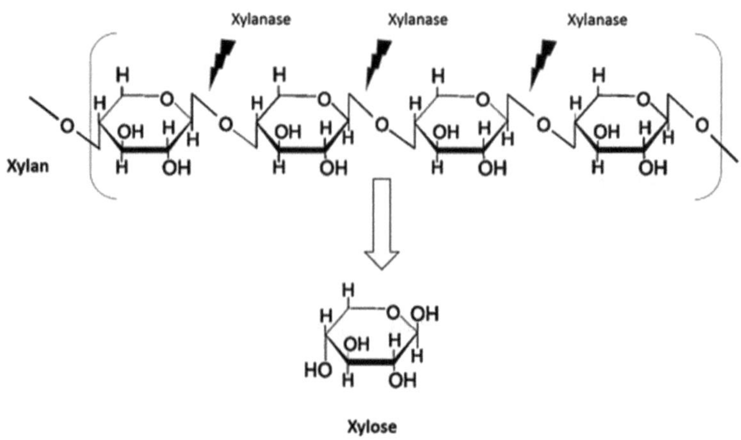

Fig. 1 Chemical structure of xylanase [6]

complex nature of xylan poses significant challenges [8] for its complete hydrolysis, necessitating the involvement of a diverse array of enzymes. Given its intricate chemical composition, the efficient degradation of xylan involves the concerted degradation of multiple hydrolytic enzymes [12].

Further these, endoxylanases substances and β-xylosidases compounds collectively known as xylanases are the primary xylanolytic substances starts reaction for xylan hydrolysis [13]. Endoxylanases cleave the homopolymeric backbone of 1,4-linked β-D-xylopyranose, generating xylo-oligomers, while β-xylosidases further hydrolyze such oligomers to release xylose [14]. Notably, endoxylanases play a crucial role in breaking glycosidic bonds, leading to the liberation of short-chain xylo-oligosaccharides [15]. Biocatalysts, particularly xylanases, have garnered increasing attention over fast reactions using catalysts of high specificity, operational efficiency following moderate pH and temperature conditions, reduced energy requirements, considered environmentally friendly nature [7].

Lignocellulosic from plant cell contributes as biomass widely acknowledged renewable resource for the sustainable [13, 16] production of bioenergy and value added fortified products, including animal fresh feed, single plant cell protein, fermented ethanol, isolated xylitol, organic acids, methane, and monosaccharides [17]. However, the efficient conversion of lignocellulosic waste into these valuable compounds presents significant challenges, such as low selectivity and high processing costs [18, 19].

Hemicelluloses are a diverse group of polysaccharides located between cellulose fibrils and lignin in the plant cell wall [11, 17]. They comprise various monosaccharides, including pentoses units of simple carbohydrates (e.g., xylose, rhamnose, and arabinose), hexoses of six unit simple sugar (e.g., glucose, mannose, and galactose), and uronic acids (e.g., 4-O-methyl glucuronic acid, D-glucuronic acid, and D-galacturonic acid) [19]. The hemicellulose backbone, which can be either homopolymeric or heteropolymeric, is primarily interlinked with β-1,4-glycosidic double bonds, with occasional β-1,3-glycosidic linkages [20, 21].

Hydrolysis of hemicelluloses into their constituent sugars achieved by following chemical and enzymatic methods [4, 21]. However, enzymatic hydrolysis is preferred over chemical hydrolysis due to its cost-effectiveness, environmental sustainability, and higher selectivity [1, 22]. The heterogeneous and complex structure of xylan, the major hemicellulosic component, presents a challenge for step by step degradation, necessitating reaction involvement of numerous hydrolytic enzymes Among them, endoxylanases and β-xylosidases collectively referred to as xylanases isolated enzymes have function for xylan degradation. Endoxylanases cleave the homopolymeric backbone of β-1,4-linked D-xylopyranose, generating xylo-oligomers, whereas β-xylosidases further hydrolyzed oligomers releases xylose monomers. Endoxylanases, in particular, facilitate the breakdown of glycosidic linkages, producing short-chain xylo-oligosaccharides.

Enzymatic biocatalysis has gained considerable attention over chemical catalysis, high reaction specificity, speedy function under moderate pH and temperature controlled conditions, lower energy requirements, following minimal environmental impact. Consequently, xylanase enzymes have been extensively explored for various biotechnological applications [23].

Xylanases are enzymes of significant industrial relevance due to their extensive applications in the pharmaceutical, food, paper, and cellulose industries. Among the various xylanases, endo-xylanases [14] and β-xylosidases are the most critical, as they play a fundamental role in the breakdown of xylan compounds from hemicellulose of plant cell [16]. Xylanase activity is typically quantified using the standard 3,5-dinitrosalicylic acid (DNS) assay, which measures energy release of reducing sugars during catalytic hydrolysis of substances [16] (expressed as xylose) resulting from xylan hydrolysis. One unit (U) enzyme xylanase reaction defined as the percentage of enzyme requires to release 1 μmol of xylose substance recorded after per minute of hydrolysis following specified assay methodology [18].

Fungal and bacterial xylanases have been extensively studied and widely produced; however, actinobacteria have also emerged as potent xylanase producers [17]. Actinobacteria constitute the largest taxonomic group within the microbial community [24]. The term "Actinobacteria" or "Actinomycetes" word derived from the Greek language "aktis" reflects lightning and "mykes" resembles to fungus, that indicate ancient grouping on the basis of intermediate groups identified in fungal and bacterial species [4]. Due to their unique physiological and metabolic characteristics, actinobacteria have been extesively investigated for their potential in diverse industrial applications worldwide [25].

2 Sources of Xylanase

Xylanase is synthesized by a diverse array of living organisms, encompassing archaea, bacteria, fungi, actinomycetes, and various eukaryotic life forms, including insects, algae, plants and immature seeds, protozoa mollusks, and nematodes [26]. These xylanase-producing organisms inhabit a wide spectrum of ecological environments, including aquatic and terrestrial ecosystems enriched by dumping decaying residues of plant biomass for xylanase isolation. The xylanases are use both for thermophilic and mesophilic habitats by following the digestive tracts of ruminant animals in feed proportions [27].

3 Inversion Mechanism of Xylan Hydrolysis

The inversion mechanism follows only single direction displacement process wherein solitary carboxylate ion facilitates the acid-catalyzed departure of a functional group. Enzymes that operate via this mechanism, particularly those belonging to glycosyl hydrolase families 8 and 43, mediate the inversion of the anomeric compounds, utilizing glutamate non nutritive sweetner and aspartate as principal catalytic substance [13]. The reaction taken place forms residues that more over serve a bases to activate a the presence of nucleophilic compound observed in water molecule, that subsequently outbreaks the anomeric carbon compound, moving towards glycosidic bond cleavage and inversion of the anomeric carbon arrangement [21]. Unlike most polysaccharide-hydrolyzing enzymes such as cellulases, cellobiohydrolases, and xylanases that retain the C1 anomeric configuration upon hydrolysis, xylan hydrolysis through the inversion mechanism follows a distinct biochemical pathway [12].

Xylanases are a diverse group of enzymes predominantly found in microorganisms, including bacteria, fungi, archaea, actinomycetes, and yeast, as well as in higher organisms such as marine algae, protozoans, mollusks, crustaceans, insects, seeds, and plants [22]. The commercial significance of xylanases has grown its marketing for manufacturing of many industrial products, getting more diversion to pulp and paper bleaching process, food and beverage processing, animal nutrition, and bioethanol production [27]. Given their robust biotechnological potential, microbial xylanases are the preferred sources for large-scale industrial use.

Microbial enzymatic complexes getting a pivotal role for conversion of plant cell wall polysaccharides in to small molecules [23], such as cellulose and hemicellulose, facilitating the release of fermentable sugars for bioconversion into value-added products at a cost-effective [22] rate Among microbial xylanases, mesophilic enzymes are widely utilized in industrial applications; however, thermostable xylanases derived from thermophilic microorganisms are increasingly favored due to their stability at high temperatures, which is essential for industrial processes [17].

Extremophilic microorganisms, known for their ability to survive in harsh environmental conditions, [18] contribute significantly to enzyme thermos tolerance, making them valuable for biotechnological applications. Moreover, bacteria, fungi, and yeasts getting more concentration extensively studied as natural sources of xylanase [22] affected by different factors of pH, temperature, carbon sources and nitrogen sources.

1. **Bacterial Sources of Xylanase**

Thermophilic xylanase-producing bacteria are isolated from diverse ecological places, including hot springs, composting sites, and deep-sea hydrothermal vents. A notable example for enzyme isolation following thermophilic Bacillus strains collected Tunisian hot springs, which were identified based on phenotypic and phylogenetic analysis of 16S rDNA sequences. These Bacillus strains exhibited significant xylanase production [3].

Another thermophilic anaerobic bacterium, *Caldicoprobacter algeriensis* TH7C1(T), isolated using hydrothermal springs reported to produce new strain of extracellular thermostable xylanase (*Xyn35*). Similarly, a thermophilic Gram-positive strain, *Thermoanaerobacterium* RXL, was identified in the scorching springs of Baoshan, Yunnan Province, China, and demonstrated efficient xylan degradation utilizing xylan, starch, and various monosaccharides and polysaccharides as substrates [15].

Additionally, xylanase genes have been successfully cloned from *Planococcus*sp. SL4, an alkaliphilic bacterial strain isolated from soda lake sediments [9]. Such alkaliphilic xylanases are particularly useful in industrial processes that require enzymatic activity under extreme pH conditions [28].

2. Fungal Sources of Xylanase

Fungal xylanases, particularly those from filamentous fungi, have been extensively used in industrial applications for over five decades due to their high stability and production efficiency [16]. Fungi belonging to lignocellulolytic genera, such as *Trichoderma reesei*, *Aspergillus niger*, followed by *Myceliophthora thermophila*, possess an extensive repertoire of xylanolytic enzymes that facilitate the breakdown of complex plant cell wall components [11].

M. thermophila, a well-established cellulolytic fungus, promising component for next future generation biofuel production possible to enhance enzymatically active ability to secrete a extensive range of thermostable lignocellulolytic enzymes [10]. Its genome encodes a high number of (glucurono)arabinoxylan-degrading enzymes, positioning it as a superior source of extracellular thermophilic xylanases compared to other fungi, such as *A. niger* and *T. reesei.* Notably, *M. thermophila* harbors 9,110 genes across seven chromosomes, including approximately 250 carbohydrate-active enzyme (CAZy) [19] genes, of which 180 are potential glycoside hydrolases. Thirteen of these genes are predicted to encode xylanases, making this organism a prolific producer of industrially relevant enzymes [29] the xylases enzymes production flowchart is given below in Fig. 2.

Other thermophilic fungi, such as *Humicola insolens*, a non-pathogenic and highly stable fungal species, also produce a diverse array of hemicellulases and cellulases.

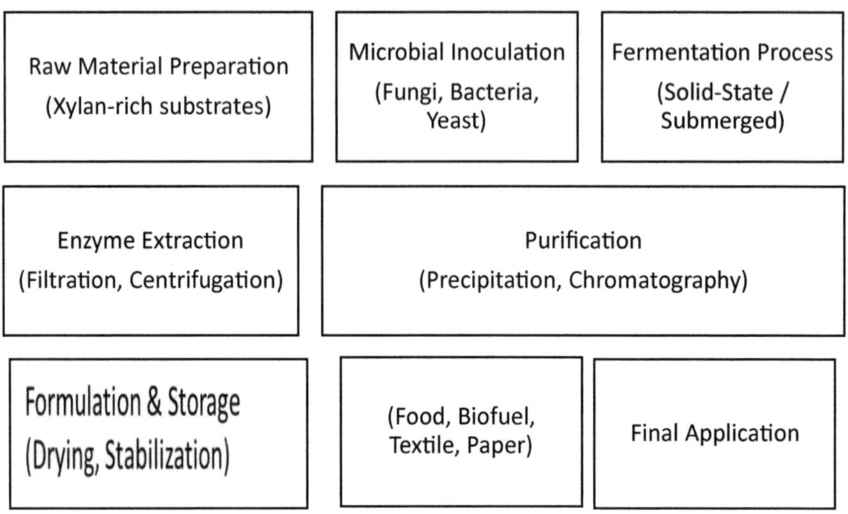

Fig. 2 Xylase enzyme production flowchart

H. insolens Y1 is a notable manufacturer of xylanolytic enzymes, including thermophilic xylanases from glycoside hydrolase families GH10 and GH11 [15]. Another promising thermophilic fungus, *Thermoascus aurantiacus*, has been recognized for its ability to synthesize thermostable cellulases and xylanases [11].

Interestingly, *T. aurantiacus* also produces antioxidant compounds as byproducts of its metabolic pathways, further enhancing its potential for the valorization of crop residues into biofuels and other sustainable bioproducts [24].

4 Industrial Applications of Xylanases

Xylanase enzymes yield gained extensive attention in recent decades due to their extensive industrial applications in various industries [8]. They play a crucial role in the pulp and paper sector, animal feed processing, beverage clarification, and biofuel production from agro-residues. Below are some of their key industrial applications [27]. Xylanases have a wide range of industrial applications due to their ability to break down xylan, a chief hemicellulosic component observed in plant cell walls [9]. These enzymes are particularly valuable in enhancing digestibility in animal feed, improving food processing, and facilitating biofuel production. Below is a more detailed discussion of their key applications [15]

1. **Pulp and Paper Industry**

The use of xylan enzyme production of high-quality white paper requires the efficient removal of lignin from pulp, a process known as bleaching. Traditionally, chemical bleaching methods utilize chlorinated agents such as chlorine dioxide (ClO_2), chlorine gas (Cl_2), and hypochlorite under high-temperature alkaline conditions (170 °C). However, these methods pose significant environmental hazards, as they generate toxic and carcinogenic byproducts [7].

In contrast, biobleaching—a sustainable alternative—utilizes microbial enzymes like xylanase to degrade hemicellulose and facilitate lignin removal. This enzymatic approach reduces the reliance on harmful chlorine-based chemicals, making the process more eco-friendly and cost-effective [7].

Xylanase treatment in the pulp and paper industry offers several advantages, including reduced chemical consumption, improved pulp brightness [8], and enhanced fiber properties. It minimizes fiber degradation, improves water retention, optimizes pulp fibrillation, enhances fiber bonding, decreases beating time, and increases fiber freeness, ultimately contributing to improved paper quality and sustainability [13].

2. **Animal Feed Processing**

Xylanase enzymes helps to break down complex plant polysaccharides, especially lignocelluloses, xylanases are essential for increasing the nutritional value and digestibility of animal feed [18]. High concentrations of lignocellulosic material, which contains vital nutrients, are found in many animal feed ingredients, such as triticale, soy-based feeds, and cereals (such as wheat, barley, and maize). In monogastric animals like pigs and poultry, the presence of non-starch polysaccharides (NSPs) such arabinoxylans makes the feed more viscous, which hinders nutrient absorption and results in ineffective digestion [18].

Xylanases aid in the release of blocked nutrients by hydrolyzing the xylan component of NSPs enzymatically, increasing the nutrients' bioavailability to animals [20]. This enzymatic pre-treatment boosts feed efficiency, improves nutrient absorption, and strengthens gut health. Furthermore, it has been demonstrated that xylanase supplementation lessens the anti-nutritional effects of arabinoxylans, improving cattle weight gain and growth performance [5].

As feed supplements, a variety of commercial microbial xylanase formulations are accessible. These come from fungal (*Aspergillus niger, Trichoderma reesei*) and bacterial (*Bacillus subtilis, Pseudomonas fluorescens*) sources, each of which is suited to a particular nutritional need [4, 29]. Because xylanase enables producers to use more economical and fiber-rich raw materials without sacrificing nutritional quality, its application in animal feed processing also lowers feed prices.

3. **Food Industry and future trends**

Xylanases are widely used in various food industries to enhance product quality and processing efficiency [7]. Their applications span across bakery products, fruit juice processing, dairy alternatives, and other food-related sectors [3].

(a) **Bakery Industry use of xylanase**

When producing bread, the enzyme xylanase is essential for enhancing the volume and texture of the dough [9]. It improves the dough's water redistribution by hydrolyzing the wheat flour's arabinoxylans and releasing soluble arabinoxylans. This leads to a softer bread crumb, better gas retention, and enhanced dough elasticity [28]. Furthermore, the flexibility of gluten networks is improved by the enzymatic breakdown of xylan, which results in improved loaf volume and overall bread structure. For the best dough conditioning, xylanases are frequently employed in conjunction with other enzymes including amylase and protease [12, 26].

(b) **Juice from fruits Explanation and application of xylanases**

To increase yield, clarity, and stability in juice processing, xylanases are commonly used in conjunction with enzymes such as cellulase and pectinase [8]. Fruit cell walls' breakdown of xylan promotes pulp liquefaction, lowers viscosity, and improves juice extraction from raw fruit material. Additionally, xylanases help to minimize sedimentation and turbidity, guaranteeing the creation of clear, superior juices. When processing apple, orange, and grape juices, where clarity is a crucial quality criterion, this enzymatic treatment is very advantageous [20].

(c) **Products Made from Dairy and Soy**

Xylanases aid in altering the fiber content and improving the solubility of plant-derived polysaccharides in the synthesis of dairy and plant-based milk [15]. Xylanase treatment during soy milk processing results in a smoother texture and better mouthfeel.

(d) **Biofuel Production**

When it comes to the enzymatic hydrolysis of lignocellulosic biomass for the production of biofuel, xylanases are crucial [21]. Cellulose, hemicellulose, lignin, and other complex polymers make up lignocellulose, a significant portion of plant biomass [18]. The production of bioethanol and other biofuels depends on the effective conversion of lignocellulose into fermentable sugars. One important stage in this process is the enzymatic degradation of xylan in hemicellulose [19]. Three primary phases are involved in the xylanase-based biofuel generation process:

(a) **Delignification and Pre-treatment**
To liberate cellulose and hemicellulose, lignin must first be broken down [9]. To increase the accessibility of polysaccharides to hydrolytic enzymes, a variety of pre-treatment techniques are employed, such as steam explosion, acid hydrolysis, and enzymatic treatment [11].
(b) **Depolymerization of Hemicellulose and Cellulose**
Hemicellulose and cellulosic components are hydrolyzed into simpler sugars, such as xylose and glucose [24], by xylanases and enzymes such xylosidase [10], ligninase, mannanase, glucanase, and glucosidase.
(c) **Fermentation and Biofuel Production**
The liberated sugars undergo microbial fermentation [7], in which they are transformed into ethanol by bacteria such *Saccharomyces cerevisiae* or *Zymomonas mobilis*. A high-value sugar alcohol that is frequently utilized in the food and pharmaceutical industries, xylitol, can also be made from sugars obtained from xylan [14]. Xylanase's use in biofuel production increases the efficiency of enzymatic hydrolysis, lowers process costs, and promotes the sustainable use of forestry and agricultural waste for the production of renewable energy [1].

5 Conclusion

Enzymes called xylanase have become important biocatalysts in a number of sectors, providing environmentally benign and sustainable substitutes for conventional chemical processes. Their importance in biotechnology and industrial applications is highlighted by their contributions to increasing the digestibility of animal feed, increasing the efficiency of food processing, and aiding the manufacture of biofuel. Innovation in sustainable bioprocessing will be fueled by the ongoing creation of microbial xylanase strains with improved substrate specificity, pH tolerance, and thermostability, which will increase their industrial value. Because of their enzymatic stability and effectiveness in harsh environments, microbial xylanases especially those originating from thermophilic bacteria and fungi—are essential in industrial applications. These enzymes' potential for application in biotechnology, biofuels, and environmentally friendly industrial processes is being increased by ongoing study into their genetic and biochemical characteristics.

References

1. Y. Bakri, M. Jawhar, M.I.E. Arabi, Improvement of xylanase production by Cochliobolus sativus in solid state fermentation. Braz. J. Microbiol. **39**(3), 602–604 (2013)
2. Bala, Singh, Sporotrichum thermophile produces biotechnologically important hydrolytic enzymes at a low cost. Bioprocess Biosyst. Eng. **39**(1), 181–191 (2016)
3. V. Juturu, J.C. Wu, Microbial xylanase: engineering, production and industrial applications. Biotechnol. Adv. **30**(6), 1219–1227 (2012)
4. B. Battan, J. Sharma, R. Kuhad, High-level xylanase production by *Bacillus pumilus* ASH under solid-state fermentation. World J. Microbiol. Biotechnol. **22**(12), 1281–1287 (2006)
5. M. Chakdar, Y. Liu, Y. Chen, Production and partial characterization of an alkaline xylanase from a novel fungus Cladosporium oxysporum. Biomed. Res. Int. **2016**, 1–7 (2016)
6. M.G. Godoy, G.M. Amorim, M.S. Barreto, D.M.G. Freire, Chapter 12—Agricultural residues as animal feed: protein enrichment and detoxification using solid-state fermentation, in *Current Developments in Biotechnology and Bioengineering*, eds. by A. Pandey, C. Larroche, and C.R. Soccol (Elsevier, 2018), pp. 235–256
7. C. Gauthier, H. Li, R. Morosol, Increase in xylanase production by *Streptomyces lividans* through simultaneous use of the sec- and tat-dependent protein export systems. Appl. Environ. Microbiol. **71**(6), 3085–3092 (2005)
8. S.D. Kshirsagar, G.D. Saratale, R.G. Saratale, S.P. Govindwar, M.K. Oh, An isolated *Amycolatopsis* sp. GDS for cellulase and xylanase production using agricultural waste biomass. J. Appl. Microbiol. **120**(1), 112–125 (2015)
9. J.L.S. Lemos, M.C.A. Fontes, N.J. Pereira, Xylanase production by *Aspergillus awamori* in solid-state fermentation and influence of different nitrogen sources. Appl. Biochem. Biotechnol. **91–93**(1–9), 681–689 (2006)
10. N. Li, P. Shi, P. Yang, Y. Wang, H. Luo, Y. Bai, Z. Zhou, B. Yao, A xylanase with high pH stability from Streptomyces sp. S27 and its carbohydrate-binding module with/without linker-region-truncated versions. Appl. Microbiol. Biotechnol. **83**, 99–107 (2009)
11. C. Li, A. Kumar, X. Luo, H. Shi, Z. Liu, G. Wu, Highly alkali-stable and cellulase-free xylanases from Fusarium sp. 21 and their application in clarification of orange juice. Int. J. Biol. Macromol. **155**, 572–580 (2020). https://doi.org/10.1016/j.ijbiomac.2020.03.249
12. E. Rosmine, N.C. Sainjan, R. Silverter, A. Alikkunju, S.A. Varghese, Statistical optimization of xylanase production by estuarine Streptomyces sp. and its application in clarification of fruit juice. J. Genetic Eng. Biotechnol. **15**, 393–401 (2017). https://doi.org/10.1016/j.jgeb.2017.06.001
13. M. Sanjivkumar, T. Silambarasan, A. Palavesam, G. Immanuel, Biosynthesis, purification and characterization of β-1, 4-xylanase from a novel mangrove associated actinobacterium Streptomyces olivaceus (MSU3) and its applications. Protein Expr. Purif. **130**, 1–12 (2017)
14. S. Parte, V. Sirisha, J. D'souza, Biotechnological applications of marine enzymes from algae, bacteria, fungi, and sponges, in *Advances in Food and Nutrition Research*, vol. 80 (Elsevier, Amsterdam, The Netherlands, 2017), pp. 75–106
15. S. Kharazmi, A. Taheri-Kafrani, A. Soozanipour, M. Nasrollahzadeh, R.S. Varma, Xylanase immobilization onto trichlorotriazine-functionalized polyethylene glycol grafted magnetic nanoparticles: a thermostable and robust nanobiocatalyst for fruit juice clarification. Int. J. Biol. Macromol. **163**, 402–413 (2020). https://doi.org/10.1016/j.ijbiomac.2020.06.273
16. Y. Xu, J. Wu, K. Zheng, D. Wu, A xylanase from Streptomyces sp. FA1: Heterologous expression, characterization, and its application in Chinese steamed bread. J. Indus. Microbiol. Biotechnol. **43**, 663–670 (2016). https://doi.org/10.1007/s10295-016-1736-8
17. J. Wu, C. Qiu, Y. Ren, R. Yan, X. Ye, G. Wang, Novel salt-tolerant xylanase from a mangrove-isolated fungus Phoma sp. MF13 and its application in Chinese steamed bread. ACS Omega **3**, 3708–3716 (2018)
18. M.D. George, Study of SmF and SSf fermentation for the production of xylanase enzyme. Jama Surg., 6–10 (2020)

19. J.P. Dos Santos, E. da Rosa Zavareze, A.R.G. Dias, N.L. Vanier, Immobilization of xylanase and xylanase–β-cyclodextrin complex in polyvinyl alcohol via electrospinning improves enzyme activity at a wide pH and temperature range. Int. J. Biol. Macromol. **118**, 1676–1684 (2018)
20. S.M. Ali, S.H. Omar, N.A. Soliman, Comanufacturing of cellulase and xylanase enzymes by way of thermophilic bacillus subtilis 276ns. Int. J. Biotechnol. Wellness. Ind. **2**(2), 65–74 (2013)
21. R. Acosta-Fernandez, T. Poerio, D. Nabarlatz, L. Giorno, R. Mazzei, Enzymatic hydrolysis of xylan from coffee parchment in membrane bioreactors. Industr. Eng. Chem. Res. **59**, 7346–7354 (2020). https://doi.org/10.1021/acs.iecr.9b06429
22. N. Bhardwaj, B. Kumar, P. Verma, A detailed overview of xylanases: an emerging biomolecule for current and future prospective. Bioresour. Bioprocess. (2019). https://doi.org/10.1186/s40643-019-0276-2
23. M.U. Maheswari, T. Chandra, Production and potential applications of a xylanase from a new strain of Streptomyces cuspidosporus. World J. Microbiol. Biotechnol. **16**, 257–263 (2000)
24. L. Zhao, Z. Yuan, N.S. Kapu, X.F. Chang, R. Beatson, H.L. Trajano, D.M. Martinez, Increasing efficiency of enzymatic hemicellulose removal from bamboo for production of high-grade dissolving pulp. Bioresour. Technol. **223**, 40–46 (2017)
25. A.D. Harris, C. Ramalingam, Xylanases and its application in food industry: a review. J. Exp. Sci. **1**(7), 1–11 (2010)
26. L. Thomas, A. Joseph, R.R. Singhania, A. Patel, A. Pandey, Industrial enzymes: xylanases, in *Current Developments in Biotechnology and Bioengineering* (Elsevier, Amsterdam, The Netherlands, 2017), pp. 127–148.
27. A. Dhillon, S. Khanna, Production of a thermostable alkalitolerant xylanase from Bacillus circulans AB16 grown on wheat straw. World J. Microb. Biotechnol. **27**(325), 327 (2000)
28. P. Parab, R. Khandeparker, U. Amberkar, V. Khodse, Enzymatic saccharification of seaweeds into fermentable sugars by xylanase from marine Bacillus sp. strain BT21. 3 Biotech **7**, 296 (2017)
29. K.-S. Hung, S.-M. Liu, W.-S. Tzou, F.-P. Lin, C.-L. Pan, T.-Y. Fang, K.-H. Sun, S.-J. Tang, Characterization of a novel GH10 thermostable, halophilic xylanase from the marine bacterium Thermoanaerobacterium saccharolyticum NTOU1. Process Biochem. **46**, 1257–1263 (2011)

19. [illegible]
20. [illegible]
21. [illegible]
22. N. Bhardwaj, B. Kumar, P. Verma, A detailed overview of xylanases: an emerging biomolecule for current and future prospective. Bioresour. Bioprocess. (2019). [illegible]
23. [illegible]
[illegible]

Chapter 2
Microorganisms Producing Xylanase

Mansoor Tariq and Imran Tarique

Abstract Xylanase, a glycoside hydrolase, is indispensable to break down xylan, a heteropolysaccharide that consists of a complex and intricate structure and constitutes the bulk of plant cell wall hemicellulose. Being the second most prolific renewable polysaccharide after cellulose, xylan consists of xylose units with attached side chains of arabinose and glucuronic acid, which leads to its heterogeneous and diverse structure. Xylanase enzymes, produced by diverse microorganisms, unlock the potential of xylan-rich biomass for industrial and environmental applications. Xylanase stands at the forefront of sustainable biotechnology, from pulp bleaching to enhancing food texture and clarity to converting agricultural waste into biofuels. This chapter gives insight into the microbial sources of xylanase, their enzymatic mechanisms, and their industrial relevance, highlighting how these natural microscopic chemists are reshaping green innovation.

Keywords Xylanase · Xylan · Microbial enzymes · Microbial sources · Biomass degradation · Biofuels production · Industrial application · Pulp and paper bio-bleaching · Food processing · Animal feed · Textile industry

1 Introduction

Xylanase, a glycoside hydrolase, is indispensable to break down xylan, a heteropolysaccharide that consists of a complex and intricate structure and constitutes the bulk of plant cell wall hemicellulose. Being the second most prolific renewable polysaccharide after cellulose, xylan consists of xylose units with attached side chains of arabinose and glucuronic acid, which leads to its heterogeneous and diverse

M. Tariq (✉)
Department of Veterinary Pathology, Faculty of Animal Husbandry and Veterinary Sciences, Sindh Agriculture University (SAU), Tando Jam, Sindh, Pakistan
e-mail: mtsamo@sau.edu.pk

I. Tarique
Healthcare Biotechnology Department, Atta-Ur-Rahman School of Applied Bio-sciences (ASAB), National University of Sciences and Technology (NUST), Islamabad, Pakistan

A. Sattar Jatoi (ed.), *Xylanase from Microorganisms*,
https://doi.org/10.1007/978-981-95-0824-2_2

structure. Through the endohydrolysis of 1,4-β-D-xylosidic bonds in xylan, xylanase propels the degradation of plant biomass. This not only aids carbon cycling within the natural environment but also provides a way to biotechnologically use lignocellulosic material for industry [27].

Industrial applicability of the enzyme spans many sectors, showing its efficiency and use in green processes. In the manufacture of pulp and paper, xylanase induces bio-bleaching, hence reducing dependence on chlorine-reagent containing chemicals, thus mitigating environmental contamination. Through hydrolysis of arabinoxylans, the quality of the product increases the product in the food industry, including mature goods and juice. A xylanase can also be used to produce biofuels, where hemicellulose is converted from lignocellulosic biomass to fermented sugar so that renewable resources can be made. Xylanase has several applications in sustainable industrial processes and performs its central role in this field [15, 25].

This chapter is given in this chapter's overview of xylanase production, action mechanisms, industrial applications, biotechnology development, adaptation strategies, obstacles, opportunities and economic and environmental impact. The role of Xylanase as a professional scientific professional and industry professional enables green biotechnology by including recent research results in informative presentations.

2 Microbial Sources of Xylanase Production

Xylanase not only helps with carbon cycling in nature, but also helps with the use of biotechnology of lignocellulosic substrates for industrial applications [28]. Xylanase for industrial applications is characterized by a wide range of microorganisms such as bacteria, fungi, yeast and seaweed. It is mainly produced by fungi and bacteria due to high yields of enzymes and desirable properties. *Aspergillus niger, Trichoderma reesei,* and *Fusarium oxysporum* are fungi that make good candidates for cellulase-free xylanases, which are crucial for applications like bio-bleaching, where cellulose conservation is crucial. Bacterial xylanases, such as those of *Bacillus subtilis, Bacillus circulans*, and *Streptomyces,* are perfect for use in extreme-temperature industries due to their high heat resistance. Instead, yeast-based xylanases, such as those of *Saccharomyces cerevisiae* and *Pichia stipitis,* are less common but still important due to their higher stability, which is useful in specific applications [12, 15].

Recent research has broadened our understanding of xylanase producers to include novel microbial resources like human intestinal flora and aquatic bacteria. For example, *Bacteroides intestinalis* secretes xylanases with different substrate specificities, and *Glaciecola mesophila* secretes a cold-active, salt-active xylanase that is perfect for low-temperature products. These results demonstrate the possibility of using unconventional resources to find xylanases with specific characteristics for industrial use [6, 12].

The microbial selection depends on some very important factors, such as enzyme yield, thermostability, ideal pH, growth ease, and genetic manipulability.

Extremophilic microorganisms, such as thermophiles and alkaliphiles, have a specific function in the production of enzymes tolerating extreme industrial environments. For the paper and pulp sectors, where high temperature is the norm, thermostable xylanases from Bacillus species are extensively applied. These factors are helpful in the choice of the most suitable sources of microbes to meet industrial needs [16].

The xylanase enzyme links through a vast range of domains, show its efficiency and adaptability in sustainable developments. Xylanase helps bio-bleaching in pulp and paper manufacture, so reducing the use of compounds including chlorine-containing agents and hence environmental pollution. By hydrolyzing arabinoxylans, xylanases improve the quality of food including baked products and fruit juices by enhancing texture and clarity. In the production of biofuels, Xylanases are also very important since they hydrolyze hemicellulose from lignocellulosic biomass into fermentable sugars, so launching the era of renewable energy sources. These several uses are evidence of the central importance of xylanase in environment friendly industrial processes [11, 26].

Up to now, major breakthroughs in xylanase research have extended its range and use. Finding new xylanases from unusual sources—such as human gut bacteria (e.g., *Bacteroides intestinalis*) and marine bacteria (e.g., *Glaciecola mesophila*)—has opened fresh avenues for enzyme manufacture to fulfil needs. Concurrent developments in enzyme engineering have given xylanases improved thermostability and substrate specificity, so better suited for industrial needs. Moreover, the importance of the enzyme in producing prebiotic xylo-oligosaccharides (XOS) has underlined considering the growing market for nutraceuticals and functional foods. Such advances show xylanase's increasing biotechnological capacity [15, 20, 27].

3 Mechanism of Xylanase Action

Xylanases are enzymes that accelerate the hydrolysis of xylan, a key hemicellulose component of plant cell walls, break the β-1,4-xylosidic bonds. This mechanism required two distinct mechanisms—retention and inversion of the anomeric configuration (based on how the enzyme is classified into specific glycoside hydrolase (GH) families). The mechanisms determine how xylanases slice the xylan backbone, their specificity, and their use in industry [1, 14, 16]. In the following sections, we explore these mechanisms briefly, including their structural base and their relevance to practical applications.

Retention Mechanism

The retention mechanism is characteristic of xylanases belonging to GH families 10 and 11. It takes place via a double displacement reaction, facilitated by two important carboxylic acid residues in the enzyme's active site:

Nucleophile Function: A single residue attacks the glycosidic bond, building a temporary covalent glycosyl-enzyme intermediate.

Acid/Base Catalyst Function: The second residue initially releases a proton to enable bond cleavage and then deprotonates a water molecule to finish hydrolysis.

This two-step reaction retains the anomeric configuration of the resulting sugar, i.e., the stereochemistry at the cleaved carbon atom is not altered [16].

GH10 Xylanases: Structurally bigger and more complex, these enzymes can degrade a broad spectrum of substituted xylans. They are versatile and can be applied to a wide variety of applications, including biofuel and food processing [1].
GH11 Xylanases: More specialized and smaller, GH11 xylanases hydrolyze unsubstituted xylans preferentially. Their substrate specificity is directed towards a specific area, which increases their effectiveness in specific degradation activities [1].

Inversion Mechanism

By contrast, the inversion mechanism, as seen in GH families like 8 and 43, is a single displacement reaction. In this, a single carboxylate ion serves two purposes [17].

Acid Catalysis: It protonates glycosidic oxygen, destabilizing the bond.
Base Activation: It deprotonates a water molecule at the same time, which attacks the sugar, resulting in hydrolysis.

This direct mechanism inverts the anomeric configuration, changing the stereochemistry of the product.

GH8 Xylanases: These are highly specific enzymes, which catalyze reactions almost solely on xylan and have little activity on other polysaccharides. Such specificity is useful in selective degradation applications.

Structural Insights and Functional Implications

The specificity and efficiency of xylanases are partially defined by their structural features, particularly the cleft used for substrate binding. Different Researchers have demonstrated the effect of variability at this site on the catalytic activity of GH10 xylanases. This information is useful in enzyme engineering, whereby researchers can design xylanases with enhanced properties such as:

Thermostability: Critical for processes like paper and pulp, where enzymes must be able to tolerate high temperatures.
Substrate Specificity: The key to reaching optimal performance in particular industrial applications [1, 17].

Researchers can construct xylanases to meet the exact standards of biotechnology by understanding these mechanisms. For instance, in the paper industry's pulp bleaching process, in alkaline and hot conditions, engineered xylanases improve pulp bleaching effectiveness while reducing expenses and the environmental impact.

4 Bacteria as Xylanase Producers

Bacteria are one of the most significant producers of xylanase enzymes, because of their high yield growth performance, inherited adaptability in environment, and competency in harsh conditions. The *Micrococcus, Paenibacillus, Bacillus, Mycobacterium, Clostridium, Streptomyces*, and *Cellulomonas* are bacterial genus that have been reported for great bacterial xylanase production [21]. This section identifies mechanisms, important bacterial species, and their enzymatic characteristics, and industrial uses.

4.1 Mechanism of Xylanase Production in Bacteria

A variety of mechanisms are utilized by bacterial xylanase producers for xylanase enzymes, including xylan-rich substrates in the agricultural residues (wheat bran and sugarcane bagasse), solid state fermentation SSF and submerged fermentation (SmF). Physicochemical factors to optimize bacterial enzyme production including pH, temperature, and carbon/nitrogen sources. These factors also employ specific media compositions and allowing the bacteria to hydrolyze xylan to accessible sugars, and have ability to break down poly-substrates [12].

4.2 Bacterial Species and Their Xylanase Profiles

Several bacterial species produce xylanase, and a special enzyme that breaks down xylan. The most important xylanase-producing bacteria and their important industrial uses are listed in the table below [18, 21]:

Bacterial species	Mode of xylanase production	Key applications
Bacillus subtilis	Extracellular	Pulp bleaching, animal feed
Paenibacillus sp.	Extracellular	Food processing, biofuel production
Clostridium thermocellum	Cell-bound	Bioethanol production
Streptomyces sp.	Extracellular	Textile industry
Cellulomonas sp.	Intracellular	Agricultural waste degradation

This diversity in production modes—whether enzymes are secreted outside the cell or retained within—directly impacts their suitability for industrial processes, offering flexibility in application design.

4.3 Industrial Applications of Bacterial Xylanases

Bacterial xylanases have many industrial applications across several sectors. Due to their strong characteristics in bacterial xylanases, they are specifically valuable in the pulp and paper industry for bio-bleaching, removing chlorine compounds, and aiding eco-friendly processes, bioethanol production, xylanases used in the hydrolysis of lignocellulosic biomass, leading to the production of fermentable sugars and enhancing overall efficiency in biofuels, animal feedstock preparation, food and feed, bakery, and beverage industries. The capability of xylanases to act against the phytopathogens for further potential usage. In animal nutrition, they hydrolyze plant polysaccharides of high complexity, increase nutrient bioavailability and digestion, and hence livestock health and productivity [18].

4.4 Genetic Engineering and Optimization

Genetic engineering has transformed the production of bacterial xylanases. With the application of recombinant DNA technology, genes of xylanases are expressed in high-expression hosts such as Escherichia coli or Bacillus subtilis, producing high enzyme titers and improving stability. Further improvement of efficiency is obtained with additional optimization methods, including metabolic engineering, enzyme immobilization, and optimized bioprocessing technologies. All these improvements enhance the commercial viability of bacterial xylanases but extend their potential into emerging applications of biotechnology [6].

5 Fungal Xylanase Producers

Fungi are extremely potent producers of xylanases owing to their tremendous capability to secrete vast amounts of extracellular enzymes. The present section is concerned with finding the major fungal species involved in xylanase production, their mechanisms of action for enzymatic activity, and their pivotal contributions in different industries [22].

5.1 Mechanism of Xylanase Production in Fungi

Fungal xylanases are mostly extracellular enzymes, secreted when fungi degrade lignocellulosic substrates. Their production is usually induced by the availability of xylan-rich substrates, which triggers a cascade of enzymatic activity specific to degrade complex hemicelluloses. These enzymes grow well in acidic to neutral pH conditions, a feature that suits well with operations in the food and beverage industries, where such conditions are dominant, rendering them very suitable for industrial applications [9].

5.2 Fungal Species and Their Xylanase Profiles

Many fungal strains are high producers of xylanase, each with unique enzymatic characteristics targeting particular applications. Below is a tabulation of key fungal xylanase producers, their production mode, and main industrial applications [8]:

Fungal species	Mode of xylanase production	Key applications
Aspergillus niger	Extracellular	Bread-making, juice clarification
Trichoderma reesei	Extracellular	Textile processing, bioethanol production
Penicillium sp.	Extracellular	Waste treatment, pulp and paper industry
Fusarium oxysporum	Extracellular	Animal feed enhancement, bioremediation
Chaetomium thermophilum	Extracellular	High-temperature enzyme processes

The extracellular nature of these xylanases allows direct contact with xylan substrates, streamlining the degradation process and boosting efficiency in industrial settings.

5.3 Industrial Applications of Fungal Xylanases

Fungal xylanases are instrumental in various industries by efficiently decomposing or reforming hemicellulose. In baking, they make dough more elastic and crumb texture better through the degradation of arabinoxylans, thereby producing better quality bread. For fruit juice clarification, these enzymes lower viscosity and turbidity due to hydrolysis of soluble hemicelluloses, producing clearer and more attractive juices. Concurrently, during the production of biofuel, fungal xylanases facilitate the hydrolysis of lignocellulosic biomass, liberating fermentable sugars needed for the synthesis of bioethanol to promote sustainable energy alternatives [10].

5.4 Enzyme Engineering for Improved Yield

Advances in the field of biotechnology have tremendously improved the production of fungal xylanases. Methods of genetic engineering by mutagenesis and process optimization of fermentation processes like submerged fermentation and solid-state fermentation have worked to maximize enzyme yield. These technologies not only increase efficiency of production but also enhance stability and activity of the enzyme, making them effective in a broad spectrum of applications in industry [9].

6 Yeast as a Source of Xylanase

Yeasts, being unicellular fungi, have become increasingly prominent in biotechnology due to their exceptional versatility for enzyme production. While filamentous fungi have long been the dominant xylanase producers—enzymes that break down xylan, a critical part of plant cell walls—certain yeasts have shown themselves to be worthy contributors in this area. Noteworthy among them are *Saccharomyces cerevisiae*, *Candida tropicalis*, and *Pichia pastoris*, which show very high xylanase activity and are used in various industries like production of bioethanol, food processing, and animal feed fortification [24]. This section explores the mechanisms of xylanase production by yeast, describes selected yeast species, and discusses their industrial importance.

6.1 Mechanism of Xylanase Production in Yeast

The yeast xylanases are secreted into the extracellular medium and thus directly find their substrate when in a xylan-rich surrounding. Their secretory process in response to xylanase gene inducers of wheat bran, agro-industrial residues, and other materials rich in xylan stimulates xylanase enzyme production. Yeast xylanases find the best activity between acidic to near neutral pH and as such will be used efficiently within many industrial activities. But certain species of yeast, like *Saccharomyces cerevisiae*, secrete intracellular xylanases, which are inside the cell and have to be recovered using specific methods [9, 24]

6.2 Key Yeast Species and Their Xylanase Profiles

Several yeast species stand out for their ability to produce xylanases, each offering unique production mechanisms and industrial benefits. The table below summarizes their characteristics [24]:

Yeast species	Mode of xylanase production	Key applications	Unique characteristics
Saccharomyces cerevisiae	Intracellular	Bioethanol production	A robust, genetically tractable yeast widely used in fermentation processes
Candida tropicalis	Extracellular	Food industry	Capable of metabolizing diverse carbon sources, enhancing their industrial flexibility
Pichia pastoris	Extracellular	Recombinant enzyme production	A preferred host for high-yield recombinant protein expression, ideal for enzyme engineering

These species highlight the diversity of yeast in xylanase production, providing options for both intracellular and extracellular enzyme systems tailored to specific industrial needs.

6.3 *Industrial Applications of Yeast Xylanases*

Yeast xylanases have key functions in various industries, and two major areas of influence are bioethanol production and food processing:

- **Production of Bioethanol**: In the biofuel industry, yeast xylanases aid in the degradation of lignocellulosic biomass, a plant-based complex matrix that is recalcitrant to degradation. By hydrolyzing xylan into easier-to-ferment sugars, these enzymes increase the yield of fermentable substrates, thereby enhancing the efficiency of ethanol production.
- **Food Industry**: Yeast xylanases are used to enhance the quality of dough by making its texture of baked food goods. They are also very helpful in clarifying the fruit juice by breaking down hemicellulose, reducing turbidity and making high-quality products. Moreover, yeast xylanases in animal nutrition, releasing more digestible nutrient accessibility through the breakdown of plant cell polysaccharides, which further improves overall animal production [23].

6.4 *Emerging Trends and Innovations*

Recent developments in genetic engineering to improve xylanase activity and stability in production of xylanase. Yeast innovations in the field of xylanase production include food, paper, and biofuel production, and utilize yeast surface display for plant protection. Pichia pastoris, an example of yeast xylanase production in the form of recombinant expression of xylanase due to its improved stability and titers. Furthermore, improved yeast strains have been created by genetic engineering, which has reduced production costs and increased large scale production. These are examples of yeast advancements, how yeast is becoming a more efficient and natural source of xylanases in modern biotechnology [9, 23].

7 Algae: A Promising Source of Xylanase

Algae have different usages in biotechnology and industry. Xylan, a major part of plant cells, is breakdown due to its enzyme. Microalgae is a novel and sustainable source of xylanase enzymes, that can be cultivated at extreme temperature for biomass yield of different types of microalgae due to genetic modification. These aquatic organisms (algae) have vast varieties exist throughout the world from freshwater ponds to saltwater oceans, everyone has their own characteristics and capability to produce the xylanase. *Chlorella vulgaris* and *Spirulina platensis* are one of them for their remarkable xylanase activity, which open the possibilities for production of nature friendly enzyme from them and their application in sustainable industry [5]. This section explores the mechanism of algae in the production of xylanase, common species of algae, and their potential in xylanase production.

7.1 Mechanism of Xylanase Production in Algae

Most algal xylanases are extracellular depending on xylan in the environment and their related ability of algae to utilize xylan as a source of carbon and energy. Xylan breaks down into small subunits called xylose, which are used by algae for metabolism. Various factors influence the xylanase production in algae, such as salinity, pH, temperature, nutrients availability, and light intensity. Algal xylanases production involves complex factors, most important are algal species and genetic regulation. The expression of xylanase genes is influenced by multiple facts such as the presence of xylan and environmental cues influence on production of xylanase enzymes in algae. [7, 8].

7.2 Key Algal Species and Their Xylanase Profiles

In the recent advancement of research, algae are less studied than in fungi or bacteria. Two well-known algal species in following table with their xylanase production processes and industrial application [4, 5]:

Algal species	Mode of xylanase production	Key applications	Distinct features
Chlorella vulgaris	Extracellular	Biofuel production	A highly productive microalga, ideal for scalable enzyme synthesis due to its robust growth
Spirulina platensis	Intracellular	Aquaculture feed	A nutrient-packed cyanobacterium, providing both xylanase and nutritional benefits

Enzyme extraction is impacted by the distinction between extracellular and intracellular production: extracellular xylanases from Chlorella vulgaris are simpler to extract, whereas intracellular enzymes from Spirulina platensis necessitate cell disruption, which adds a step to the procedure.

7.3 Potential Applications of Algal Xylanases

Significant application of xylanase in industries by algae xylanases offer variable solutions. The following are main categories of algae xylanase usage:

- **Biofuel Production**: Algal xylanases can break down lignocellulosic materials, such as agricultural residues, into simpler sugars, boosting the efficiency of bioethanol production and supporting renewable energy goals.
- **Wastewater Treatment**: By degrading organic waste in water systems, these enzymes provide an environmentally friendly approach to bioremediation, reducing pollution naturally.
- **In Aquaculture Feed**: potential application of algal xylanases to the fish diets increases FCR in fishes, lowers feed waste, and increases nutrient digestibility. Microalgae have great potential substitute of fish meal and oil in aquatic farming [5, 7].

7.4 Challenges and Opportunities Ahead

Although algae xylanase has great potential, it also has various challenges including optimizing enzymes production, sustaining its ability for commercialization, needs effective extraction methods and scaling up its use in industry. While its opportunities on the other hand have a lot of potential for production of biofuel (bioethanol and biogas), bioremediation for heavy metals and organic pollutants, and industrial applications (textile, pulp and paper production) [4].

8 Protozoa and Their Role in Xylanase Production

Protozoa belong to the various groups of single-cell organisms that play an important role in the xylanase production, especially in compound stomach rumen. Where it breaks down the plant hemicellulose, is facilitated by protozoa. Besides other microorganisms, protozoa influence the overall microbial colonies in the rumen, and their production of xylanase by digestion and plant material by ruminants. The most important protozoan species include *Entodinium* and *Diplodinium,* along with other microbes are vital for the digestion of plant cell walls in ruminants for extraction of nutrients from plants [20]. This portion highlights the role of protozoa in xylanase production, specific species of protozoa related to xylanase production and find out their industrial applications.

8.1 Role of Protozoa in Xylan Degradation

Protozoa are hosts of rumen ecosystem and often work with bacteria and fungi in the rumen to degrade plant cell walls. Ruminal protozoa significantly contribute to all fibrinolytic activity. Rumen protozoa also contribute to metabolic products in the form of xylan degradation and production of volatile fatty acids (VFAs) and gases, which ultimately are an important source of energy for animals and come microbes. Some protozoa like *Neocallimastix frontalis* enhance the degradation of xylan, on other hand, *Entodinium* species co-work with other microbes of rumen to break down fibrous plant and increase chances of nutrient utilization and digestion. Likewise, *Diplodinium* species helps termites for degradation of cellulose and hemicellulose, so they can utilize the wood as food. Overall symbiotics effects of protozoa to access the nutritional capacity of plant, the protozoa play an important role in the rumen ecosystem along with other microbes [4, 20].

The following are the main protozoan species, their roles, and practical applications [19]:

Protozoan species	Symbiotic role	Key applications	Unique contributions
Entodinium spp.	Hosts xylanase-producing microbes	Animal nutrition	Boosts rumen digestion, improving feed efficiency in livestock
Diplodinium spp.	Aids cellulose and hemicellulose breakdown	Bioenergy research	Supports biomass degradation studies for biofuel production

8.2 *Industrial and Agricultural Implications*

The protozoal xylanase production presents below in the field of animal nutrition and renewable energy:

- **Animal Nutrition**: Protozoa are one of the most important agents in the production of xylanases in ruminant by promoting the degradation of hemicellulose, which improves animal nutrient absorption ability. This process gives a model for the creation of enzyme additives that may maximize the use of feed, possibly reducing costs and enhancing sustainability in livestock sector.
- **Bioenergy Source**: Protozoa and their symbionts' capability to come into bioenergy science. They play a significant role in different bioenergy processes like anaerobic digestion and wastewater treatment. Protozoa also contribute to the process of recycling plant waste to produce biofuels like biogas or bioethanol for biological ecosystem [2].

Future advancement in the xylanase synthesis from protozoan symbionts holds significant potential for this novel enzyme. Though the challenges are present in the form of a complex natural environment.

9 Comparative Analysis of Xylanase Production in Different Microorganisms

Xylanases are enzymes that hydrolyze xylan to lower-molecular-weight sugars, play important roles in several industrial processes. Here, these enzymes are made by a range of microorganisms that differ from each other in characteristics influencing their enzyme activity, stability, and industrial applications. This section presents a detailed comparison of these microbial sources and highlights their advantages, disadvantages, and prospects in biotechnology applications (Fig. 1).

9.1 *Overview of Xylanase Production*

The following table offers an overview of xylanase production in various microorganisms, including a brief summary of their efficiency, main industrial uses, and notable characteristics or difficulties [2, 5, 20]:

Microorganism type	Xylanase efficiency	Key industrial use	Unique features/Challenges
Bacteria	High	Pulp and paper	Thermostable, easily genetically modified
Fungi	Very high	Bioethanol, textiles	Abundant enzyme output, pH-specific activity

(continued)

(continued)

Microorganism type	Xylanase efficiency	Key industrial use	Unique features/Challenges
Yeast	Moderate	Food processing	Intracellular enzymes, engineering potential
Algae	Low	Biofuels	Extracellular production, scalability hurdles
Protozoa	Indirect	Livestock digestion	Symbiotic enzyme sources, extraction difficulty

9.2 *In-Depth Comparison*

9.2.1 Bacteria

Bacterial xylanases are most valued for their efficiency, because of the fast growth rate of such microbes and their ease of adaptation to genetic alteration. These enzymes tend to be thermostable, and therefore ideal for application in the pulp and paper segment, where they assist in bio-bleaching by hydrolyzing hemicellulose at high temperatures. This cuts down on chemical usage, promoting eco-friendliness. Besides this, bacterial xylanases are promising candidates in animal feed and biofuel, due to their robustness and versatility.

9.2.2 Fungi

Fungi are the most efficient xylanase producers, and they secrete vast quantities of extracellular enzymes. Their xylanases work optimally under acidic to neutral pH conditions, which is ideally suited for applications such as bioethanol production—where they break down lignocellulosic biomass—and textile processing, where they improve fabrics by hemicellulose removal. Yet, their susceptibility to harsh conditions, including elevated temperatures or pH changes, can limit their application in harsher industrial environments.

9.2.3 Yeast

Xylanases are moderately efficiently synthesized by yeast, a unicellular fungal species, usually intracellularly. The extraction process is difficult due to intracellular production, which prevents industrial use. Yeast xylanases, on the other hand, are better for food processing because they enhance dough consistency and juice clarity. Because of their genetic engineering suitability, they are also intriguing candidates

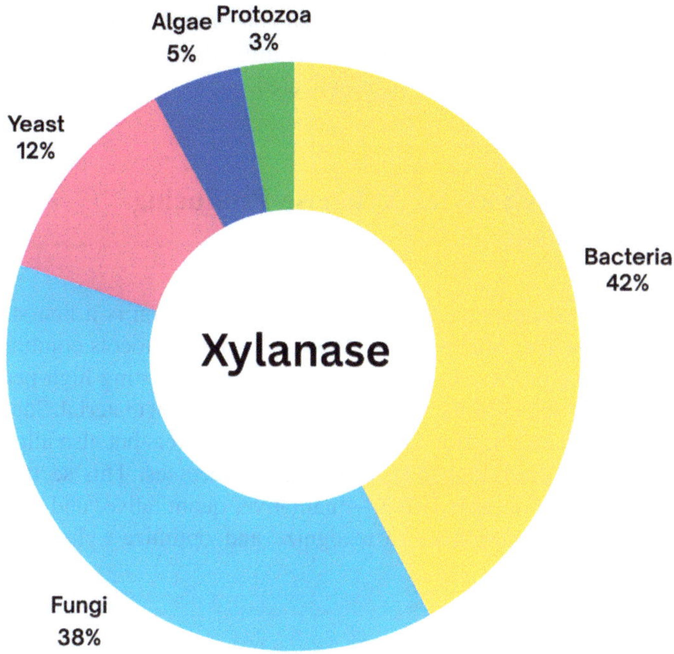

Fig. 1 Based on current data, it is the estimated percentage contribution of each microorganism to total xylanase production

to produce particular enzymes, particularly bioethanol, where higher yields could make them a crucial industrial function.

9.2.4 Algae

Currently, algae are a novel source of xylanases with low efficiency. Their extracellular enzymes, which can withstand a variety of environmental conditions, have the potential to break down plant biomass and produce biofuel. Although algae are a renewable resource, the difficulty of expanding cultivation and optimizing enzyme production limits their industrial adoption. Research is still being done to break down the obstacles and realize their full potential.

9.2.5 Protozoa

Through the accommodation of symbiotic bacteria and fungi in their systems, protozoa indirectly aid in the production of xylanase. These symbionts break down plant matter in termites' and ruminants' digestive tracts by secreting xylanases, which aid in the absorption of nutrients. This function has significance in livestock digestion and stimulates bioenergy research on microbial consortia. Unfortunately, harvesting protozoan-associated xylanases industrially is difficult because isolating enzymes from these symbiotic relationships is complicated.

10 Screening Methods for Xylanase-Producing Microorganisms

Detection of microorganisms with high xylanase production is a first step toward realizing their value for industrial processes. Screening protocols combine qualitative and quantitative methodologies for isolating and evaluating high-performance strains from varied sources like soil, compost, and rotten plant material. Such methods not only enable the identification of new xylanase producers but also allow selected strains to meet the severe demands of industrial processes. This section summarizes the major screening approaches—qualitative, quantitative, and sophisticated molecular techniques—employed to recognize and optimize xylanase-producing microorganisms [21].

10.1 Qualitative Screening

Qualitative screening is normally the initial phase for xylanase-producing microorganisms' detection, based on simple plate-based assays. In this approach, microbial isolates are plated onto xylan-supplemented agar plates (e.g., birchwood or oat spelt xylan) as the sole carbon source. After incubation, xylanase activity is detected as clear areas around colonies, a visual sign of xylan hydrolysis. To enhance detection, dyes like Congo red are usually utilized; this dye adsorbs on intact xylan, colorizing it red, while hydrolyzed areas turn clear after sodium chloride destaining. The procedure is prized due to its simplicity, cost, and capability for rapid screening of large populations of microbes. Still, it gives only a preliminary evaluation of xylanase production, and a further quantitative assay must be made to verify the findings [7].

10.2 Quantitative Screening

Quantitative screening techniques are crucial for a more accurate assessment of xylanase activity, especially when choosing strains for industrial use. The Dinitro salicylic acid (DNS) method, which measures the reducing sugars (like xylose) released during xylan hydrolysis, is commonly accepted as a standard. This assay involves mixing a xylan substrate with a microbial culture supernatant or crude enzyme extract. The resulting reducing sugars then react with DNS reagent to produce a colour change that can be measured using spectrophotometry at 540 nm. The quantity of xylanase enzyme that releases 1 μmol of reducing sugar per minute under normal conditions is known as its activity, and it is measured in units (U). Reducing sugars are also quantified by complementary methods like the Somogyi-Nelson assay, and hydrolysis products can be thoroughly and precisely analyzed using High-Performance Liquid Chromatography (HPLC). These methods have proven successful in assessing high-yielding strains that exhibit their industrial potential, such as *Aspergillus flavus* (493.33 U/mg) and *Bacillus pumilus* (388.82 U/mg) [3, 13].

10.3 Advanced Screening Techniques

Higher molecular and genetic techniques help identify and optimize xylanase producers in addition to traditional assays. PCR and gene sequencing make it possible to identify xylanase-encoding genes, like xynA in Bacillus species, which reveal details about a strain's genetic potential. To increase the efficiency of enzyme production, recombinant DNA technology can also be used to overexpress these genes in host organisms such as Escherichia coli. These cutting-edge techniques facilitate the identification of novel xylanase producers and expedite their transition from environmental isolates to industrial accumulation [13, 15].

11 Future Perspectives and Biotechnological Applications of Xylanase

Xylanases enzymes, which break down xylan, the main structural polysaccharide found in plant cell walls, are essential for a variety of biotechnological and industrial processes, from pulp processing to the production of biofuel and food. New advancements in xylanase research continue to create opportunities for their use as sustainability and efficiency gain more attention globally [10]. The new fields of study and development could improve xylanase applications even more and offer creative answers to today's problems. Recombinant technology, genetic engineering, novel microbial discoveries, and environmentally friendly bioprocessing techniques are the foundations of these developments.

11.1 Emerging Research Areas

11.1.1 Recombinant Xylanase Production

Large-scale recombinant synthesis in host organisms such as *Escherichia coli, Pichia pastoris,* or *Bacillus subtilis* has been made possible by the development of recombinant DNA technology, which has revolutionized the production of xylanase. The selection of hosts is based on how well they express foreign genes and how easily they can grow. Researchers can produce more xylanases at a reduced cost through recombinant production, increasing the enzymes' availability for industrial uses. Additionally, this technique allows for the customization of enzyme properties, such as increased stability or activity, to improve their performance in applications like textile processing or biomass saccharification [17, 22].

11.1.2 Genetic Modifications for Enhanced Performance

In order to meet industrial demands, xylanase functionality is being improved through targeted genetic engineering. Researchers can alter enzyme genes to improve characteristics like thermostability, pH tolerance, or substrate specificity using techniques like site-directed mutagenesis and directed evolution. For instance, it might be possible to make a fungal-derived gene-modified xylanase that is suitable for the harsh conditions of biofuel production to stay active at high temperatures [5]. In line with industry sustainability goals, these developments optimize process efficiency and reduce reliance on chemical additives.

11.1.3 Novel Microbial Sources

Superior xylanases are being produced by unlocking untapped microbial diversity. Microbes from harsh environments, such as deep-sea sediments, volcanic hot springs, and hypersaline lakes, can produce enzymes that are adapted to harsh circumstances. Extremozymes enable xylanases to perform at their best in a wider range of circumstances by exhibiting characteristics like salt resistance or thermostability. For example, a thermophilic bacterium might produce a xylanase that enables high-temperature pulp bleaching, process simplification, and energy conservation [11].

11.2 Future Trends in Xylanase Biotechnology

11.2.1 Metagenomics for Novel Xylanases

Enzyme discovery is being revolutionized by metagenomics, the study of genetic material is directly isolated from environmental samples, which eliminates the need to cultivate microorganisms. By using this technique, researchers can access the vast genetic potential of uncultivated microorganisms and find xylanase genes with novel or enhanced characteristics. By identifying enzymes with unique catalytic efficiencies or resistance to commercial inhibitors, metagenomic library screening may increase the toolkit available for uses like the breakdown of lignocellulose in the production of bioethanol [17].

11.2.2 Synthetic Biology for Enzyme Optimization

By producing enzymes with specific functions, synthetic biology holds the potential to completely transform the design of xylanase. Through laboratory evolution and computational resources, researchers can create xylanases that perform best for a particular application. In biorefineries, for example, a synthetically enhanced xylanase would work better with cellulases to improve the breakdown of complex plant biomass. By decreasing resource consumption and increasing efficiency, precision engineering makes industrial processes more environmentally and economically sustainable [1].

11.2.3 Eco-friendly Bioprocessing Techniques

The incorporation of xylanases into green bioprocessing is a major trend propelling their future use. In the paper and pulp sector, xylanases minimize the use of chlorine-based bleaching chemicals, reducing toxic waste. In biofuel manufacturing, they improve enzymatic liberation of fermentable sugars from biomass, reducing dependence on energy-hungry chemical pretreatments [26]. In the future, xylanases may be integral to next-generation, environmentally friendly systems—zero-waste biorefineries, for example—where they cooperate with other biocatalysts to optimize resource utilization and minimize footprint.

12 Conclusion

Xylanase production is highly diverse among microorganisms, each with specific advantages suited to specific industrial applications. Bacteria and fungi currently predominate, with their high efficiency and stable enzyme titres, but yeast and algae offer potential for the breakthroughs inherent in genetic and cultivation developments. Protozoa, as indirect participants, highlight the contribution of symbiotic systems to the study of enzymes. Choosing the best microbial source depends on recognizing these relative strengths and weaknesses, providing compatibility with the requirements of temperature, pH, and scalability in biotechnological operations. The qualitative, quantitative, and high-level screening procedures form a firm platform for selecting microorganisms with superior xylanase production capacity. Coupling the ease of plate assays, the specificity of biochemical analyses, and the versatility of molecular tools allows researchers to select and streamline strains. The path of xylanase biotechnology is characterized by swift developments in production methods, enzyme engineering, and eco-friendly applications. This volume presents a comprehensive insight into xylanase-producing microbes, their enzymatic activity, and their existing industry applications, giving a solid groundwork for future developments. Equipped with such information, scientists and biotechnologists are suitably positioned to harness these budding trends, formulating next-generation xylanases that will propel industries towards more efficiency, sustainability, and technology.

References

1. E. Abedi, F. Fatemi, Y. Sefidbakht, S.E.R. Siadat, Development and characterization of a thermostable GH11/GH10 xylan degrading chimeric enzyme. Enzyme Microb. Technol. **149**, 109854 (2021). https://doi.org/10.1016/j.enzmictec.2021.109854
2. T. Abena, A. Simachew, A review on xylanase sources, classification, mode of action, fermentation processes, and applications as a promising biocatalyst. Biotechnologia **105**(3), 273–285 (2024). https://doi.org/10.5114/bta.2024.141806
3. H.F. Alves-Prado, F.C. Pavezzi, R.S.R. Leite, V.M. de Oliveira, L.D. Sette, R. Dasilva, Screening and production study of microbial xylanase producers from Brazilian Cerrado. Appl. Biochem. Biotechnol. **161**(1–8), 333–346 (2010). https://doi.org/10.1007/s12010-009-8823-5
4. N. Bhardwaj, K. Agrawal, P. Verma, Xylanases: an overview of its diverse function in the field of biorefinery, in *Bioenergy Research: Commercial Opportunities & Challenges*, ed. by M. Srivastava, N. Srivastava, R. Singh (Springer, 2021), pp. 295–317. https://doi.org/10.1007/978-981-16-1190-2_10
5. S. Bhushan, U. Jayakrishnan, B. Shree, P. Bhatt, S. Eshkabilov, H. Simsek, Biological pretreatment for algal biomass feedstock for biofuel production. J. Environ. Chem. Eng. **11**(3), 109870 (2023). https://doi.org/10.1016/j.jece.2023.109870
6. L. Cao, R. Zhang, J. Zhou, Z. Huang, Biotechnological aspects of salt-tolerant xylanases: a review. J. Agric. Food Chem. **69**(31), 8610–8624 (2021). https://doi.org/10.1021/acs.jafc.1c03192
7. R. Chaudhary, T. Kuthiala, G. Singh, S. Rarotra, A. Kaur, S.K. Arya, P. Kumar, Current status of xylanase for biofuel production: a review on classification and characterization. Biomass Convers. Biorefinery, 1–19 (2021)
8. R. Chaudhary, T. Kuthiala, G. Singh, S. Rarotra, A. Kaur, S.K. Arya, P. Kumar, Current status of xylanase for biofuel production: a review on classification and characterization. Biomass Convers. Biorefinery **13**(10), 8773–8791 (2023). https://doi.org/10.1007/s13399-021-01948-2
9. S. Dahiya, A. Rapoport, B. Singh, Biotechnological potential of lignocellulosic biomass as substrates for fungal xylanases and its bioconversion into useful products: a review. Fermentation. **10**(2), 82 (2024)

10. F.M. Dar, P.M. Dar, Fungal xylanases for different industrial applications, in *Industrially Important Fungi for Sustainable Development: Volume 2: Bioprospecting for Biomolecules*, ed. by A.M. Abdel-Azeem, A.N. Yadav, N. Yadav, M. Sharma (Springer International Publishing, 2021), pp. 515–539. https://doi.org/10.1007/978-3-030-85603-8_14
11. A. Dukare, K. Sharma, S. Kautkar, J. Dhakane-Lad, R. Yadav, V. Nadanathangam, S. Saxena, Microbial xylanase aided biobleaching effect on multiple components of lignocelluloses biomass based pulp and paper: a review. Nord. Pulp Pap. Res. J. **38**(3), 459–480 (2023). https://doi.org/10.1515/npprj-2023-0005
12. H. Fernandes de Souza, L. Aguiar Borges, D. Di Próspero, V. Gonçalves, J. Vitor dos Santos, M. Sousa Bessa, M. Fronja Carosia, M. Vieira de Carvalho, I. Viana Brandi, E. Setsuko Kamimura, Recent advances in the application of xylanases in the food industry and production by actinobacteria: a review. Food Res. Int. **162**, 112103 (2022). https://doi.org/10.1016/j.foodres.2022.112103
13. B. Kalim, N.M. Ali, Optimization of fermentation media and growth conditions for microbial xylanase production. 3 Biotech. **6**(2), 122 (2016)
14. M. Kataoka, F. Akita, Y. Maeno, B. Inoue, H. Inoue, K. Ishikawa, Crystal structure of Talaromyces cellulolyticus (formerly known as Acremonium cellulolyticus) GH family 11 xylanase. Appl. Biochem. Biotechnol. **174**(4), 1599–1612 (2014). https://doi.org/10.1007/s12010-014-1130-9
15. D. Kaur, A. Joshi, V. Sharma, N. Batra, A.K. Sharma, An insight into microbial sources, classification, and industrial applications of xylanases: a rapid review. Biotechnol. Appl. Biochem. **70**(4), 1489–1503 (2023). https://doi.org/10.1002/bab.2469
16. B.S. Kim, I.J. Kim, Strategies and recent trends in engineering thermostable GH11 xylanases. Catalysts **15**(4), Article 4 (2025). https://doi.org/10.3390/catal15040317
17. N. Li, R. Zhang, J. Zhou, Z. Huang, Structures, biochemical characteristics, and functions of β-xylosidases. J. Agric. Food Chem. **71**(21), 7961–7976 (2023). https://doi.org/10.1021/acs.jafc.3c01425
18. D. Mu, P. Li, T. Ma, D. Wei, M. Montalbán-López, Y. Ai, X. Wu, Y. Wang, X. Li, X. Li, Advances in the understanding of the production, modification and applications of xylanases in the food industry. Enzyme Microb. Technol. **179**, 110473 (2024). https://doi.org/10.1016/j.enzmictec.2024.110473
19. T. Mujtaba, H. Mohsin, A review on microbial xylanases and their industrial applications. J. Biol. Allied Health Sci. **1**, 1 (2021)
20. R. Pan, J. Sun, F. Xin, W. Jiang, M. Jiang, Enzymes for biomass pretreatment: a comprehensive review (Springer, 2025), pp. 1–19. https://doi.org/10.1007/10_2025_275
21. M. Phuyal, U. Budhathoki, D. Bista, S. Shakya, R. Shrestha, A.K. Shrestha, Xylanase-producing microbes and their real-world application. Int. J. Chem. Eng. **2023**(1), 3593035 (2023)
22. A. Sarangi, H. Thatoi, Xylanase as a promising biocatalyst: a review on its production, purification and biotechnological applications. Proc. Natl. Acad. Sci. India Sect. B: Biol. Sci., 1–16 (2024)
23. M. Sohail, N. Barzkar, P. Michaud, S. Tamadoni Jahromi, O. Babich, S. Sukhikh, R. Nahavandi et al., Cellulolytic and xylanolytic enzymes from yeasts: properties and industrial applications. Molecules **27**(12), 3783 (2022)
24. K. Šuchová, C. Fehér, J.L. Ravn, S. Bedő, P. Biely, C. Geijer, Cellulose-and xylan-degrading yeasts: enzymes, applications and biotechnological potential. Biotechnol. Adv. **59**, 107981 (2022)
25. S. Tundo, G. Mandalà, L. Sella, F. Favaron, R. Bedre, R.M. Kalunke, Xylanase inhibitors: defense players in plant immunity with implications in agro-industrial processing. Int. J. Mole. Sci. **23**(23), Article 23 (2022). https://doi.org/10.3390/ijms232314994
26. A. Walia, S. Guleria, P. Mehta, A. Chauhan, J. Parkash, Microbial xylanases and their industrial application in pulp and paper biobleaching: a review. 3 Biotech. **7**(1), 11 (2017). https://doi.org/10.1007/s13205-016-0584-6

27. S. Yegin, Microbial xylanases in xylooligosaccharide production from lignocellulosic feedstocks. Biomass Convers. Biorefinery **13**(5), 3619–3658 (2023). https://doi.org/10.1007/s13399-022-03190-w
28. G. Kaur, M.S. Taggar, A. Kalia, M. Krishania, A. Singh, Fungal secretomes of Aspergillus terreus repertoires cultivated on native and acid/alkali treated paddy straw for cellulase and xylanase production. BioEnergy Res. **17**(1), 145–159 (2024)

Chapter 3
Xylanase Production Techniques

Abdul Sattar Jatoi

Abstract Xylanase enzymes have become important for industry because they break down xylan into xylose while finding applications in pulp and paper manufacturing along with textile processing and food production and animal feed applications and biofuel manufacturing. The evaluation of xylanase product techniques follows a detailed framework that includes discussions about microbial strains as well as fermentation strategies and genetic modifications and process enhancement methods. Research has focused on fungal and bacterial strains for studying their productivity of xylanase when grown under various cultivation conditions. The discussion includes both solid-state and submerged fermentation techniques which demonstrate their industrial-scale properties through detailed assessments of benefits and constraints. The research evaluates contemporary developments in recombinant DNA technology and strain improvement methods to improve enzyme output and stability. This work critically examines how each physicochemical parameter including pH, temperature, carbon and nitrogen sources as well as inducers affects the production of xylanase. The assessment ends by analyzing present obstacles and analyzing possible paths for future sustainable development of xylanase production at affordable costs.

Keywords Xylanase · Production · Techniques

1 Introduction

The hydrolytic enzyme Xylanase functions to degrade xylan, which represents the major substance in plant cell walls, as it breaks down xylan into xylose alongside xylo-oligosaccharides [1]. Xylan exists in plants at a level reaching 35% of plant mass while standing as the second most prevalent polysaccharide after cellulose so xylanase functions as an indispensable enzyme for dismantling lignocellulosic biomass [2]. During the last three decades, xylanase has experienced exponential growth because various industries use it in pulp and paper bleaching and

A. S. Jatoi (✉)
Department of Chemical Engineering, Dawood University of Engineering and Technology, Karachi, Pakistan
e-mail: abdul.sattar@duet.edu.pk

A. Sattar Jatoi (ed.), *Xylanase from Microorganisms*,
https://doi.org/10.1007/978-981-95-0824-2_3

textile processing and animal feed production and food and beverage manufacturing and biofuel manufacturing processes [3]. The pulp and paper industry employs xylanases to run bio-bleaching processes at lower chlorine usage levels for producing brighter quality papers [4]. Registration of xylanases in animal nutrition helps reduce anti-nutritional factors within feed while enhancing the uptake of nutrients [5]. The food industry relies on xylanases for fruit juice clarification and clarification tasks in addition to improving dough processing and enhancing bread quality and extracting coffee oils and plant oils. Xylanases represent a major component of bioethanol production through agricultural residue conversion to fermentable sugars to help establish sustainable, renewable energy systems [6]. Biochemical analysis of xylanase enzymes places them into glycoside hydrolase families (GH10 and GH11) through sequence comparison and structural analysis, and their catalytic methodology assessment. Different family types possess unique mechanisms of action alongside different enzyme specificities that direct their use to industrial applications [7]. GH10 xylanases demonstrate a wide range of substrate recognition that permits them to modify highly branched xylans but GH11 enzymes display specific reaction pathways leading to elongated xylan chains. The growing global emphasis on sustainability and waste valorization together with biotechnology, requires efficient and cost-efficient methods to produce xylanases [8]. Xylanase industrial production through natural microorganism screening has evolved to include DNA recombinant methods and substrate innovation with agricultural waste material to minimize production expenses and environmental repercussions[9]. Our knowledge about multiple ways to generate xylanase serves essential functions for both academic progress and industrial deployment because this enzyme holds crucial economic value, together with environmental significance [10]. The chapter covers complete xylanase production techniques through analysis of microbial diversity alongside submerged and solid-state and semi-solid fermentation methods, substrate requirements, production parameter enhancements and recombinant methodology, purification and scale-up considerations. The sections present advanced enzyme biotechnology approaches along with scientific findings, which will direct upcoming revolutionary advancements while optimizing biomass deployment for different biotechnological applications. The production techniques for xylanase solutions maintain an essential connection between microbiological research and biochemical engineering with industrial sustainability goals.

2 Microbial Sources of Xylanase

The successful industrial xylanase production at large scale depends on microbial selection because microbes serve as the main producers of industrial enzymes because they grow fast and they are genetically manageable and able to break down wide ranges of substrates [10]. The three major microbial domains that produce xylanase are bacteria, fungi and actinomycetes [11]. The enzyme production capacity of fungi remains highest when using filamentous types like *Aspergillus niger*, *Aspergillus oryzae*, *Trichoderma reesei* and *Penicillium chrysogenum* under solid-state fermentation conditions for extracellular enzyme secretion particularly xylanase [12]. Fungal xylanases function best in food and beverage as well as textile industries because their pH tolerance spans from acidic to neutral. Initial production of enzymes by fungi happens best on lignocellulosic waste materials because fungi show excellent performance in this environment [13]. *Bacillus subtilis* together with *Bacillus pumilus* represent the most extensively researched bacterial genera for enzyme production along with *Streptomyces* spp., *Thermomonospora*, and *Paenibacillus*. These bacterial species offer efficient fermentation cycles and quick bacterial growth to industry [14]. The thermostable and alkaline pH-active characteristics of bacterial xylanases make them suitable for paper pulp bleaching and textile processing applications under hot and basic conditions [15]. The xylanases from thermophilic bacteria residing in hot springs and compost piles possess substantial stability that enables their utilization in industrial applications [16]. The ability of actinomycete Streptomyces species to degrade complex polysaccharides together with the production of endo- and exo-xylanases makes them effective in performing complete xylan hydrolysis [17]. Actinomycetes also generate highly stable enzymes that maintain their activity within extensive pH conditions and heightened temperatures [18]. Scientific interest in extremophiles isolated from harsh environmental spaces such as deep-sea vents and heavy salinity environments and polar environments has grown significant during recent years [18]. The extremozyme-producing capabilities of these microbes create xylanases that thrive under harsh industrial situations due to their ability to function in extreme pH conditions and extreme temperatures and high salt concentrations [19]. The xylanases produced by halophilic archaea maintain high efficiency in high-saline processes especially during fish-processing wastewater treatment. Academic institutions and researchers investigate psychrophilic xylanases originating from freezing environments for applications as cold wash detergents and low-temperature food processing. The expression of heterologous xylanase in yeast strains including *Candida* and *Pichia pastoris* has gained importance because these strains perform post-translational modifications while producing active recombinant enzymes [20]. The quick growth and well-developed genetic tools make *E. coli* the best choice for xylanase gene cloning yet it faces industrial limitations due to protein inclusion bodies and the absence of glycosylation. Eukaryotic organism expression systems consisting of *Pichia pastoris* and *Saccharomyces cerevisiae* successfully tackle the expression problems by enabling appropriate enzyme folding and correct secretion. Metagenomic studies enable scientists to locate new xylanase genes by directly studying environmental DNA without cultivating these genes. The research method has enabled scientists to discover entirely new enzymes among the xylanases which demonstrate distinctive catalytic abilities. Synthetic biology along with CRISPR-Cas9 technology allows scientists to precisely modify microbe genomes for better xylanase production and optimized metabolic pathways which results in better enzyme synthesis efficiency. The various microbial communities produce xylanase enzymes which establish a wide range of structural and functional features [21]. The selection of microbial enzymes depends mostly on several factors including required properties of pH, temperature, stability together with substrate specificities and desired industrial use. Cost-efficient microorganism fermentation systems combined with improved strain selection methods together with new bioprospecting discoveries continue to broaden the number of microbes that produce xylanases [22]. Modern-day biotechnological methods with traditional microbial selection create promising opportunities to develop specific xylanase production systems for biofuel and pharmaceutical industries' needs. Untapped microbial diversity and metabolic potential will enable the optimization of xylanase production methods that conform to principles of green chemistry and circular economy [21].

3 Fermentation Techniques

Industrial biotechnology depends on fermentation as a fundamental method to produce enzymes with xylanase synthesis as its primary function through microorganism-controlled environments for efficient enzyme secretion [23]. Three methods serve for xylanase production through fermentation: submerged fermentation (SmF), solid-state fermentation (SSF) and semi-solid or semi-submerged fermentation [24]. The three production methods demonstrate unique operating conditions alongside separate advantages and disadvantages which determine their application range for specific microbes and industrial uses [25]. Commercial enzyme operations choose submerged fermentation (SmF) as their primary fermentation technology because it provides convenient upscaled production along with high process parameter control and friendly integration with automated bioreactor devices [26]. SmF involves growing microorganisms in a water-based nutrient solution which enables proper mixing and oxygen supply and pH maintenance [27]. Bacteria and yeasts demonstrate excellent suitability for SmF since these microorganisms thrive under natural conditions in aqueous solutions [28]. The fermentation process requires detailed control of temperature along with pH values and dissolution oxygen levels as well as agitation rates to achieve optimal enzyme releases from microbial agents [29]. The liquid broth obtained from SmF proves highly beneficial for continuous production of xylanase because it enables straightforward filtration or centrifugation to collect the extracellular enzyme [30]. Production costs rise additionally and the enzyme purification becomes more difficult because SmF requires significant water usage along with increased energy needs which result in enzyme concentration dilution.

The cultivation of microorganisms on solid materials occurs through solid-state fermentation (SSF) while no or minimal free water exists during this process [31]. Many fungi and actinomycetes naturally grow by utilizing decaying plant matter and other organic substrates and this method effectively imitates such natural environments. SSF demonstrates natural affinity toward filamentous fungi including Aspergillus, Trichoderma and Penicillium because these microorganisms effectively penetrate solid substrates while producing significant extracellular xylanase amounts

[32]. SSF demonstrates its main benefit through the effective use of low-cost agro-industrial residue materials including wheat bran, corn cobs, rice straw along with sugarcane bagasse and sawdust as production substrates [33]. These lignocellulosic materials provide both physical growth support to microorganisms and natural xylan content that acts as an enzyme production inducer. Volumetric productivity within SSF systems remains high while the process requires little moisture and uses minimal water and energy [34]. Enzyme stability and activity levels tend to be superior when produced through SSF than in SmF because microbes develop stress adaptation mechanisms when operating under low-water medium. The successful implementation of SSF fermentation requires specialized bioreactor technology including tray fermenters and packed-bed reactors together with rotary drum fermenters because it faces obstacles related to heat and mass transfer process and control parameter instability and scale-up issues.

Semi-solid or semi-submerged fermentation introduces a blend of SmF and SSF operational features through its intermediate medium condition. The solid substrate wetting level in this approach exceeds SSF but stops short of creating a complete liquid environment. The manufacturing method seeks to achieve the high enzyme production capabilities of SSF with the operational benefits of SmF. The combination of better nutrient and oxygen availability in semi-solid systems along with solid substrate advantages distinguishes them from SSF operations. The production of xylanases and other enzymes through this method becomes increasingly popular whenever biological organisms prove sensitive to moisture and when economic production substrates are deployed.

Enzyme quality alongside high yields can only be attained when process parameters are optimized, despite which fermentation method is employed [35]. During SmF-based enzyme production, the specific quantities of agitation speed along with aeration rate together with inoculum size, and nutrient formulation profoundly influence the enzyme output [36]. When using SSF method the critical production variables include substrate type and particle size, together with initial moisture content and temperature along with incubation time and microbial inoculum density [37]. Fermentation technique selection happens through a combination of microbial strain selection along with substrate physicochemical properties and target xylanase application purposes. A thermophilic bacterial strain under SmF would provide optimal results when aiming to produce bio-pulping xylanase, but food-grade xylanase from fungi often needs fungal production through SSF with agro-waste substrates [38].

Modern bioprocess engineering research has targeted the advancement of more efficient large-scale fermentation systems. The research on SmF includes the implementation of fed-batch and continuous fermentation methods which help lengthen production times while increasing yield levels. Aeration distribution and temperature control improvements in SSF bioreactor systems attempt to solve large-scale application limitations in moisture distribution and aeration and temperature control [39]. Monitoring tools that incorporate online sensors for measuring pH and dissolved oxygen and CO_2 levels enable real-time fermentation environment optimization in both SmF and SSF systems. Researchers are using computational modeling and artificial intelligence to develop programs that predict optimal fermentation operational conditions by simulating fermentation kinetics [40].

The production of xylanase depends heavily on fermentation methods since these methods control the yield and activity levels together with economic efficiency levels [41]. Submerged fermentation operations provide precise control and scalable production while solid-state fermentation performs under renewable biomass-based green chemistry with reduced environmental effects. Semi-solid systems enable operators to combine the advantages of SmF and SSF by delivering an intermediate operation method. Technical selection for xylanase fermentation requires precise adaptation to biological needs and industrial parameters because technological improvements will result in improved efficiency and sustainability along with multiple industrial applications [42].

4 Substrates for Xylanase Production

Xylanase production heavily relies on substrates because they function as microbial growth matrices and carbon fuels particularly under solid-state fermentation conditions. Xylanase production yield together with its activity levels and production costs depend heavily on which substrates scientists choose and how they process them [43]. Xylanase production strategies now focus heavily on agro-industrial residues and lignocellulosic biomass because these substances help lower production costs while sustaining high activity levels and environmental sustainability. Many parts of the world have access to inexpensive agro-industrial residues which include wheat bran, corn cobs, rice husks, sugarcane bagasse, sawdust, groundnut shells, palm press fiber, and soybean hulls as hemicellulose-rich substrates that contain large amounts of xylan [44]. The material demonstrates economic viability and sustainability along with pollution reduction that stems from its conversion into value-added products out of renewable waste. The research community has extensively investigated wheat bran as a production substrate since it contains high amounts of hemicellulose and features a loose structural design enabling microbial colonization together with enzyme secretion in both submerged and solid-state processes [45]. The combination of corn cobs with rice straw works well as fermentation substrates because their xylan components allow oxygen penetration and microbial penetration [46]. Raw plant biomass with a recalcitrant combination of lignin along with cellulose and hemicellulose maintains low xylanase production yields because these compounds create difficulties for direct microbial processing. The production of xylanase requires pretreatment methods that enhance substrate accessibility in order to improve the inducible xylanase output [47].

The main approaches for pretreatment include physical methods combined with chemical ones together with physico-chemical operations and biological approaches [48]. Through physical methods including milling, grinding and steam explosion plant cell walls become more accessible by breaking down larger plant cell sizes and altering structural formations [49]. The xylan substance becomes more accessible to microbial enzymes when chemicals like acids and alkaline solutions (NaOH and $Ca(OH)_2$) and organosolv compounds break the lignin barrier and dissolve hemicellulose components. The exposure of xylan increases best with alkaline pretreatment methods when the cellulose material remains intact. The xylan breakdown through acid hydrolysis produces monomeric sugars that could either serve as strong xylanase inducers or fail to do so due to differences in microbial systems. The physico-chemical treatment methods AFEX and hydrothermal treatment merge physical substrate breaking with chemical modification to develop a more accessible substrate. Organisms capable of degrading lignin can perform biological pretreatment while remaining environmentally friendly although their process times extend beyond other methods. The selection process of pretreatment depends on biomass type along with microbial system choice and downstream processing requirements while striving to balance operational efficiency with expenditure and environmental sustainability factors.

The chemical composition of the substrate including its contents of xylan and lignin as well as inhibitors establishes a vital role in how microorganisms metabolize this material and activate enzymes. Natural compounds present in some substrates act as both expression suppressors for enzymes together with growth obstruction factors for bacteria. The selection of appropriate substrates requires additional attention to substrate concentrations because high loading leads to substrate inhibition while affecting aeration and increasing the viscosity of submerged systems. The activity levels of microorganisms in solid-state fermentation decrease when moisture content falls below optimal levels and excessive moisture leads to bacterial contamination and decreased porosity. Enzyme production ends up reaching its highest point when the right substrate-to-moisture ratio becomes established.

Xylanase production becomes more efficient through the combination of multiple agro-residues as substrate mixtures which balance nutrients and generate beneficial interactions. The combination of wheat bran with rice straw and sugarcane bagasse improves xylanase fermentation performance due to better carbon-to-nitrogen ratio balance as well as improved aeration and wider spectrum of xylan structures which enhance broad-spectrum xylanase activity. Scientists conduct studies using soluble birchwood xylan as a pure xylanase inducer in submerged fermentation but avoid its industrial application due to high production costs. When used in purified form xylan allows researchers to precisely control carbon source inputs yet fails to show substrates that exist within natural environments.

Xylanase gene expression reacts strongly to the way substrates promote induction. Microbes tend to activate xylanase gene transcription when exposed to xylan-rich substrates instead of repressing xylanase production through carbon catabolite repression from carbon sources such as glucose or cellulose. Process optimization requires the combination of xylan or xylo-oligosaccharides inducers with complex biomass to achieve higher enzyme titer levels. Scientists modify genetic systems to create strains which avoid carbon source repression to make xylanase production possible by default or through regulated promoters.

Substrate sustainability demonstrates substantial importance in all its aspects. Bio-digesting agricultural waste products eliminates landfill doubts and emission problems and creates rural economic benefits by joining the bio-product value chain. Industrial enzyme manufacturing benefits from local agro-waste materials used in xylanase production because this approach supports the circular economy concepts and minimizes the carbon dioxide emissions from manufacturing operations. The tropical regions use oil palm empty fruit bunches along with cassava peels whereas temperate areas have more availability of barley husks and beet pulp as substrates.

The selection along with optimization of substrates directly influences every aspect of economical xylanase production including operational efficiency and environmental sustainability. Proper pretreatment of agro-industrial wastes transforms them into environmentally sustainable bases which strongly enhance enzyme production from microbial activity. All parameters regarding substrate relationship with the microbe together with its chemical structure and physical texture along with how it affects fermentation kinetics must receive thorough assessment. Through extensive knowledge of substrates combined with contemporary blending and pretreatment techniques the production of xylanase can be optimized while lowering manufacturing expenses and sustaining green industrial bioprocessing practices.

Study	Microorganism	Xylanase yield (Units/ml or U/g)	Optimal temperature (°C)	Optimal pH	Carbon source	Fermentation time (hrs)	Effect of inoculum size	Other key findings
Gupta et al.	*Aspergillus niger*	55 U/ml	30	5.5	Xylan	48	3% inoculum size	Higher yields achieved with xylan as carbon source
Joo et al.	*Trichoderma reesei*	120 U/g	35	5.0	Xylose	72	5% inoculum size	High xylanase activity with xylose; optimal enzyme at pH 5.0
Singh et al.	*Penicillium oxalicum*	200 U/ml	37	6.0	Xylan	72	4% inoculum size	High thermostability at 50 °C; excellent for industrial processes
Kaur et al.	*Aspergillus terreus*	150 U/g	30	5.5	Xylan	48	3% inoculum size	Xylanase production enhanced with pre-treated lignocellulosic biomass
Souza et al.	*Trichoderma reesei*	250 U/ml	30	5.0	Xylan	96	4% inoculum size	High enzyme production in submerged fermentation
Vasanthan et al.	*Bacillus subtilis*	95 U/ml	37	7.0	Glucose	48	2% inoculum size	Best results with glucose as carbon source for faster enzyme production
Kaur et al.	*Aspergillus niger*	120 U/g	35	5.0	Xylan	72	5% inoculum size	Strong xylanase activity with xylan as the sole carbon source
Nisa et al.	*Trichoderma reesei*	250 U/ml	40	5.5	Xylan	72	3% inoculum size	Metal ions like Ca^{2+} enhance xylanase stability

(continued)

(continued)

Study	Microorganism	Xylanase yield (Units/ml or U/g)	Optimal temperature (°C)	Optimal pH	Carbon source	Fermentation time (hrs)	Effect of inoculum size	Other key findings
Chand et al.	*Aspergillus niger*	100 U/ml	30	5.5	Xylan	48	4% inoculum size	Enhanced xylanase activity in pulp and paper industry applications
Mondal et al.	*Trichoderma viride*	110 U/ml	35	6.0	Xylan	72	3% inoculum size	Successful in biopolishing and textile processing
Dong et al.	*Aspergillus niger*	200 U/g	40	5.0	Xylan	72	5% inoculum size	Recombinant xylanase showed improved thermal stability
Kaur et al.	*Penicillium* sp.	80 U/ml	30	6.0	Xylan	48	3% inoculum size	Higher enzyme activity under controlled agitation rates

5 Effect and Optimization Production Parameters for Xylanse Production from Microorganism

The production parameters require optimization to maximize both the yield output and activity level of xylanase during microbial fermentation operation. A detailed adjustment of physical and chemical parameters that directly affect microbial growth and enzyme synthesis and secretion process takes place. The optimization process in industrial biotechnology targeting maximum enzyme activity also involves minimizing production expenses while shortening process durations and securing predictable results whenever other industrial scales are applied. Temperature together with pH values represent the main operational variables that affect fermentation conditions along with the incubation period and aeration controls and agitation intensity and inoculation volume and nutritional elements including carbon and nitrogen components while making use of inducer substrates. These parameters generate complex interactions which need specific optimal values due to the combination of particular microbial strains with fermentation approaches (submerged or solid-state) and the various substrate types used in the process. Enzyme biosynthesis depends heavily on the temperature factor because it influences how fast the microorganism grows while also determining how stable the produced enzyme remains. Mesophilic bacteria along with fungi function optimally at temperatures within the range of 30–40 °C yet *Thermomonospora* and *Bacillus stearothermophilus* bacteria perform excellent xylanase activity at temperatures between 50 and 60 °C. Higher operating temperatures enhance the solubility of xylan substances while simultaneously lowering contamination risks through better substrate penetrative capabilities. Extremely high or low temperatures will denature proteins or prevent microbial processes from taking place which requires strict temperature regulation.

Microbial physiology together with enzyme expression undergoes change because pH regulates the ionization of nutrients and determines the microbial cell surface electrostatic properties. Fungal strains excel at pH ranges between 5.0 and 7.0 when producing maximum yields however bacterial strains work best within a pH scope of 7.5 to 9.0. The stability of enzymes depends on pH levels therefore operators must keep fermentation pH within ideal limits. The production process for industrial bioreactors uses pH sensors together with automated solution dosing systems to achieve pH stability through acid and base solutions. The duration of microbial incubation controls which stage of growth the microorganisms will produce maximum xylanase synthesis. Xylanase enzyme formation reaches its maximum levels during the late exponential growth phase and early stages of stationary growth. The fermentation timeline spans between one day and five days according to both the used microorganism and the established production parameters. The process of prolonged incubation results in depletion of important nutrients while accumulating harmful substances along with degrading the enzyme therefore requires regular activity testing to determine the optimal collection moment.

Drastic factors in submerged fermentation relate to oxygen transfer through agitation and aeration because these techniques affect nutrient mixing and heat distribution and oxygen transfer. The enzyme production efficiency of Bacillus or Aspergillus species depends on adequate oxygen availability during their growth processes. The synthesis of xylanase by fungi becomes limited when insufficient oxygen reaches the culture while too much agitation causes physical harm to the fungal cell structures. The right level of agitation maintains uniform media circulation through the tank space while keeping cell integrity intact. Aeration remains challenging during solid-state fermentation since the process lacks free-flowing liquid which typically requires specialized bioreactors or manual turning for air circulation and temperature control management.

The production of enzymes depends on two variables: the inoculum size as well as its age. The production of lower enzyme activity and prolonged microbial growth happens when inoculum is set too low while too high a dose leads to premature nutrient exhaustion and unbalanced microbial spread. Research reveals that bacterial xylanase production reaches its best results when using a 5–10% v/v inoculum volume of 24-h-old culture but specific conditions differ between organisms. The fermentation output depends heavily on the inoculum's current phase as it develops through lag then log and finally reaches stationary phase because actively growing cultures produce superior results.

Xylanase production heavily depends on the specific carbon and nitrogen elements used in combination with their corresponding amounts in the growth medium composition. The use of xylan-containing substrates including birchwood xylan and wheat bran and corn cob powder maintains xylanase expression and functions as carbon source but simple sugars like glucose and sucrose create enzyme suppression through carbon catabolite repression. The problem can be addressed by using multiple carbon sources together with controlled-release organic matter. Organic nitrogen components like yeast extract, peptone, soybean meal together with inorganic nitrogen compounds including ammonium sulfate, ammonium nitrate have been utilized as nutrient sources. Medium selection depends on the chosen microorganism alongside the cost criteria of the medium composition. Higher yields occur with organic nitrogen sources because they deliver extra growth factors yet they possess a higher price tag. The optimal microbial metabolism and enzyme secretion process needs the presence of both micronutrients phosphates along with magnesium, calcium, iron and trace elements and calcium.

Researchers use traditional and advanced optimization approaches to validate the best possible combination and levels of these parameters. Asian Design Technique stands as one of the traditional optimization methods that manipulates one variable independently while maintaining other components at steady values. This basic optimization method remains straightforward but fails to detect variable interactions and requires large amounts of workforce and time. Modern optimization methods use the combination of Response Surface Methodology (RSM) with Taguchi Design of Experiments (DOE) as statistical techniques. RSM depends on regression models to conduct multiple-factor analysis along with interaction examination while predicting the best operation points through streamlined experimentation methods. Taguchi methods provide techniques to identify superior operating conditions when operating under environments with variable characteristics by focusing on robustness and noise tolerance. Artificial intelligence tools including genetic algorithms and neural networks together with machine learning models are currently applied to forecast enzyme production volumes and enhance difficult non-linear systems that rely on extensive datasets.

The transfer of laboratory-optimized results into industrial production demands process scale-up implementation although it causes new problems including inefficient mixing and irregular thermal zones and damaging mechanical forces. The validation of optimized procedures requires pilot-scale trials because they enable testing under genuine production conditions. Real-time monitoring systems connected to process control software through examples like pH along with temperature and CO_2 evolution and dissolved oxygen measurements enable constant optimization of consistency in multiple production batches.

Parameter	Effect on xylanase production	References
Temperature	Xylanase production is highly temperature-dependent. Higher temperatures (30–40 °C) generally enhance enzyme activity and yield, but extreme temperatures can cause denaturation of the enzyme	Gupta et al.
pH	The optimal pH for xylanase production varies by microorganism, but it generally ranges from 4.5 to 6.5. pH affects the enzyme's stability and its ability to hydrolyze xylan	Souza et al., Singh et al.
Carbon source	Carbon sources like xylan, glucose, and xylose promote high xylanase production. Xylan is typically the best carbon source for inducing xylanase production, but excessive glucose may repress enzyme production	Kaur et al., Joo et al.
Nitrogen source	Organic nitrogen sources (e.g., yeast extract, peptone) typically promote higher enzyme production compared to inorganic sources (e.g., ammonium nitrate). Nitrogen limitation can enhance enzyme synthesis	Gupta et al., Singh et al.
Agitation speed	Agitation speed can influence oxygen transfer rate and mixing, which are crucial for microbial growth and enzyme production. Too high agitation speed may shear cells and decrease yield	Mondal et al.
Inoculum size	Higher inoculum sizes can lead to faster growth rates and enzyme production, but too large an inoculum can lead to overcrowding and resource depletion	Vasanthan et al.
Fermentation time	Longer fermentation times can lead to increased enzyme production, although prolonged times may lead to enzyme degradation. Typically, optimal enzyme production is observed between 48 and 72 h of fermentation	Chand et al.
Salinity (salt concentration)	Moderate salt concentrations can improve microbial growth and enzyme production by stabilizing the enzyme structure. Excessive salt may inhibit microbial growth	Kaur et al.

(continued)

(continued)

Parameter	Effect on xylanase production	References
Oxygen availability	Xylanase production in submerged fermentation is highly oxygen-dependent. Insufficient oxygen can result in lower yields, while excess oxygen can sometimes lead to oxidative stress	Dong et al.
Metal ions (co-factors)	Certain metal ions such as Ca^{2+}, Mn^{2+}, and Fe^{2+} can act as co-factors, enhancing xylanase stability and activity. However, excessive metal ions can lead to enzyme inhibition	Nisa et al., Kaur et al.
Pre-treatment of substrate	Pre-treatment of the lignocellulosic substrate (e.g., by steam explosion, alkali, or acid treatment) enhances xylan availability, thus increasing xylanase production and its efficiency in biofuel production	Kaur et al., Chand et al.
Cationic surfactants	The addition of cationic surfactants, such as cetyltrimethylammonium bromide (CTAB), can improve enzyme activity by stabilizing the enzyme structure and enhancing its solubility	Mondal et al.
Substrate concentration	Higher substrate concentrations can increase enzyme production, but high concentrations can also lead to substrate inhibition. The optimal concentration depends on the microbial strain	Joo et al.
Inhibitors (e.g., glucose)	High concentrations of glucose can inhibit xylanase production by catabolite repression, where the microorganism preferentially metabolizes glucose over xylan	Kaur et al., Souza et al.

6 Genetic and Metabolic Engineering for Enhanced Xylanase Production

Genetic and metabolic engineering techniques have become powerful methods to boost xylanase production by enabling scientists to handle DNA sequences and metabolic routes of microorganisms in directed ways. Precision through genetic engineering allows scientists to raise enzyme yields by using rational approaches for modifying enzyme properties while also broadening substrate ranges. The genetic engineering process of xylanase production involves gene expression optimization along with bottleneck removal and defines better secretion methods coupled with regulatory network engineering until gene transfer into sturdy host systems occurs. Metabolic engineering takes a systematic approach for redesigning enzyme biosynthesis pathways that improves production flux and reduces competitive production pathways. High-throughput screening coupled with systems biology when used together with these approaches provides exceptional control over microbial physiology to support production of xylanase for multiple industrial purposes.

Scientists typically achieve xylanase overproduction and cloning through genetic manipulation of high-yield organisms which get placed into *E. coli, B. subtilis, P. pastoris,* and *A. niger* host systems. Scientists have used Thermotoga maritima's thermophilic bacteria genes and Bacillus's alkaliphilic strain genes to clone them into mesophilic hosts for making thermostable or alkali-tolerant xylanases. The application of recombinant systems under T7, AOX1 and gpdA promoters leads to enhanced transcription levels by using powerful dedicated promoters. Xylanase gene codon optimization facilitates effective translation in foreign host organisms particularly within bacterial-fungal genetic combinations. The use of recombinant production enables easier purification through the incorporation of affinity tags like His-tags that helps industries that produce enzyme formulations.

The field of genetic engineering targets an increase in secretion efficiency of xylanase in Bacillus and filamentous fungi because these microorganisms demonstrate strong extracellular enzyme production capabilities. Xylanase gene secretion is enhanced by using signal peptide engineering techniques to add either native or external signal sequences to xylanase genes thus achieving better membrane transport to extracellular targets. The use of the signal peptides amyE from *Bacillus subtilis* and glucoamylase from *Aspergillus niger* successfully enhances xylanase secretion levels. Modifying cell wall or membrane proteins helps increase enzyme yield by reducing secretion bottlenecks while simultaneously lowering protease activity that exists in the medium.

The control of regulatory gene systems provides major pathways to tackle metabolic barriers in production processes. Researchers can modify transcription factors and repressors involved in carbon catabolite repression through knockouts or mutations of CRE1 in *Trichoderma reesei* and CcpA in *Bacillus subtilis* to keep the xylanase expression active despite glucose and other easy sugars being present. The modification or permanent activation of activators such as XlnR in Aspergillus results in continuous high-level enzyme production of xylanase. Xylanase enzyme production can be sustained through modifications to xylan metabolism pathways such as increasing pentose sugar transporter activity and enhancing xylo-oligosaccharide degradation inside the cell.

Microbial production of xylanase through enhanced metabolic designs relies on genome-scale models and flux balance analysis together with transcriptomics and proteomics and metabolomics results for designing pathway optimization. Changing the carbon flow from central metabolism (such as glycolysis) to xylanase production pathways raises the limited availability of precursor molecules (including amino acids alongside ATP and NADPH) essential for protein formation. Operation of redox balance systems alongside optimized ATP production becomes essential when dealing with high-cell-density fermentations because metabolic constraints naturally reduce enzyme outputs. The invention of CRISPR/Cas9 by synthetic biology has made it possible to modify microbial genomes efficiently and accurately thus supporting the integration of complex biosynthetic pathways and simultaneous genomic editing. Engineers integrate various components including inducible promoters with feedback systems and biosensors to achieve dynamic control mechanisms for xylanase expression through internal or external indications.

The development of custom microbial consortia through co-culture experiments allows different engineered strains to perform specialized functionality where one strain produces xylanase and another deals with lignin break down and metabolizes leftover products. The separation of biological tasks between different strains decreases metabolic stress within single microorganisms while making enzyme production more flexible and expandable. The production of enzymes through synthetic microbial communities becomes achievable by using quorum sensing and signaling modules which enable the coordination of enzyme production based on population density or environmental conditions.

The combination between adaptive laboratory evolution and genetic engineering allows industries to develop strong xylanase-producing strains that tolerate stress conditions which include elevated temperatures and extreme pH levels and heavy substrate concentrations. The process of genome shuffling uses protoplast fusions between mutant strains to generate improved xylanase-producing strains that display higher production rates together with increased resistance to inhibitory conditions.

Engineered microbial strains need to fulfill extensive safety examinations and characterizations both prior to their commercial implementation and when used for food and pharmaceutical applications. The regulation of genetically modified microorganisms (GMMs) through biosafety measures necessitates them to stay free of detrimental metabolites and allergens. The implementation of food-grade selection markers and chromosomal integration and self-cloning with genes from related species assists businesses to fulfill regulatory standards.

Genetic engineering facilitates the modification of xylanase properties through which industry can fulfill their technical demands. The optimization of xylanase properties through site-directed mutagenesis and domain swapping and directed evolution provides increases in thermal stability and pH tolerance and specific activity as well as proteolytic resistance. Scientists modify surface amino acids by adding charged residues to improve both enzyme solubility and thermostability and the introduction of disulfide bonds strengthens structural rigidity. Xylanase fusion with carbohydrate-binding modules (CBMs) gives the enzyme better substrate binding properties for improving its efficiency on insoluble xylan and lignocellulosic biomass.

7 Fermentation Strategies for Xylanase Production

Xylanase production relies on fermentation strategies to establish operational processes which control microbial growth and biosynthesis and effectiveness of product recovery. Selection of the correct fermentation process remains vital because it enables both maximum enzyme output and low-cost manufacturing alongside confident product reliability during industrial applications. The production of xylanase utilizes two prevailing fermentation techniques which include Submerged Fermentation (SmF) and Solid-State Fermentation (SSF) but demand specific microbial strain selection and considered the required enzyme characteristics alongside final processing requirements. Xylanase production has been enhanced through hybrid process combinations and fed-batch culture and continuous fermenter systems and updated bioreactor technologies which now enable commercial-scale production at affordable prices. Improvements in bioprocess monitoring together with automation techniques and statistical optimization procedures have enabled manufacturers to achieve greater control of fermentation variables which results in enhanced productivity levels.

The dominant industrial method in biotechnology utilizes Submerged Fermentation (SmF) because it works with standardized bioreactor systems easily and operators can regulate process parameters effectively. Culture of microorganisms in liquid nutrient environments under submerged fermentation allows dissolution or suspension of the substrate which typically involves xylan-containing or wheat bran and oat spelt xylan and agricultural hydrolysates. Thêm *Bacillus subtilis and Bacillus licheniformis and Streptomyces* species grow under aerobic conditions inside stirred-tank bioreactors that use continuous agitation and aeration to distribute oxygen and promote mixing homogeneity. The cultivation of fungi such as *Aspergillus niger* and *Trichoderma reesei* happens successfully under submerged conditions but operators need to manage foaming and viscosity as well as shear sensitivity. SmF provides several benefits because it can be efficiently sterilized while offering real-time monitoring of dissolved oxygen and pH and temperature levels and benefiting from simple downstream processing through its liquid broth nature. The process technique demands substantial amounts of water while needing complex equipment and results in less concentrated enzyme outputs particularly throughout the initial stages of developing the manufacturing process.

Solid-State Fermentation (SSF) operates by processing fungi in solid substrates which contain low water amounts to follow the natural ecological conditions of many filamentous fungi. SSF has become a preferred method for manufacturing xylanase from agro-industrial residues which include wheat bran and corn cob alongside sugarcane bagasse and rice straw and sawdust. Microbial growth takes place on these substrates which simultaneously act as carbon material and inducer substances and microbial growth base. *Aspergillus awamori* together with *Penicillium chrysogenum and Trichoderma viride* demonstrate suitability for SSF operations because they exhibit excellent performance growing on solid media while secreting substantial quantities of extracellular enzymes. The SSF process produces stronger concentrations of products while minimizing both pollutant risks and wastewaters as well as decreasing the need for capital investment. SSF encounters problems because different regions of the system can become unevenly moistened while heat can build up and oxygen supply faces difficulties and the technology faces challenges when trying to scale up production rates. Current improvements in SSF technology using tray fermenters and rotating drum reactors along with packed-bed bioreactors increase scalability but monitoring and controlling the process presents limited precision when compared to SmF methods.

Fed-batch fermentation represents an enhanced version of SmF by adding fresh nutrients stepwise to the culture without removing any broth. Using this approach prevents substrate inhibition while preserving cell viability which consequently extends the productive fermentation period. Fed-batch systems enable effective but controlled xylan or xylose inducer addition which supports enzyme production through avoidance of catabolite repression emanating from glucose or other sugars. The implementation of feeding strategies provides customizable control mechanisms that base microbial metabolic processes on oxygen consumption levels and pH values and gas analysis data. Enzyme production through continuous fermentation systems provides operators with operational stability and shorter production interruptions and reliably consistent end products while remaining less widespread in the industry. The method includes maintaining an active flow of fresh medium while extracting culture broth continuously which supports prolonged manufacturing operations. Long-term applications of xylanase production in the industry remain limited due to the difficulties of maintaining stable strains together with blocking microbiologic contamination.

The design parameters of bioreactors together with their structural configuration significantly affect strategy performance in fermentations. The standard reactor for SmF production remains stirred-tank systems because they combine flexibility with effective oxygen maintenance. The optimization process of parameters including impeller type as well as agitation speed and aeration rate ensures both high oxygen transfer rates and limited damage from shear forces. Two different bioreactor setup options can be utilized for SSF including rotating drums paired with static tray devices to manage substrate compaction. Advanced bioreactor systems such as fluidized bed and packed-bed offer enhanced mass and heat transfer ability and currently get explored for both research-scale and industrial SSF applications. Immobilized cell or enzyme usage through fixed or fluidized bed bioreactors increases both the stability and reusability as well as streamlines downstream separation processes.

A new strategy combines co-culture fermentation that entails raising multiple microbial strains together to benefit from their beneficial cooperative interactions. Each microbial strain has its niche due to distinct specialization since one breaks down complex xylan fibers first while the second specializes in producing xylanases. The combined microbial operation benefits intricate biomass feedstocks by eliminating the requirement for pretreatment procedures. Monitoring and designing the strain balance together with medium optimization and competitive inhibition prevention demands thorough management.

The implementation of modern fermentation systems includes process analytical technologies (PAT) as well as automated control systems. The system utilizes sensors to measure pH and temperature together with oxygen and carbon dioxide levels and biomass density so fermentation conditions stay optimal. Bioreactors of higher complexity feature monitoring systems that use sensor information to adjust the flow of nutrients and machine operations and gaseous inputs. 이러한 시스템能够实现数据驱动的酶合成参数优化仪器下表示 Response Surface Methodology (RSM) 而个体和 Artificial Neural Networks (ANN) 或者 Design of Experiments (DoE) 等统计表元件. A combination of omics-based information with bioprocess design features allows scientists to optimize fermentation conditions because they can track gene expression using transcriptomics and monitor metabolic fluxes using metabolomics for maximizing enzyme production.

Fermentation performance is directly affected by the methods used to treat substrates together with the selection of specific substrates. The accessibility of xylan in lignocellulosic biomass needs physical and chemical pretreatment or enzymatic procedures to become possible. Processors can apply steam explosion or dilute acid hydrolysis or alkali treatment to break down complex structures thereby inducing better enzyme production. A combination of trace nutrients with surfactant addition helps enhance microbial enzyme secretion function alongside medium oxygenation efficiency. Adding Tween-80 alongside polyethylene glycol (PEG) improves fungal growth alongside the activity of xylanase in SSF systems.

The production of high-titer xylanase regarding downstream processing becomes simpler because of low-viscosity fermentations. Different methods of concentration including ultrafiltration and precipitation and chromatography and lyophilization can be used for enzyme extracts depending on their intended applications and purity needs. The production of xylanase through SmF systems simplifies the filtration and separation tasks although SSF systems generate denser enzymes but demand elaborate processes to remove enzymes from solid substrates. Submerged fermentation provides exact control and expandability whereas solid-state fermentation generates economic benefits and large quantities of enzymes when using agricultural waste materials. The combination of hybrid fermentation approaches with monitoring technology advancements and statistical methods and omics applications has boosted xylanase production while making it more efficient along with being sustainably managed. The advancement of bioprocessing technologies will lead to automated fermentation approaches with increased energy efficiency that supports processing different feedstocks which will create sustainable industrial enzyme production systems.

8 Recovery and Purification of Xylanase

The process of xylanase recovery and purification stands as an essential requirement for achieving high purity along with activity levels in industrial research applications of enzymes. The recovery process alongside purification techniques establish economic viability as well as performance quality attributes of xylanase leading to its end activity and operational stability. Microbial cells need to be harvested in recovering xylanase enzymes through extraction from fermentation broths based on different fermentation procedures. During submerged fermentation microorganisms release the produced xylanase into solution through which the enzyme can easily separate from cells by using filtration or centrifugation methods. The microorganism separation through sonication or homogenization and enzymatic lysis is less necessary for xylanase production due to its extracellular secretion habits. The process of solid-state fermentation (SSF) needs extra extraction measures because xylanase becomes bound to solid substrates during this technique. The process of enzyme recovery in solid matrix conditions includes solid matrix dissolution through buffer solution washing combined with high-speed mixing techniques. The extraction process for enzymes depends on substrate nature and moisture content along with solid matrix properties affecting effectiveness and requires solid material filtering at the end. Concentration of the enzyme stands as the second production hurdlefollowing extraction procedures. Ultrafiltration utilizes selective protein size retention to extract enzyme solution concentration when it removes small molecules and salts along with other impurities in addition to concentrating xylanase. The xylanase concentration process includes ammonium sulfate precipitation as a cost-effective scalable method which allows protein precipitation through high salt solubility. Xylanase precipitation occurs after adding ammonium sulfate because the method selectively traps the enzyme while allowing other solution contents to remain in the supernatant. The reduction of enzyme solution volume through this step makes the solution ready for subsequent purification steps. Several purification methods follow concentration to obtain highly pure xylanase. Affinity chromatography together with ion-exchange chromatography and size-exclusion chromatography provide the most common approaches to separate xylanase from other proteins and contaminants through distinct operational characteristics. Affinity chromatography represents the most efficient way to purify xylanase through the enzyme's specific binding to xylan ligands or their related compounds. The xylanase binds to a xylan-containing resin before its elution happens by adjusting pH levels and ionic strength or including competitive ligands. The process produces xylanase of high purity in one step which eliminates additional processing steps. The separation process in ion-exchange chromatography functions through net charge discrimination of proteins. The xylanase retention or elution occurs during chromatographic runs through charged resins following adjustments to the mobile phase pH or ionic strength levels. The process proves effective for isolating xylanase when it exhibits specific charge-to-mass characteristics compared to other proteins found in the solution. The enzyme purification process incorporates ion-exchange chromatography with other methods like affinity chromatography along with ammonium sulfate precipitation. Proteins separate through Size-exclusion chromatography (SEC) also termed gel-filtration chromatography according to their size dimensions as large proteins exit first followed by prolonged retention of smaller proteins. The method functions effectively to extract low-molecular-weight contaminants and validate the cleaned state of purified xylanase during intermediate chromatographic separation procedures. Enzyme preparations reach their final stage of purity during SEC since this method eliminates all small contaminants which could disrupt enzyme activity. Hydrophobic interaction chromatography (HIC) serves as another purification method for xylanase by means of its hydrophobic regions which bond to hydrophobic resins during high salt solutions. A decreasing salt gradient allows xylanase to become easily recoverable from the column. Xylanase purification through this method becomes effective because its hydrophobic characteristics enable separation from other protein components. SDS-PAGE constitutes the primary method for evaluating enzyme purity during purification through electrophoresis especially when employing sodium dodecyl sulfate–polyacrylamide gel electrophoresis (SDS-PAGE). SDS-PAGE analysis determines xylanase molecular weight and verifies the single protein band specific to the target enzyme within the preparation. SDS-PAGE functions as a main diagnostic tool which cannot operate independently as a purification method yet remains vital for evaluating the success achieved through other purification techniques. Xylanase activity needs special attention during purification procedures to ensure its operational capacity throughout each procedure step. Activity assays include the step of enzyme–substrate incubation using xylan-derived substrates followed by measuring reaction end products which signal xylan degradation ability of the enzyme. The measurement of enzyme activity per protein concentration defines Specific Activity which serves as an important assessment tool to evaluate purification method success. A high level of specific activity demonstrates that most proteins within the sample represent functional xylanase whereas minimal specific activity displays poor purification of the enzyme. Storage and application formulation of xylanase requires additional work because the ingredient needs to maintain stability throughout these processes. Enzyme stability faces influences from different conditions including environmental temperature range and pH situations as well as salt content level and stabilizer substances or cofactors. The enzyme stability tests involve storing the protein under different conditions to track its activity levels throughout the testing period. Manufacturers use various formulation techniques to maintain stability of xylanase throughout its storage period and application processes. Long-term stability of enzyme solutions becomes possible through addition of stabilizing agents including glycerol, trehalose and mannitol. Xylanase becomes available as a dry powder after lyophilization for maintaining stability at ambient storage temperatures. Freeze-drying xylanase through lyophilization process enables the enzyme to survive for extended periods which enhances market applicability. These applications utilize xylanase immobilization processes by attaching the enzyme to agarose beads or ceramic particles which make the enzyme stable enough for continuous process use. The expanded recovery operation encounters specific difficulties when it deals with processing large amounts of culture broth efficiently. Industrial and pilot macro-scale purifications of enzyme products demand adjustments from small laboratory methods to meet requirements for column dimensions together with operational expenses and flow amount specifications. Protocol adjustments of chromatography and ultrafiltration methods will become necessary to process rising volumes of enzyme extracts while maintaining product purity alongside enzymatic activity during volume expansion. Comprehensible xylanase production at commercial levels depends on effective recovery and purification methods especially when aimed at industrial uses that include feed supplement additives and detergent products and biofuels and food processing applications. Xylanase production economics heavily depends on the selection of purification methods because advanced equipment and complex procedures accompany the production of highly pure enzymes. Widespread application requires both active purity and economical availability of huge quantities of xylanase. The laboratory technique for obtaining purified xylanase includes serial operations starting from enzyme extraction followed by concentration through various chromatographic means to optimize both recovery rate and product purity. The advancement of bioprocess engineering combined with purification technology development enables the scalable economical production of commercial xylanase which allows numerous applications. Moving ahead the commercial scale of xylanase production requires better recovery and purification strategies to reduce operational costs while enhancing general availability across different biotechnological industries.

9 Applications of Xylanase

Xylanase serves as a biocatalyst for xylan conversion into xylose because of its ability to degrade hemicellulose within lignocellulosic materials. This makes the enzyme applicable across multiple industrial sectors. The enzyme enhances industrial operations by breaking down tough polysaccharides into basic sugars that enable applications in biofuel production, food processing, animal nutrition, paper production and manufacturing and environmental projects. The biofuel industry makes use of xylanase to facilitate decomposition of lignocellulosic biomass since bioethanol production encounters significant challenges because of this material. The pretreatment effectiveness of xylanase depends on its ability to convert xylan present in agricultural residues and wood and grass materials into fermentable sugars. The breaking down of fermentable sugars through microorganisms produces ethanol from renewable biological material. The implementation of xylanase during the pretreatment of lignocellulosic biomass creates a major advancement toward biofuel production through increased efficiency in bioethanol manufacturing for substituting fossil fuels. The food industry utilizes xylanase enzymes for enhancing baked goods quality by producing superior texture in various items such as bread and cakes as well as cookies. Xylanase enzymes split xylan components in flour so water bonds release making dough processing easier and yielding superior quality end products. The enzymatic properties of Xylanase aid in decreasing dough viscosity and improve bubble formation within the end product. The application of xylanase also functions as dough conditionern to extend the product lifespan of bakery items through its ability to slow down staling processes. The enzyme functions at two levels in fruit juice processing where it helps both juice clarification and turbidity reduction. Broken-down xylan in fruit pulp decreases juice viscosity which lets manufacturers extract liquid more efficiently and obtain clearer end products. Xylanase functions within the animal feed sector because it enhances the breakdown of feed products. Animals specifically monogastric types including poultry and pigs lack sufficient ability to decompose complex plant compounds like xylan yet the integration of xylanase to animal feed promotes xylan decomposition which generates additional fermentable sugars improving feed use efficiency. The improved feed utilization through xylanase use results in better animal growth together with more efficient feed conversion ratio and a more sustainable animal agriculture due to decreased need for extra feed supplements. Through its use in animal feed xylanase reduces environmental impact since it prevents polysaccharides from passing undigested through animals to end up as manure pollutants of nitrogen and phosphorus. Xylanase conducts improvements in both bleaching operations and paper quality throughout the paper and pulp industry. Xylanase enables pulp manufacturers to decrease their reliance on strong bleaching chemicals because it decomposes pulp-bound xylan material during the processing stage. Through enzymatic processing the product achieves greater paper quality standards that both increase its brightness level and decrease the need for chlorine bleaching chemicals which reduces environmental harm. Xylanase shows great effect in kraft pulping by breaking down hemicellulose residuals and making chemical treatment more efficient. The enzyme operates as an active agent in textile manufacturing by performing biopolishing of fabrics. The application of xylanase on cotton textiles and other plant materials leads to pectin and hemicellulosic components' breakdown which produces fabrics that feel softer and maintain color quality better while becoming smoother. The enzyme provides textile manufacturers with a tool to eliminate strong chemicals thereby supporting sustainable environmentally friendly production methods. The environmental sector utilizes xylanase to decompose agricultural wastes and various organic waste products which have high xylan content. The biodegradation and composting process receives an enhanced decomposition rate through the application of xylanase which speeds up how plant residues break down thus resulting in improved compost quality. Industrial application of xylanase includes bioremediation by degrading hard-to-remove organic pollutants in both water and soil through its enzymatic activity. Research has evaluated the potential usage of this enzyme for marine oil spill clean-up operations that involve integrating xylan-rich plant materials into bioremediation approaches for contaminated sites. Pharmaceutical experts have started to explore the utilization of xylanase for developing drug delivery systems. Engineers utilize the enzyme to build pharmaceutically activated carrier systems that enhance medicine delivery systems particularly for treating gastrointestinal conditions. The selectivity of Xylanase allows it to break down gastrointestinal tract mucosal polysaccharides to create sustained therapeutic drug delivery through specific control mechanisms that improve drug performance and minimize unwanted effects. Makers of cosmetic products utilize xylanase as a vital ingredient in exfoliants because its structure-breaking effects on outer skin polysaccharides enable the removal of dead skin cells. The enzymatic action of xylanase offers a gentle option for skin care products instead of harsh abrasive exfoliants thus making it appropriate for sensitive skin applications. Xylanase offers biotechnological capabilities for extracting liquid chemicals from sustainable raw materials. Biological technology uses xylanase to create bio-based chemical products like organic acids and sugars as well as alcohols which result from breaking down hemicellulose. The production of biodegradable plastics for the plastics industry together with environmentally friendly solvents and lubricants and surfactants are applications of these chemicals. Xylanase-based production of bio-based chemicals remains important for industries because it enables fossil fuel reduction and environmental pollution minimization. Further advancements in research about xylanase lead to an ongoing expansion of biotechnological and industrial applications because of this versatile enzyme. Engineering progress in enzyme development has optimized xylanase properties for stability together with improved activity levels and selective action which has created new industrial potential. Better properties from recombinant xylanases have boosted production efficiencies as well as created additional commercial potential. The genetic modification of microorganisms to produce xylanase with enhanced stability and modified substrate orientation leads to better biomass conversion for biofuel production and improved food and feed sector processing. Genetic engineering makes xylanase more suitable for operation in demanding industrial settings which includes high temperature and extreme pH conditions thus improving its functional capabilities across different applications. Xylanase-based technologies in development will create major disruptive changes across different industrial sectors like agriculture and food processing and energy production and environmental sustainability practices. Xylanase stands set to become a fundamental tool which promotes environmentally conscious and efficient operation in numerous applications including renewable energy production along with waste management and bioremediation because industries actively strive for greener sustainable practices. The enzyme provides value because of its ability to break down complex organic materials through its versatile performance while generating minimal environmental impact thus supporting sustainable technologies and circular bioeconomy development.

Topic	Author(s)	Year	Title	Journal/ Conference	Key findings/ Conclusions
Xylanase production by fungi	Gupta R et al.	2003	Xylanases: Microbial Sources, Production, Purification, and Applications	*Food Technology and Biotechnology*	Discussed various fungal sources of xylanase production, particularly focusing on *Aspergillus niger* and *Trichoderma reesei*
Genetic engineering of xylanase	Joo H-S et al.	2015	Cloning, Expression, and Characterization of Xylanase Genes from *Trichoderma reesei*	*Applied Biochemistry and Biotechnology*	Examined the cloning and expression of xylanase genes from *T. reesei*, highlighting the use of recombinant systems for improved enzyme production
Thermostable xylanases for industrial use	Singh S et al.	2018	Thermostable Xylanases: A Review on Production and Applications	*Bioresource Technology*	Reviewed various thermostable xylanases, including their potential applications in high-temperature industrial processes such as pulp and paper

(continued)

(continued)

Topic	Author(s)	Year	Title	Journal/ Conference	Key findings/ Conclusions
Xylanase in bioethanol production	Kaur A et al.	2019	Role of Xylanase in the Bioconversion of Lignocellulosic Biomass to Bioethanol	*Renewable and Sustainable Energy Reviews*	Investigated the role of xylanase in breaking down lignocellulosic materials, improving the yield of bioethanol production
Recombinant xylanases in industrial applications	Souza R et al.	2014	Recombinant Xylanases for Industrial Applications: The Road Ahead	*Biotechnology Advances*	Discussed the advancement of recombinant DNA technology in producing xylanases with enhanced stability and activity for various industrial uses
Xylanase in animal feed	Vasanthan T et al.	2017	Use of Xylanases in Animal Feed: A Review	*Animal Feed Science and Technology*	Reviewed the impact of xylanase in improving feed efficiency and digestibility in livestock, particularly for monogastric animals
Xylanase in textile industry	Mondal M et al.	2016	Application of Xylanase in Textile Processing: Current Trends and Future Prospects	*Textile Research Journal*	Analyzed the use of xylanase in biopolishing and fabric softening, highlighting its environmental benefits over traditional chemical methods
Xylanase in pulp and paper industry	Chand A et al.	2012	Application of Xylanase in the Pulp and Paper Industry	*Journal of Biotechnology*	Described the use of xylanase in the bleaching process in the paper industry, reducing the need for chemical treatments and improving paper quality

(continued)

(continued)

Topic	Author(s)	Year	Title	Journal/ Conference	Key findings/ Conclusions
Xylanase in bioremediation	Nisa S et al.	2021	Xylanase as a Bioremediation Tool: Degradation of Lignocellulosic Waste	*Environmental Science and Pollution Research*	Explored the potential of xylanase in bioremediation, particularly for the degradation of xylan-rich agricultural residues
Xylanase in food industry	Kaur A et al.	2020	Application of Xylanase in Food Processing	*Food Research International*	Focused on the use of xylanase to improve the texture of baked goods and the clarification of fruit juices
Xylanase engineering for enhanced activity	Dong S et al.	2017	Protein Engineering of Xylanase: Strategies for Improved Activity and Stability	*Journal of Molecular Catalysis B: Enzymatic*	Reviewed the protein engineering approaches used to enhance the activity and stability of xylanases, including mutagenesis and directed evolution

10 Xylanase in Biotechnology and Genetic Engineering

Xylanase generates intense interest throughout biotechnology research and genetic engineering because it functions in flexible industrial applications and allows engineers to develop multiple research-oriented benefits. The biotech applications of xylanase depend mostly on its capacity to convert xylan components found in plant cell walls into marketable sugar products. The various applications of xylanase have established its essential position across multiple sectors including biofuel manufacturing and food manufacturing and waste management systems. Xylanase biotechnology advances exist because scientists thoroughly understand the composition of its molecules together with its enzymatic activities and genetic elements controlling production. Scientists initially used genetic engineering to enhance microorganism xylanase output by studying fermentation environments and altering the genetics of their host organisms specifically fungi and bacteria. Research scientists increased production yields and industrial enzyme stability by modifying microorganism genetics through modifications of Aspergillus sp., Trichoderma sp., and Bacillus sp. The selection process for optimal microbial hosts constitutes a key factor within genetic engineering for xylanase production. Fungal microorganisms

Aspergillus niger and *Trichoderma reesei* have become standard for xylanase production because they operate efficient secretion mechanisms to release extracellular enzymes outside culture dishes. Scientists choose *Bacillus subtilis* bacterial cells for their fast reproduction speed together with their manageable genetic characteristics. The advancement of recombinant DNA technology allows researchers to clone and express xylanase genes within heterologous organisms thus creating extra microorganisms that can produce this enzyme. Research in xylanase biotechnology reached a major milestone through enzyme gene cloning that allowed expression in high-production biological systems featuring *Escherichia coli* and yeast strains which resulted in recombinant enzyme generation. The industrial demand for enzymatic biofuel production benefits from large-scale production capabilities of recombinant xylanase. The production of recombinant xylanase starts with gene insertion into plasmid vectors after vector plasmids are introduced to the right microbial host. The controlled expression of the enzyme becomes possible through genetic manipulation which enhances the process of gene transcription and protein translation. Researches can utilize site-directed mutagenesis or protein engineering methods to modify the cloned xylanase gene so it obtains better properties including improved stability together with enhanced specificity along with better tolerance toward pH and temperature conditions. Protein engineering methods significantly contribute to developing xylanases which perform with enhanced capabilities and special applications. Changing the active site amino acid sequence enables researchers to develop enzymes with specific substrate profiles which improves their aptness for industrial usage. Modification through protein engineering allows xylanases to specialize their activity toward hardwood or softwood xylan which produces improved performance in paper and pulp bleaching and biofuel production. Researchers use directed evolution methods to develop stable versions of xylanases which aftermarket they can apply in industrial processes operating at elevated temperatures such as pulp and paper mills or textile processing operations because enzymes frequently encounter harsh environments. Enzyme makers focus on generating blended xylanase products to achieve synergistic effects as one main biotechnology application of xylanase technology. Xylanases typically function with additional enzymes including cellulases, pectinases and lignases for optimizing complex lignocellulosic material degradation. Enzyme cocktails show particular value in biofuel production since the total sugar production from plant matter needs multiple enzymes that can break down each part of the plant wall structure. The genetic modification of microorganisms to generate enzyme combinations that contain xylanase combined with other related enzymes increases biomass breakdown efficiency leading to better biofuel production specifically ethanol yields. The combination of biofuel production enhancement with waste agricultural waste conversion into useful products constitutes one of the key outcomes of this approach. Genetic engineering techniques developed new production methods for xylanase in alternative microorganisms thus creating additional possibilities for enzyme manufacturing. Scientists use *Pichia pastoris* methylotrophic yeast as a production host for recombinant xylanase through its successful expression system. The microbial expression system *P. pastoris* provides multiple benefits that make it a superior option than conventional microbial systems due to its high-density cell culture capabilities and simplified genetic alterations and excellent recombinant protein secretion capacity. Synthetic biology used in recent development produced engineered microbial consortia that enhance production efficiency of xylanase alongside other enzymes. Genetically modified microorganisms integrated into engineered consortia cooperate to enhance whole enzyme manufacturing by effectively using resources and breaking down challenging substrates. The industrial use of genetic engineering methods led to substantial improvements in both the economical aspects and production scale of xylanase manufacturing. Before recent times xylanase producers utilized submerged fermentation processes to create their products using big bioreactors and large amounts of growth media. The application of genetically modified microorganisms made the production more efficient through elevated enzyme outputs and decreased manufacturing expenses. Genetic modifications optimize xylanase stability in industrial environments which consist of high temperature or acidic or alkaline pH or chemical-containing operational settings. Through genetic modification of the enzyme scientists can achieve durability during long processing runs to support industries where time-sensitive applications exist such as paper and pulp production and textiles manufacture as well as animal feed manufacturing. The genetic techniques have enhanced xylanase secretion levels in extracellular production systems because industries require efficient enzyme recovery for large-scale production. The optimization of microbial host cells for better xylanase production efficiency depends on metabolic engineering methods used to optimize carbon and nitrogen utilization pathways. Genetic manipulation of host strains allows researchers to increase production rates of xylanase without adding more nutrients because it makes precursor metabolites for enzyme synthesis more accessible. The expression of genes related to secondary metabolism can be normalized through genetic engineering to avoid competing with the production of preferred xylanase enzymes. The conversion of plant biomass receives particular promise from xylanase implementation in plant biotechnology research. Genetic modification of plants with Xylanase enables the breakdown of biomass making it more suitable as a biofuel production resource. Scientists created genetically modified crops through their genetic modification with xylanase genes to achieve enhanced xylanase production levels that boost plant cell wall breakdown efficiency in biomass conversion processes. Biofuel production efficiency can be increased through genetically modified crops which eliminate the requirement for external enzyme supplements which consequently makes biofuels more financially viable than fossil fuels. Scientists are currently investigating the agricultural utilization of xylanase as a means to boost agricultural production output. These biochemical changes allow presumable increased access to plant tissue nutrients as xylanase breaks down plant cell wall xylan compounds thus promoting agricultural productivity and enhancing crop yields. Xylan degradation by this enzyme demonstrates usefulness as an agent to enhance soil structure since it breaks down organic matter and creates better nutrient availability for crops. Union between xylanase application with plant growth-promoting elements holds potential to transform sustainable agriculture through yield improvement together with minimized chemical fertilizer requirements. The development of bioreactors for commercial enzyme manufacturing demonstrates potential as a new field for xylanase-based genetic engineering. The production of xylanase through engineered microorganisms occurs in solid-state fermentation (SSF) systems which use agricultural waste as the substrate. Using waste materials as substrates in this approach provides an environmentally beneficial method that lowers costs associated with liquid growth media. Genetically modified organisms used in SSF-based production enhance both enzyme output rates and financial efficiency which enables large-scale commercial xylanase manufacturing. New progress in synthetic biology and its creation of artificial biological systems have helped scientists better understand xylanase production regulation systems. The controlled modification of regulatory processes in host microbes enables researchers to enhance xylanase gene expression rates which results in lower production costs and decreased synthesis times. Xylanase production and optimization together with its applications have dramatically progressed through the combination of genetic engineering and biotechnological techniques. Through strain efficiency improvements and enzyme stability advancements along with delivery methodology development genetic engineering enables broad industrial application of xylanase. Future progress in xylanase biotechnology research will create novel applications of this enzyme that extend across renewable energy production and biofuels technology alongside agriculture and environmental sustainability advances and other unspecified industrial sectors.

References

1. M. Phuyal, U. Budhathoki, D. Bista, S. Shakya, R. Shrestha, A.K. Shrestha, Xylanase-producing microbes and their real-world application. Int. J. Chem. Eng. **2023**, 3593035 (2023)
2. R. Chaudhary, T. Kuthiala, G. Singh, S. Rarotra, A. Kaur, S.K. Arya et al., Current status of xylanase for biofuel production: a review on classification and characterization. Biomass Convers. Biorefinery **13**, 8773–8791 (2023)
3. F. Güler, F. Özçelik, Screening of xylanase producing Bacillus species and optimization of xylanase process parameters in submerged fermentation. Biocatal. Agric. Biotechnol. **51**, 102801 (2023)
4. T.B. Machado, L.C.S. Corrêa Junior, M.V.C.d.V. de Mattos, G.V. Gautério, S.J. Kalil, Sequential alkaline and ultrasound pretreatments of oat hulls improve xylanase production by Aureobasidium pullulans in submerged cultivation. Waste Biomass Valor. **12**, 5991–6004 (2021)
5. G.S. Barbieri, H.B.S. Bento, F. de Oliveira, F.P. Picheli, L.M. Dias, F. Masarin et al., Xylanase production by talaromyces amestolkiae valuing agroindustrial byproducts. Biotech **11**, 15 (2022)
6. G. Anand, M. Leibman-Markus, D. Elkabetz, M. Bar, Method for the production and purification of plant immuno-active xylanase from trichoderma. Int. J. Mol. Sci. **22**, 4214 (2021)
7. F.S. Ire, I.J. Chima, V. Ezebuiro, Enhanced xylanase production from UV-mutated Aspergillus niger grown on corn cob and sawdust. Biocatal. Agric. Biotechnol. **31**, 101869 (2021)

8. Y. Zhang, Z. Jiang, Y. Li, Z. Feng, X. Zhang, R. Zhou et al., The combined cultivation of feruloyl esterase-producing strains with CMCase and xylanase-producing strains increases the release of ferulic acid. Microorganisms **10**, 1889 (2022)
9. J. Kaushal, M. Khatri, G. Singh, S.K. Arya, A multifaceted enzyme conspicuous in fruit juice clarification: an elaborate review on xylanase. Int. J. Biol. Macromol. **193**, 1350–1361 (2021)
10. M.G. Morán-Aguilar, M. Calderón-Santoyo, J.M. Domínguez, M.G. Aguilar-Uscanga, Optimization of cellulase and xylanase production by Aspergillus niger CECT 2700 using brewery spent grain based on Taguchi design. Biomass Convers. Biorefinery **13**, 7983–7991 (2023)
11. A. Sarangi, H. Thatoi, Xylanase as a promising biocatalyst: a review on its production, purification and biotechnological applications. Proc Natl Acad Sci India Sect B: Biol Sci (2024)
12. H. Fernandes de Souza, L. Aguiar Borges, V. Dédalo Di Próspero Gonçalves, J. Vitor dos Santos, M. Sousa Bessa, M. Fronja Carosia, et al., Recent advances in the application of xylanases in the food industry and production by actinobacteria: a review. Food Res. Int. **162**, 112103 (2022)
13. G. Malhotra, S.S. Chapadgaonkar, Production of thermo-alkali-stable xylanase from Bacillus licheniformis isolated from natural hot water geyser. J. Microsc. Ultrastructure **9**, 131–135 (2021)
14. Y.F. Abdelaliem, M.H. Mahmoud, N.A. Elkassem, S.M. Mansour, M.F. Ramadan, A.A.A. Mohdaly, Utilization of agro-industrial biowastes to produce xylanase using Aspergillus niger AUMC 14230: optimization of production parameters. Rendiconti Lincei. Scienze Fisiche e Naturali **34**, 941–951 (2023)
15. A. Pasalari, A. Homaei, Isolation and molecular identification of xylanase-producing bacteria from Ulva flexuosa of the Persian Gulf. Processes **10**, 1834 (2022)
16. R. Siwach, S. Sharma, A.A. Khan, A. Kumar, S. Agrawal, Optimization of xylanase production by Bacillus sp. MCC2212 under solid-state fermentation using response surface methodology. Biocatal. Agric. Biotechnol. **57**, 103085 (2024)
17. K. Min, H. Kim, H.J. Park, S. Lee, Y.J. Jung, J.H. Yoon et al., Improving the catalytic performance of xylanase from Bacillus circulans through structure-based rational design. Biores. Technol. **340**, 125737 (2021)
18. J. Kaushal, A. Raina, G. Singh, M. Khatri, S.K. Arya, N. Karmegam et al., Methodical study implicating the effectiveness of Microbial treatment over xylanase enzymatic treatment for pulp bio-bleaching. Environ. Technol. Innov. **28**, 102731 (2022)
19. J. Zhang, S. Liu, H. Sun, Z. Jiang, Z. Zhou, X. Han et al., Enzyme production potential of Penicillium oxalicum M1816 and its application in ferulic acid production. Foods **10**, 2577 (2021)
20. A. da S. Pereira, C.P.L. Souza, R.C.B. Franson, T.F. Ferreira, P.F.F. Amaral, From agri-food wastes to enzyme production: a systematic review with methodi ordination. Waste Biomass Valor. **15**, 5843–5870 (2024)
21. H. Ramírez-Lagunes, M.G. Aguilar-Uscanga, M.I. Infanzón-Rodríguez, B. Sachman-Ruiz, J. Gómez-Rodríguez, C. Nolasco-Hipólito et al., Optimization of xylanase production from Aspergillus tamarii SCBH2 using response surface methodology. Biomass Convers. Biorefinery **13**, 12213–12223 (2023)
22. F. Salzano, M. Aulitto, G. Fiorentino, D. Cannella, E. Peeters, D. Limauro, A novel endo-1,4-β-xylanase from Alicyclobacillus mali FL18: Biochemical characterization and its synergistic action with β-xylosidase in hemicellulose deconstruction. Int. J. Biol. Macromol. **264**, 130550 (2024)
23. C. Espinoza-Abundis, C. Soltero-Sánchez, E. Romero-Borbón, J. Córdova, Cellulase and xylanase production by a newly isolated Penicillium crustosum strain under solid-state fermentation, using water hyacinth biomass as support, substrate, and inducer. Fermentation **9**, 660 (2023)
24. M.G. Moran-Aguilar, I. Costa-Trigo, M. Calderón-Santoyo, J.M. Domínguez, M.G. Aguilar-Uscanga, Production of cellulases and xylanases in solid-state fermentation by different strains of Aspergillus niger using sugarcane bagasse and brewery spent grain. Biochem. Eng. J. **172**, 108060 (2021)

25. F. Ameen, Purification and characterization of xylanase produced by Aspergillus fumigatus isolated from the Northern Border region of Saudi Arabia. Fermentation **9**, 595 (2023)
26. M.M. Bakry, S.S. Salem, H.M. Atta, M.S. El-Gamal, A. Fouda, Xylanase from thermotolerant Bacillus haynesii strain, synthesis, characterization, optimization using Box-Behnken Design, and biobleaching activity. Biomass Convers. Biorefinery **14**, 9779–9792 (2024)
27. S.A. Belorkar, H. Kausar, Valorization parameters to determine the fermentative applicability of selected fruit peels for xylanase production. Waste Biomass Valor. **14**, 185–196 (2023)
28. D.I. Desai, B.D. Iyer, Optimization of medium composition for cellulase-free xylanase production by solid-state fermentation on corn cob waste by Aspergillus niger DX-23. Biomass Convers. Biorefinery **12**, 1153–1165 (2022)
29. M.S. Poomani, I. Mariappan, K. Muthan, V. Subramanian, A thermotolerant yeast from cow's rumen utilize lignocellulosic biomass from wheat straw for xylanase production and fermentation to ethanol. Biocatal. Agric. Biotechnol. **50**, 102741 (2023)
30. A. Khaswal, S.K. Mishra, N. Chaturvedi, S. Saini, B. Pletschke, R.C. Kuhad, Microbial enzyme production: unlocking the potential of agricultural and food waste through solid-state fermentation. Bioresource Technol. Reports **27**, 101880 (2024)
31. L.P.S. Vandenberghe, A. Pandey, J.C. Carvalho, L.A.J. Letti, A.L. Woiciechowski, S.G. Karp et al., Solid-state fermentation technology and innovation for the production of agricultural and animal feed bioproducts. Syst. Microbiol. Biomanuf. **1**, 142–165 (2021)
32. N. Oiza, J. Moral-Vico, A. Sánchez, E.R. Oviedo, T. Gea, Solid-state fermentation from organic wastes: a new generation of bioproducts. Processes **10**, 2675 (2022)
33. D. Chmelová, B. Legerská, J. Kunstová, M. Ondrejovič, S. Miertuš, The production of laccases by white-rot fungi under solid-state fermentation conditions. World J. Microbiol. Biotechnol. **38**, 21 (2022)
34. J. Wang, Z. Huang, Q. Jiang, H. Roubík, Q. Xu, A. Gharsallaoui et al., Fungal solid-state fermentation of crops and their by-products to obtain protein resources: the next frontier of food industry. Trends Food Sci. Technol. **138**, 628–644 (2023)
35. N.D. Cerda-Cejudo, J.J. Buenrostro-Figueroa, L. Sepúlveda-Torre, C. Torres-León, M.L. Chávez-González, J.A. Ascacio-Valdés et al., Solid-state fermentation for the recovery of phenolic compounds from agro-wastes. Resources **12**, 36 (2023)
36. Y. Cui, S. Peng, D. Deng, M. Yu, Z. Tian, M. Song et al., Solid-state fermentation improves the quality of chrysanthemum waste as an alternative feed ingredient. J. Environ. Manage. **330**, 117060 (2023)
37. A.M. Bonilla Loaiza, R.M. Rodríguez-Jasso, R. Belmares, C.M. López-Badillo, R.G. Araújo, C.N. Aguilar et al., Fungal proteins from Sargassum spp. using solid-state fermentation as a green bioprocess strategy. Molecules **27**, 3887 (2022)
38. G.-A. Martău, P. Unger, R. Schneider, J. Venus, D.C. Vodnar, J.P. López-Gómez, Integration of solid state and submerged fermentations for the valorization of organic municipal solid waste. J. Fungi **7**, 766 (2021)
39. C.R. Piecha, T.C. Alves, M.L.d.O. Zanini, C.d.P.L. Corrêa, F.P.L. Leite, V. Galli et al., Application of the solid-state fermentation process and its variations in PHA production: a review. Arch. Microbiol. **205**, 11 (2022)
40. F. Karimi, D. Mazaheri, M. Saei Moghaddam, A. Mataei Moghaddam, A.L. Sanati, Y. Orooji, Solid-state fermentation as an alternative technology for cost-effective production of bioethanol as useful renewable energy: a review. Biomass Convers. Biorefinery (2021)
41. G. Jin, Y. Zhao, S. Xin, T. Li, Y. Xu, Solid-state fermentation engineering of traditional Chinese fermented food. Foods **13**, 3003 (2024)
42. F. Lu, E.A. Alenyorege, N. Ouyang, A. Zhou, H. Ma, Simulated natural and high temperature solid-state fermentation of soybean meal: a comparative study regarding microorganisms, functional properties and structural characteristics. LWT **159**, 113125 (2022)
43. A. Mattedi, E. Sabbi, B. Farda, R. Djebaili, D. Mitra, C. Ercole et al., Solid-state fermentation: applications and future perspectives for biostimulant and biopesticides production. Microorganisms **11**, 1408 (2023)

44. F. Bibi, N. Ilyas, M. Saeed, S. Shabir, A.A. Shati, M.Y. Alfaifi et al., Innovative production of value-added products using agro-industrial wastes via solid-state fermentation. Environ. Sci. Pollut. Res. **30**, 125197–125213 (2023)
45. J.P. López-Gómez, J. Venus, Potential role of sequential solid-state and submerged-liquid fermentations in a circular bioeconomy. Fermentation **7**, 76 (2021)
46. T.M. Rocha, P.R.F. Marcelino, R.A.M. Da Costa, D. Rubio-Ribeaux, F.G. Barbosa, S.S. da Silva, Agricultural bioinputs obtained by solid-state fermentation: from production in biorefineries to sustainable agriculture. Sustainability **16**, 1076 (2024)
47. L.H.S. de Menezes, P.C. Oliveira, E.L. do Espírito Santo, M.S. Gonçalves, M. Bilal, H.A. Ruiz et al., Solid-state fermentation as a green technology for biomass valorization: optimization techniques for bioprocess—an overview. BioEnergy Res. **17**, 42–58 (2024)
48. O. Martínez-Avila, J. Llimós, S. Ponsá, Integrated solid-state enzymatic hydrolysis and solid-state fermentation for producing sustainable polyhydroxyalkanoates from low-cost agro-industrial residues. Food Bioprod. Process. **126**, 334–344 (2021)
49. R. Chablé-Villacis, E. Olguin-Maciel, T. Toledano-Thompson, L. Alzate-Gaviria, H.A. Ruiz, R. Tapia-Tussell, Enzymatic hydrolysis assisted with ligninocellulolytic enzymes from Trametes hirsuta produced by pineapple leaf waste bioconversion in solid-state fermentation. Biomass Convers. Biorefinery **13**, 9095–9106 (2023)

Chapter 4
Biochemical Properties of Xylanase

Owais Bag, Abdul Sattar Jatoi, Bazil Owais Nasir, and Saim Akram

Abstract Xylanases serve to hydrolyze xylan, an important plant hemicellulose and break it into Xylo oligosaccharides and xylose. The way they work, depending on the optimal pH and temperature ranges which substances they use and their rate of action, mostly determines their use in industries. Depending on their source and the way they are built, xylanases differ remarkably in their weights, where they are charged and how they hold up at high temperatures. Glycosylation and similar processes greatly improve an enzyme's strength and solubility, while also protecting it from being destroyed by other enzymes. Also, hydrolysis intermediates such as xylooligosaccharides can interfere with enzyme activity by xylooligosaccharides competing with the main product, making xylooligosaccharides block the enzyme or by reducing how much of the main product is produced despite the presence of plenty of enzyme. It is necessary to understand these biochemical characteristics to make xylanase applications more effective in bioconversion, biofuel development, food processing and the paper and pulp industry. By making xylanases more effective with enzyme engineering and optimization, researchers make them useful for sustainable applications in biotechnology.

Keywords Biochemical · Xylanase · Food

1 Introduction

The main function of xylanases is to help break down xylan, a major component of plant cell walls considered a hemicellulosic polysaccharide [1]. They cut the xylan chains by breaking the β-1,4-xylosidic bonds, making shorter oligomers and releasing xylose. Many different bacteria, fungi and actinomycetes produce xylanases and their chemical structure depends on which microorganisms produced them and where they lived. Industrial processes can use xylanase better when their properties such as

O. Bag · A. S. Jatoi (✉) · B. O. Nasir · S. Akram
Department of Chemical Engineering, Dawood University of Engineering and Technology, Karachi, Pakistan
e-mail: abdul.sattar@duet.edu.pk

A. Sattar Jatoi (ed.), *Xylanase from Microorganisms*,
https://doi.org/10.1007/978-981-95-0824-2_4

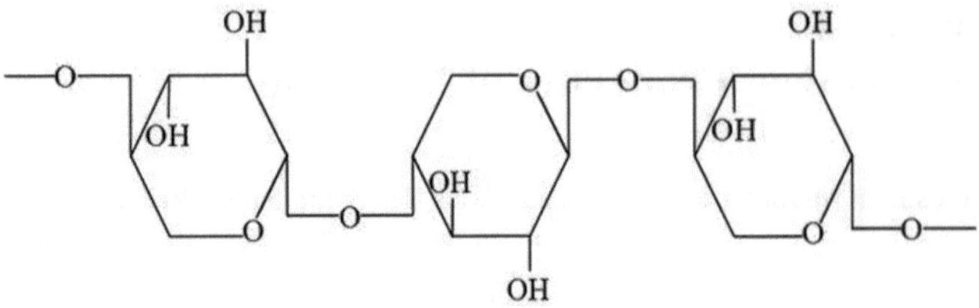

Fig. 1 Hemicellulose (xylan) molecular structure [5]

optimum temperature and pH, important kinetics, substrate specificity and molecular weight, are understood. Under different processing situations, the performance of xylanases is determined by these properties which are essential for using them in biofuel production, food and feed, paper and pulp and textile industries [2].

Apart from their catalytic functions, xylanases can be better stabilized and improved in solubility and defense against proteolytic breakdown by post-translational changes like glycosylation, phosphorylation and disulfide bond formation. Besides, certain intermediates blocking products such as xylobiose and other xylooligosaccharides, have a major impact on enzyme performance and overall yield. Interactions between how xylanases work and the conditions in which they function make it important to do detailed biochemical study of these enzymes [3]. Thanks to new developments in molecular biology and enzyme engineering, candidates for xylanases are being created with better biochemical properties that promote efficient and environmentally friendly bioprocesses.

Three main structural polysaccharides present in plants are cellulose, lignin and hemicellulose. More than 60% of the hemicellulose in cereal crops is xylan, approximately 20–25% in softwood and 30% in hardwood. The addition of various D-xylopyranosyl molecules and short β-1,4 side chains makes plant tissues more complex and alters both their strength and rate of water intake [4].

It can be observed in Fig. 1 that xylan consists of D-xylopyranosyl units glued together with β-1,4 links. Most of the hemicellulose in plant cell walls makes the walls tough and stretchy, thanks to this polymer. Since the structure of xylan is complex, it protects plant cell walls and forms part of biomass.

2 Xylanase: Nature's Molecular Scissors

The xylanase (EC 3.2.1.8) found in glycoside hydrolases splits the β-1,4-xylosidic bonds in xylan, producing shorter xylooligosaccharides and D-xylose. Plant-calcium-dependent enzymes are needed for the breakdown of plants and have major effects on how carbon cycles around the world. The use of xylanases in commercial bioprocessing minimizes the environmental burden of producing pulp and paper, animal feed additives and biofuels. A variety of organisms in nature, including bacteria, fungi, protozoa and marine algae, host xylanases [6]. That these enzymes are found in many living things reflects their great importance for both ecology and evolution. To get the energy stored in plant cell walls, xylanases—useful enzymes produced by microorganisms—decompose plant substances. Xylanases cause glycosidic bonds to be broken by stepping through the bonds twice. Typically, this reaction depends on two catalytic part in the enzyme: a nucleophile and a proton donor. The reaction first creates a link called a glycosyl-enzyme intermediate which is then cleaved to give the product [7]

To categorize xylanases, various GH families are chosen because of their structural and amino acid sequence comparisons. Even though xylanases are located in families 5, 8, 30 and 43, the highest number are found in families 10 and 11. Because each family of enzymes has different structural features and activities, they show the many ways they have evolved. GH10 xylanases tend to be larger and have a $(\beta/\alpha)8$ barrel structure, while GH11 xylanases are little and adopt a β-jelly role. Due to different structures, enzymes have different types of substrates and work differently. Unlike GH11 xylanases, GH10 xylanases can hydrolyze xylan with substituted or unsubstituted groups.

Due to their special abilities in chemistry, xylanases find many applications in industry. Xylanases are applied in pulp and paper production to improve paper quality and cut the use of strong chemicals. They help to digest animal feed better and produce important food ingredients such as Xylo oligo saccharides in the food and feed businesses [8].

Extraction of sugars from lignocellulosic biomass is enhanced in biofuels because of xylanases. Such enzymes help break up xylan, making cellulose easier for cellulases to reach and boost the outcomes of biomass conversion. The different types of xylanases have encouraged scientists to use enzyme engineering to adapt them for industrial purposes. Engineering techniques such as directed evolution and rational design have contributed to enhancing the catalytic strength, pH tolerance and heat resistance of xylanases [9]. Thanks to this work, new xylanase variants have been formed that are more useful in different industrial applications.

Thanks to recent discoveries in these areas, we now understand in detail how xylanases operate. By using both molecular dynamics and detailed crystal structures, it is now much simpler to see how enzymes attach to and work on their substrates. Due to this understanding, xylanases can now be made with new substrate preferences and better enzyme performance. Biologists now have a better idea of how plant–microbe relationships form and how enzymes involved in the breakdown of plant cell walls developed from studying xylanases. Because of the appearance of microbes and plants, bacteria found the need for many forms of xylanases, while plants started making intricate xylan structures [10]. As a result of the ongoing arms race, xylanases exist in many different and impressive forms in nature.

3 A Historical Overview and Evolution of Xylanase Research

3.1 Advancements in Enzyme Biotechnology

There has been great development in studying xylanase over the past century, with most important results emerging in the last decades of the twentieth century. When xylanase-producing fungi were noticed in the 1950s and 1980s, it caused scientists to pursue research into purifying and improving enzymes [11]. In this investigation, Aspergillus niger, Trichoderma reesei and Penicillium chrysogenum were critical species to use. In the 1990s, the introduction of recombinant DNA technology led to major changes in research on xylanase. Thanks to this method, researchers could clone the xylanase genes and then create opportunities for making bigger batches and modifying enzymes. Researchers made enzymes that work better, withstand changes in pH and heat, because they could alter the genes of xylanase. Because of this technology, xylanases could now be adapted to suit specific steps in the paper and pulp, food and feed and biofuel industries [5].

In the last several years, drug design has improved thanks to combining testing large numbers of materials, engineering proteins and using advanced computing. As a result, we can now discover new xylanases from many sources much faster and make it easier to create enzymes with specific properties. The use of metagenomics has now enabled scientists to investigate nonculturable bacteria, introducing new xylanases for use in industry [12].

3.2 Early Discoveries

There has been great development in studying xylanase over the past century, with most important results emerging in the last decades of the twentieth century. When xylanase-producing fungi were noticed in the 1950s and 1980s, it caused scientists to pursue research into purifying and improving enzymes. In this investigation, Aspergillus niger, Trichoderma reesei and Penicillium chrysogenum were critical species to use [13].

In the 1990s, the introduction of recombinant DNA technology led to major changes in research on xylanase. Thanks to this method, researchers could clone the xylanase genes and then create opportunities for making bigger batches and modifying enzymes. Researchers made enzymes that work better, withstand changes in pH and heat, because they could alter the genes of xylanase. Because of this technology, xylanases could now be adapted to suit specific steps in the paper and pulp, food and feed and biofuel industries.

In the last several years, drug design has improved thanks to combining testing large numbers of materials, engineering proteins and using advanced computing. As a result, we can now discover new xylanases from many sources much faster and make it easier to create enzymes with specific properties. The use of metagenomics has now enabled scientists to investigate nonculturable bacteria, introducing new xylanases for use in industry [14].

4 Molecular and Structural Biochemistry of Xylanase

4.1 Amino Acid Composition and Active Site Residues

The mode of conservation in the catalytic domains of xylanases highlights their key part in how the enzyme functions. Basic to the action of xylanases are the amino acid residues Asp and Glu. For the hydrolysis of xylan, it is necessary to have these residues.

- Glutamic acid (Glu) offers protons that take part in the enzyme's functions. The enzyme helps to separate the xsylvania chain by giving the oxygen in the glycosidic bond a proton. As a result of protonation, the glycosidic link between monomers is less stable and so can be broken more easily.
- Asptatic acid (Asp) is essential as both a base and a nucleophilic catalyst. The xanthan structure triggers a water molecule to target the anomeric carbon in the xylose unit. Asp draws electrons during hydrolysis, helping to stabilize the intermediate that arises during that step. A stable structure in the catalyst allows it to lower its energy activation and function efficiently.

The active site of the enzyme requires these catalytic residues to be positioned perfectly for it to work properly [15]. They have the right form for catalytic reactions and are most often found at the tips of β-strands in the enzyme's catalytic section. The location of amino acids in the active pocket of the enzyme impacts how it distinguishes xylan from other substances. They are necessary to guide the substrate into position, connect and detect it.

Other conserved amino acids in addition to the catalytic ones also affect the general structure and function of xylanases. They include remainders that assist in substrate binding, help maintain the structure of the enzyme and might be part of allosteric effects or between proteins [16].

4.2 Three-Dimensional Structural Configuration

The enhanced stability, substrate selection and catalytic powers of xylanases are due to their three-dimensional makeup. There are several GH families of xylanases and of these, GH10 and GH11 are the most thoroughly understood. This structure comprises eight parallel β-strands in the center, surrounded by eight α-helices and the catalytic site is placed at the C-terminal tip of those β-strands, providing a place for the substrate to interact [17]. In contrast, GH11 Xylanases are folded in a β-jelly roll form consisting of two large beta sheets joined in a "right-hand" manner, have a deeply nested pocket for substrates and are more compact with a key loop or thumb, that participates in substrate interactions and product release. The ways these enzymes are built affect how well they function: GH10 enzymes can act on many xylan structures, including highly modified ones, while GH11 xylanases are better equipped for simpler xylans. Furthermore, GH11 xylanases are more stable in extremes and so are simpler to produce. Understanding the differences between structure and function allows for smarter engineering of xylanases, resulting in their improved stability in industrial processes and improved efficiency on various substrates [18].

5 Biochemical Properties of Xylanases: A Detailed Exploration

Plant cell walls contain the complex polysaccharide xylan, which is hydrolyzed by a class of glycoside hydrolases known as xylanases (EC 3.2.1.8). Their diverse biological properties (Table 1) enable them to function and influence their potential applications across several industries [19].

5.1 Substrate Specificity and Mode of Action

Xylanases show different substrate specificity due to the parent organism and distinctive form of the xylan substrate. Even though some xylanases can work on different links, others are highly specific and target the four-carbon-xylose bonds in xylan. This describe important traits that distinguish various endo-xylanases. The groups placed on the xylose molecules need to be permitted or are strongly preferred by them. These mechanisms include endo-acting which removes connections inside the xylan molecule and exo-acting which cuts off the sugars from the ends of the chain. Common endo-xylanases provide Xylo oligosaccharides that differ in terms of length. In their study, Zhang et al. found that xylans decreased cellulase functioning which means xylanases are needed to break down biomass.

Xylanases work and are applied based on how specific they are to their substrates. The enzyme's ability to react only with certain xylans is influenced by the unique shape of its active site [20]. Additions of acetyl groups, arabinose and glucuronic acid in the side chains of xylan elements may alter how the enzyme can access the joints it needs to break. Although some xylanases only breakdown specific xylan groups, others can hydrolyze several different types of xylan.

The process by which xylanases act is another important feature of their biology. The main endo-xylanases within the fungi break the backbone of xylan to form different-length xylo-oligosaccharides. The disorder in the way the substrate

Table 1 Overview of biochemical properties of xylanases

Category	Property	Description
Basic characteristics	Optimal pH	Typically ranges from 4.5 to 6.5; varies with microbial source (acidophilic to neutral enzymes)
	Optimal temperature	Commonly 40–70 °C; thermostable types exceed 80 °C
	Molecular weight	Usually 20–50 kDa; GH11 enzymes are smaller than GH10 enzymes
	Isoelectric point (pI)	Generally ranges from 4.0 to 6.0, depending on the source
	Enzyme families (GH)	Mainly GH10 and GH11; GH10 has a TIM barrel, GH11 a β-jelly roll fold
Catalytic properties	Substrate specificity	GH10: broad specificity, can cleave substituted xylan; GH11: narrow specificity
	Catalytic efficiency (Vmax)	Higher for simpler xylan substrates (especially GH11)
	Michaelis constant (Km)	Varies widely (e.g., 1–10 mg/mL); indicates substrate affinity
	Mode of action	Endo-acting; cleaves internal β-1,4-xylosidic linkages in xylan
Stability factors	Thermal stability	Enhanced by glycosylation and disulfide bonds; GH11 generally more stable
	pH stability	Broad pH range in some enzymes; important for industrial applications
	Protease resistance	Improved by glycosylation and structural compaction
Regulation and modulation	Inhibitors	Competitive: xylose, xylobiose; Non-competitive: metal ions, plant proteins (e.g., TAXI, XIP)
	Activators	Allosteric activators, reducing agents (e.g., DTT), and certain oligosaccharides
Post-translational modifications	Glycosylation	Enhances thermal stability, solubility, and protease resistance
	Phosphorylation	May modulate enzyme activity and localization

(continued)

Table 1 (continued)

Category	Property	Description
	Disulfide bond formation	Stabilizes enzyme structure and activity
	Proteolytic processing	Activation from precursor forms to mature enzyme

hydrolyzes makes the material quickly lose its viscosity and molecular weight. On the contrary, brief oligosaccharides or individual xylose molecules are produced when exo-xylanases act on the ends of the chain. Most of the time, complete xylan breakdown works best when different xylanases and some auxiliary enzymes are used together.

Recent research has explained how xylanase works in detail. An example is Vardakou et al., who clarified the structurally important role in xylanase activity using advanced molecular dynamics simulations and X-ray crystallographic analysis. Research uncovered which amino acid residues in the active area of the enzyme are necessary for binding the substrate and carrying out its tasks [21].

In addition, xylanases matter not only because they only break down xylan. Zhang et al. discovered that if xylan is around, cellulase activity goes down. This strengthens the idea that xylanases help convert lignocellulosic biomass. Many are now interested in developing enzyme cocktails with xylanases to boost and reduce the costs of making biofuel by relying on several enzyme classes for breaking down biomass (Fig. 2).

5.2 *pH Optima and Stability*

You should also learn what the pH optimum of xylanase is, since it tells you when xylanase is most effective. A source organism's worth depends greatly on its location in nature. Optimum pH for several xylanases falls in the range of 4.0 to 9.0, according to Singh and Alokika. When using enzymes in industry, their ability to work well over a range of pH values is very important. Since pH levels in industry can change, recognizing how well an enzyme holds up at different pH ranges is very important.

An xylanase's stability and the correct pH are both necessary for excellent industrial results. The best pH for xylanase activity is typically the same as the environmental pH where the organism was found. Some bacterial xylanases are most active in slightly alkaline conditions (pH 6.0–8.0), but fungal xylanases are typically more active in acidic pH (4.0–6.0). Finding that special pH-active xylanases are made by extremophiles means there is a larger range of where these enzymes can be effective.

Keeping pH consistent is just as important, mainly in industrial cases where the enzyme faces differing pH values. A number of xylanases will act the same over a broad pH range, while others are strongly affected by small changes in pH. Verma and Satyanarayana have found that an xylanase from Geobacillus thermoleovorans

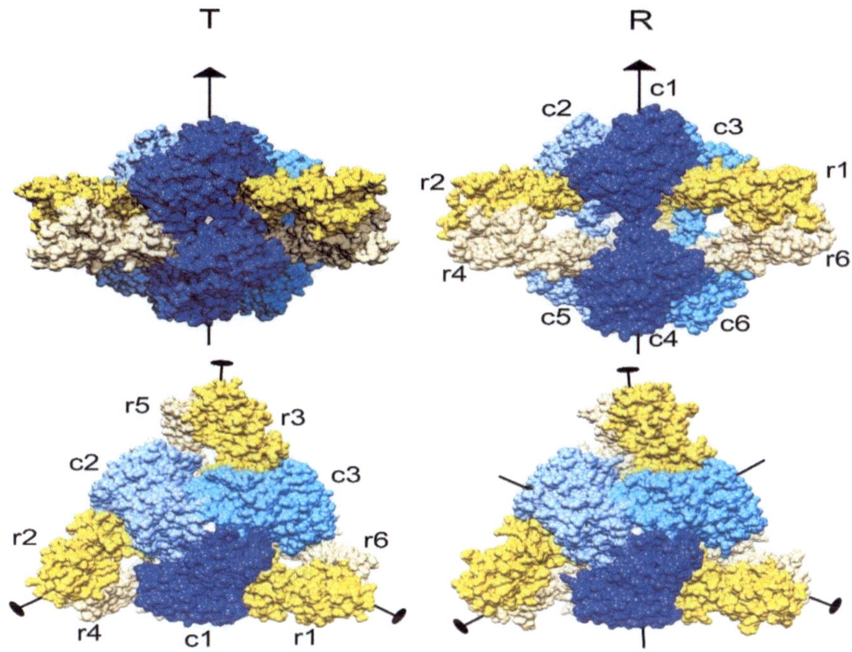

Fig. 2 Quaternary structure of ATCase in the T (left) and R (right) states with the molecular threefold axis vertical (top) and viewed down the molecular threefold axis (bottom). The molecule expands 11 Å along the threefold axis during the allosteric transition. During the T to R transition the regulatory dimers rotate ± 6° around their respective twofold axes and the catalytic trimer rotate ± 7.5° around the threefold axis. The catalytic chains are shown in shades of blue and the regulatory chains are shown in yellow and tan [22, 23]

is active across a pH range of 5.0 to 9.0 which allows it to be used efficiently in many industrial processes [24].

A lot of effort has been made to study the molecular reason behind pH stability of xylanases. Studies have pointed out that disulfide bonds, where charged amino acids are located on the enzyme surface and specific structural segments are all important for maintaining the enzyme's activity across different pH levels. Improving the pH stability of xylanases with protein engineering is now possible thanks to our knowledge of these factors.

Xylanases with improved pH stability have been created with advances in rational design and controlled evolution. Yang et al. added improved alkaline stability to xylanase which is applied in pulp and paper, by carrying out site-directed mutagenesis and error-prone PCR. Similarly, Xu et al. used computational methods to find out which residues have the biggest impact on pH stability and enhanced the pH tolerance of their xylanase through that approach [24].

Although the efficiency of the enzyme counts, so do pH stability and optimization. How stable the enzyme is, if it works well with other process pieces and the total cost of the process all depend on these characteristics. In fruit juice clarification,

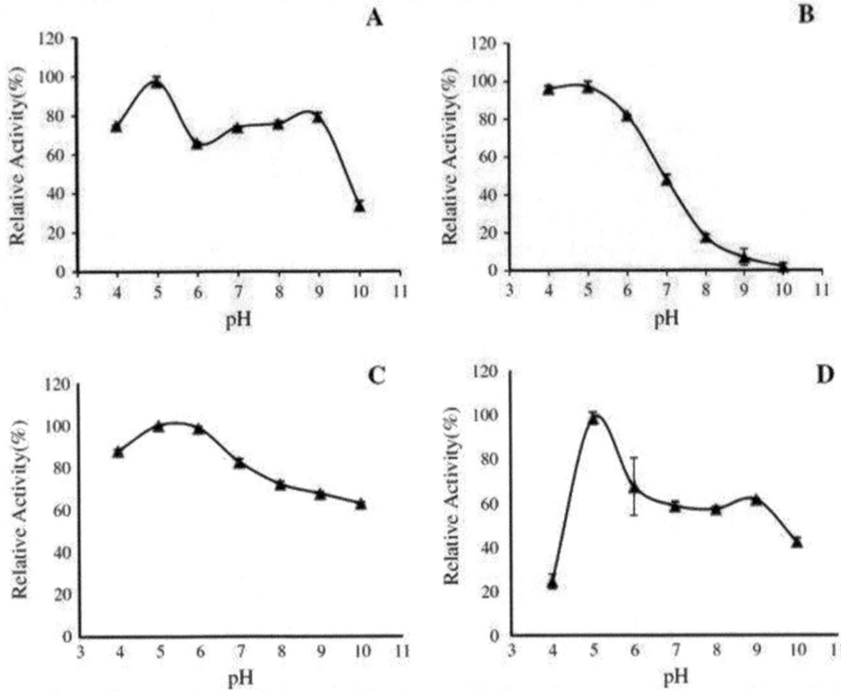

Fig. 3 pH optima and pH stability curve of xylanase of S. obclavatum from SSF and SF. Panel **a** pH optima curve of xylanase from SSF. Panel **b** pH optima curve of xylanase from SF. Panel **c** pH stability curve of xylanase from SSF. Panel **d** pH stability curve of xylanase from SF. Relative enzyme activities (% of maximum) were plotted against pH. All the results were expressed in mean ± SD from n = 3 [24]

food producers use mildly acid-stable xylanases, but for detergent work, they prefer xylanases that are stable in an alkaline environment (Fig. 3).

5.3 *Temperature Optima and Thermostability*

Xylanases are active in a particular temperature range in the same way pH is. They found in 2019 that the typical temperature lurks between 30 and 60 °C. Industry often wants materials that are able to operate normally despite being exposed to high temperatures. Kumar et al. and Wu et al. notice the benefits of using thermostable xylanases in the pulp and paper industry. Developing xylanases that can function well in hot industrial processes is an active field of research.

Industrial applications of xylanases are strongly dependent on their optimal working temperature and their stability at those temperatures. The temperature that supports maximal xylanase activity can be different for each source, while mesophilic xylanases usually work best around 30–50 °C and thermophilic versions can work best at temperatures exceeding 80 °C. The different temperatures that xylanases can work at make it possible to choose the right one for different manufacturing tasks, including low-temperature processes in the food sector and high-temperature uses in biofuels. A key trait in commercial xylanases is being able to keep their shape and function when subjected to high temperatures. Among the benefits of thermostatable xylanases are improved reaction speeds, a reduced risk of food spoilage and better dissolving of the structure of the substrate. Thermostable xylanases for bio-bleaching are particularly appreciated among the pulp and paper companies according to Wu et al. and Kumar et al. Lots of research has been done into how xylanaseists become thermostable chemically. An increase in salt bridges, hydrogen bonds, hydrophobic interactions, disulfide bonds and binding sites for metals help a protein stand up to heat. Structure comparisons between mesophilic and thermophilic xylanases have revealed the types of amino acids and changes in folding that make it possible for them to function under high temperatures. New developments in protein engineering have led to scientist making more resistant xylanases. Rational design allowed Liu et al. to increase the number of salt bridges in a mesophilic xylanase, making its new version much more thermostable. In much the same way, Zheng and colleagues revealed that xylanase can be applied in severe industrial contexts thanks to directed evolution which led to the creation of a xylanase that operates at high temperatures up to 100 °C [24]. Not only can they operate in heat, but thermostable xylanases are also significant. Unlike many other enzymes that destabilize at high pH, in organic solvents or after proteolysis, these enzymes usually have increased stability. This extra stability means that the shelf life is longer, you can use lower enzyme amounts and the process economics will improve in industrial use. Thermostable xylanases have given rise to additional opportunities in biomass conversion processes. Xylanases that function at high temperatures can improve the pretreatment part of lignocellulosic ethanol production and simplify the overall process. Further research is ongoing to develop new techniques to increase the temperature stability of xylanases. Among the listed solutions are investigating heat-tolerant microbes, finding efficient ways to immobilize enzymes and using expected mutations suggested by blind software (Fig. 4).

5.4 *Molecular Weight and Structure*

The type of enzymes and the substances they act on are chosen by the structure and the amino groups of a protein. The significance lies in the fact that this allows us to study the structure and duties of xylanases to update their features for protein engineering.

Using structural information and amino acid sequence, xylanases are grouped into different GH families. Xylanases from the GH10 and GH11 families are more commonly found than other families. Xylanases from the GH10 group have molecular weights between 30 and 50 kDa and a characteristic (β/α) eightfold structure,

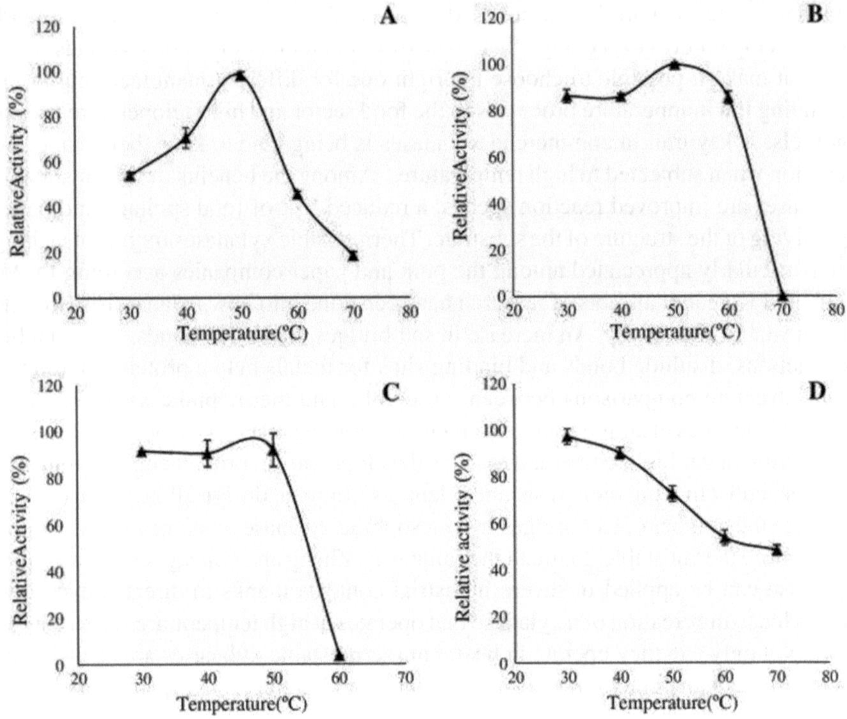

Fig. 4 Temperature optima and thermal stability curve of xylanase ofS. obclavatumfrom SSF and SF. Panel **a** Temperature optima of xylanase from SSF. Panel **b** Temperature optima of xylanase from SF. Panel **c** Thermal stability of the xylanase from SSF. Panel **d** Thermal stability of the xylanase from SF. Relative enzyme activities (% of maximum) were plotted against temperature. All the results were expressed in mean ± SD from n = 3[24]

making them normally larger. At the same time, GH11 xylanases are organized in a smaller β-jelly roll shape, generally weighing 20–30 kDa. Xylanases need conserved glutamate residues in their catalytic region to carry out hydrolysis. CBMs found on xylanases make it easier for them to join with the xylans in the substrate. The organization and features of the structural components can decide how stable, selective and active an enzyme becomes. Understanding how the structure affects the function of xylanases allows us to customize them for particular industrial uses. Techniques called protein engineering may be applied to change the efficiency of using a catalyst, its sensitivity to pH or how resistant the protein is to heat. These changes may produce xylanases that function better and adapt well to many sectors, including biofuel, food and feed and paper and pulp (Fig. 5).

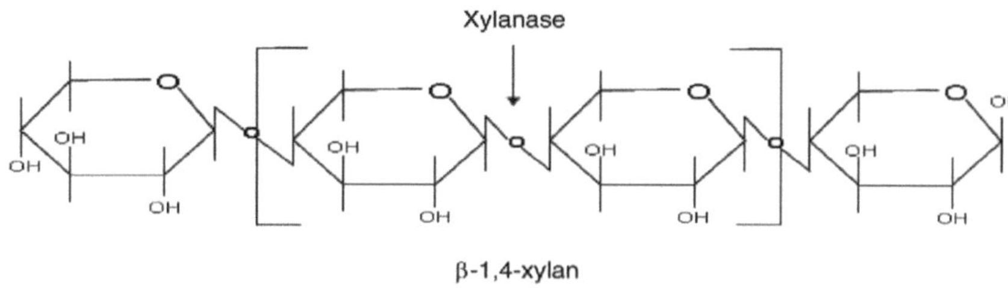

Fig. 5 Structure of xylan, xylopyranosyl residues linked through b-1,4 linkages; point of xylanase action is shown [25]

5.5 *Metal Ion Requirements and Effects*

In some cases, xylanases function and are more stable when calcium or magnesium is present. These ions can enhance both the shape and function of the enzyme. From time to time, metal ions in the mixture can decrease how well xylanase works. In some cases, xylanases use metal ions to improve the way they catalyze reactions. As an illustration, calcium ion binding improves the catalytic ability in several xylanases by preserving the structure of their active site. In a number of xylanase forms, involvement by magnesium ions in binding the substrate or releasing the product is possible. The addition of metal ions does not often have a beneficial effect on xylanase. It is possible for higher amounts of specific metal ions to sometimes lessen the function of xylanase. For instance, by linking to some important amino acids or producing alterations that block the enzyme, heavy links between low temperature, errors and low enzyme activity. There are many ways in which metal ions relate with xylanases, each depending on how the enzyme is created. Certain xylanases can carry out their activities independently of metal ions, whereas others only function with the presence of specific metals. Improving xylanase activity in industrial processes and designing ways to enhance enzyme stability in different environments require a clear picture of the role of metals.

5.6 *Inhibitors and Activators*

In both natural settings and industry, enzyme performance is regulated by chemicals called xylanase inhibitors or activators that have strong effects on enzymes. Xylanase inhibitors are broken into two categories: competitive which compete with the substrate for active site binding and non-competitive which grab onto other sites and affect the action of the enzyme, for example by certain metal ions. On the other hand, many plants produce protein-based xylanase inhibitors such as XIP and TAXI which are part of their defenses against microbial organisms, so it is difficult for industrial biomass processing to deal with this resistance. Alternatively, xylanase activators make enzymes more effective by allosteric regulation and chemical modification of essential amino acids and also by altering substrate-enzyme or product-enzyme interactions. Since xylanase efficiency changes depend on different factors, it is essential to understand modulators to optimize these industries. In addition, exploring natural inhibitors of xylanase may support strategies for making crops resistant to pests and help create medicine to treat fungal infections.

5.7 Hydrolysis Products and Product Inhibition

What a xylanase does and the structure of xylooligosaccharide determine the types of end products created by hydrolyzing xylan enzymatically. Usually, hydrolysis results in several forms of xylooligosaccharide oligomers and on occasion one or more xylases. If hydrolysis products build up, they could become an inhibitor in the same way that regular product inhibition can stop the activity of the enzyme. Most of the time, bacterial xylanases break the β-1,4-glycosidic units in the xylan backbone, producing a mix of xylooligosaccharides. Examples of these are xylobiose, xylotriose and xylotetraose. Often, xylose monomers result if more xylanases or longer reaction periods, are present. The different products formed during enzymatic treatment have a major impact on the whole process. Over time, an accumulation of these products through product inhibition can decrease how well xylanase functions. The catalytic power of enzymes is lowered when the end products of the reaction bind to their active site or allosteric parts.

The presence of hydrolysis products, mainly xylooligosaccharides, inhibits xylanases by multiple methods and thus decreases their efficiency. Xylobiose and other short xylooligosaccharides halt substrate binding by competing for the active site, hence decreasing the rate of the reaction. Certain end products of an enzyme inhibit catalytic action, since they connect with allosteric sites and cause changes in the enzyme structure. Additionally, when xylooligosaccharides are very long and found in large amounts, they might compete with xylan as the substrate. Because product inhibition varies with the selected xylanase, product levels and experimental conditions, it is an important issue in xylanase use in industry. Methods to prevent hydrolysis inhibition involve repeatedly removing hydrolysis products, creating specialized enzymes, using several enzymes in a system and adjusting pH, temperature and concentration. Managing the interaction between products and xylanase leads to improvements in all applications involving xylanases in industries such as biofuel, business and food development (Fig. 6).

5.8 Glycosylation and Post-translational Modifications

Most xylanases are modified by glycosylation which adds polysaccharide chains to the protein and affecs their stability, performance and ability to bind certain substrates. In xylanases, the two major types of glycosylation include N-linked,

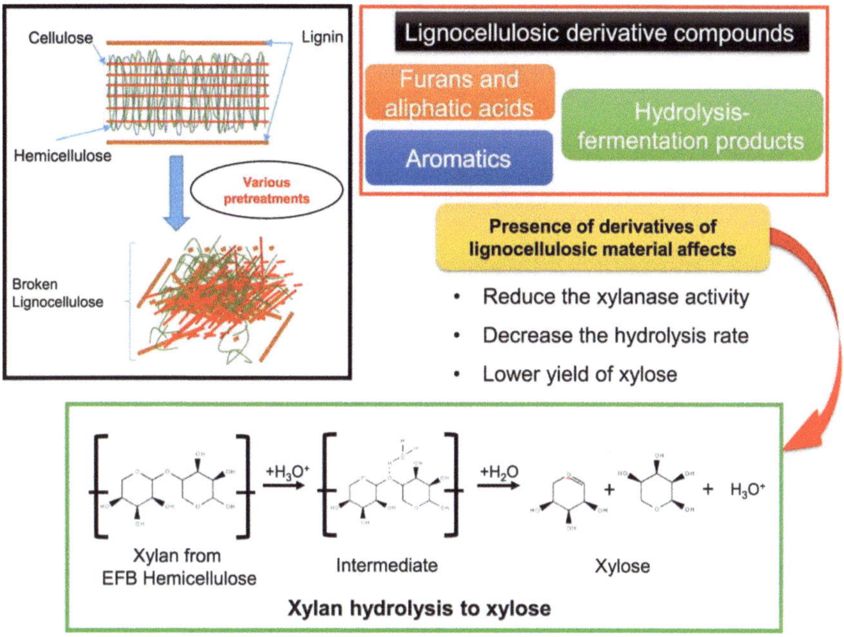

Fig. 6 Xylanase inhibition by the derivatives of lignocellulosic material [26]

in which sugars bind to asparagine residues following an Asn-X-Ser/Thr, without proline rule and O-linked, in which sugars attach to the hydroxyl part of serine or threonine. Because of their many forms, ranging from basic monosaccharides to elaborate oligosaccharides, these glycan chains make xylanases function in many different ways. Glycosylation makes enzymes more resistant to temperature and low or high pH, prevents their breakdown by proteases, improves how easily they dissolve in fluids and can increase, reduce or not affect their ability to react with a substrate—it is different for every enzyme variant. Apart from glycosylation, phosphorylation, acetylation, the formation of disulfide bonds and proteolytic processing also influence how xylanase functions. Studying these changes is key for making recombinant enzymes work better, for designing better enzymes, for clarifying how the structure of xylanases influences their function, for choosing xylanases for industrial purposes and for searching for adaptations among different kinds of enzymes (Fig. 7).

6 Substratum Explicitness and Creation of the Product

Due to both the source organism and environmental conditions, the substrate preferences of xylanases depend on the structure of xylan they are decomposing. A number of xylanases can cut bonds in arabinoxylan as well, but many appear to be almost exclusively focused on breaking the beta-1,4-xylosidic connection found in the main

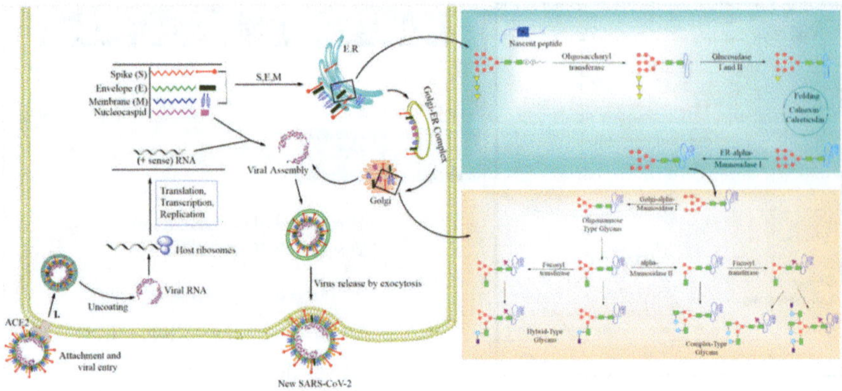

Fig. 7 Role of glycosylation in post-translational modifications (PTMs) of SARS-CoV-2 proteins [27]

part of xylan. Because the structure of its active site includes several modified xylose units, the enzyme is able to function. Determining xylan substitution and certain side chains is crucial for correctly identifying the substrate for binding. Depending on whether xylan is consumed alone or with replacements, xylanases are sorted into two classes.

Eighty-five to ninety-five percent of the breakdown products formed during xylan hydrolysis are sphere-shaped xylooligosaccharides, with rest being free xylose. What these cleavage products look like is affected by the type of action (endo or exo) by the enzyme, the degree of substitution within the substrate and the situation of the reaction. Activation of endo-xylanases leads to segments in the chain and the resulting xyloligosaccharides have different degrees of polymerization. Exo-xylanases help by chopping xylose chains into two. How many xylooligosaccharides are produced has a strong effect on their value and usage as prebiotics. To make prebiotic XOS, scientists depend on using enzymes called endo-xylanases, as Linares-Pastén et al. stated.

7 Environmental and Ecological Significance

As xylanase aids in breaking down xylan which is a big part of plant hemicellulose, it greatly supports both global carbon and organic matter cycles. Microorganisms, fungi and certain bacteria produce xylanase in soil, helping to break down plant remains and give off sugars that nourish others as well as help improve soil fertility. Enzymes in forests and farms speed up decomposition of plant parts which supports nutrient use and helps limit the buildup of lignocellulosic waste. Additionally, the action of xylanase enables herbivores and termites to convert fibrous plant matter into energy which supports their productive relationships with intestinal bacteria. Since some members of the group make xylanase, they also help plants

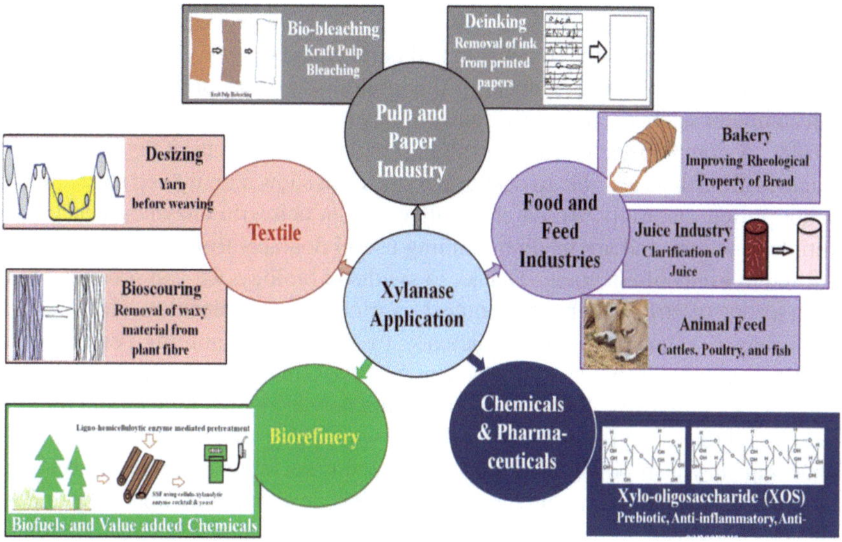

Fig. 8 Xylanase as a greener tool in different industries [28]

by competing with disease-causing microorganisms or causing plants to increase their defensive measures. The ecological purpose of xylanase matters for its use in sustainable farming, waste treatment and maintaining ecosystems that can withstand climate change (Fig. 8).

8 Future Perspectives and Research Trends

It offers a great opportunity to make xylanases fit specific industrial applications. The overall structure of enzymes is often resolved to ensure higher catalytic activity, reliable action with particular chemicals, proper working in various pH levels and strong temperature resistance. These tools are being used to give xylanases stronger and better features: direct evolution, rational design and semi-rational methods. Adding the ability to work at higher temperatures is an important focus in the research of xylanases. Scientists are trying out different ways to make enzymes more heat resistant such as using disulfide bonds, carefully arranging the charge spread on the protein surface and modifying areas called loops. The stability of xylanases over a range of pHs is also very important, since they are commonly used in situations that are strongly acidic or alkaline. The aim is to change essential amino acid residues that regulate the pH of the enzyme. Different xylan structures in biomass sources require efforts to make xylanases that work with a wider variety of substances. Protein engineering methods are applied to make xylanases function more quickly in catalysis. In other words, the process changes the main structure of the enzyme and updates its active site to help speed up reactions and make the process of extracting the product easier. It is possible to develop and build entirely new xylanases by making use of synthetic biology. Scientists in this multidisciplinary domain use concepts from computer science, engineering and biology to engineer biological systems with set roles. A common technique in synthetic biology is computing designs for new xylanases. Scientists are able to design enzymes that show activity and structure that are not present naturally. Researchers often design chimeric enzymes which have multiple functions and are made by joining useful domains from various xylanases or different kinds of enzymes. Thanks to synthetic biology, we can produce effective cascades of enzymes that decomposes various xylan molecules. The objective is to enhance xylan breakdown efficiency by either programming several xylan digesting enzymes into a single organism or by assembling the different enzymes into enzyme complexes. Teams of scientists are now trying to engineer microbes to produce and release xylanases that work very well, using modern synthetic biology. Enhancing enzyme production requires constructing secretion pathways, developing new promoters and maximizing the efficiency of gene expression tools. Demand for xylanases in industries supports more sustainable development. These enzymes help reduce pollution, increase the use of natural energy and need fewer dangerous chemical reagents. The use of xylanases supports the circle bioeconomy because they help to benefit from forestry and agriculture waste. Thanks to xylanases, biofuels, biochemicals and biomaterials can be produced by helping break down tough lignocellulosic components. Xylanases help make many industrial activities more environmentally friendly. Xylanase adds value to pulp and paper bleaching by enabling more sustainable processes and lower gas emissions. Further research may be directed at improving xylanases so that they operate well under very strong pH conditions, both extremely high and low. As a result such materials would bring greater value to industrial processes and help develop technology that is safe for the environment and sustainable. Research into xylanases is growing in healthcare and nanotechnology. Such enzymes can be applied in drug delivery approaches and in making nanocellulose.

9 Conclusion

Because xylanases are useful in many sectors and environments, they have grown in importance. They hold great value in biotechnology, agriculture and other parts of industry thanks to the wide range of their commercial uses and biochemical traits. Since xylanases break down a key substance in plant cell walls, called xylan, they are now considered major contributors to making an economy more bio-based and sustainable. Xylanases have the potential to control environmental risks and improve the process of biofuel production. The presence of xylanases in food production has proved highly useful, playing a part in making juice and baked goods both more effectively and with better results. It is hoped that applying xylanases in the feed industry will enhance animal nutrition and benefit the environment. As we advance in the study of enzymes, we are seeing the growth of much more effective and focused xylanases. While boosting existing results, these new achievements are also providing chances to discover new areas where xylanases can be useful. Work to boost both the stability and performance of xylanase enzymes, as well as their ability to digest a larger range of substances, makes the field's prospects bright. Using both multifunctional enzymes and xylanases in complex industrial methods should greatly improve several industries. It is hard to say enough about the important role that xylanases play in making development more sustainable. By staying essential, these enzymes help industries generate less waste, require fewer chemicals and become more sustainable as the world moves toward using more renewable resources. In short, xylanases are an exciting area of enzyme technology that can still develop in many directions. The role of these useful enzymes in building a better and greener industry will only grow as research and development continue to produce new discoveries.

References

1. E. Schmitz, J. Francis, K. Gutke, E. Nordberg Karlsson, P. Adlercreutz, M. Paulsson, Chemical and biochemical bleaching of oat hulls: the effect of hydrogen peroxide, laccase, xylanase and sonication on optical properties and chemical composition. Biotechnol. Reports **30**, e00624 (2021)
2. L.O. Souza, A.R. de Brito, R.C.F. Bonomo, N.B. Santana, J.L.d. Almeida Antunes Ferraz, E. Aguiar-Oliveira et al., Comparison of the biochemical properties between the xylanases of Thermomyces lanuginosus (Sigma®) and excreted by Penicillium roqueforti ATCC 10110 during the solid state fermentation of sugarcane bagasse. Biocatalysis Agric. Biotechnol. **16**, 277–284 (2018)
3. G.V. Gautério, L.G.G. da Silva, T. Hübner, T.d.R. Ribeiro, S.J. Kalil, Xylooligosaccharides production by crude and partially purified xylanase from Aureobasidium pullulans: biochemical and thermodynamic properties of the enzymes and their application in xylan hydrolysis. Process Biochem. **104**, 161–170 (2021)
4. S. Wang, H. Han, J. Dong, M. Li, Q. Wang, L. Wang et al., A novel GH11 β-1,4-xylanase from Fusarium verticillioides: its eukaryotic expression, biochemical characterization and synergistic effect with cellulase on lignocellulosic biomass degradation. Int. J. Biol. Macromole. **305**, 141169 (2025)
5. I.A. Olopoda, O.T. Lawal, O.V. Omotoyinbo, A.N. Kolawole, D.M. Sanni, Biochemical characterization of a thermally stable, acidophilic and surfactant-tolerant xylanase from Aspergillus awamori AFE1 and hydrolytic efficiency of its immobilized form. Process Biochem. **121**, 45–55 (2022)
6. C.C. Terrone, C.d. Freitas, C.R.F. Terrasan, A.F.d. Almeida, E.C. Carmona, Agroindustrial biomass for xylanase production by Penicillium chrysogenum: purification, biochemical properties and hydrolysis of hemicelluloses. Electron. J. Biotechnol. **33**, 39–45 (2018)
7. X. Qi, X. Liu, Y. Zhang, J. Hao, H. Rao, D. Zhao et al., A novel wheat arabino-xylanase from Trichoderma gamsii: study on biochemical characterization, substrate recognition mechanism and its application in wheat flour dough and Chinese bread. LWT **219**, 117550 (2025)
8. O.R. Ghaffar, A. Abdollahi, The effects of corncob and xylanase supplementation on performance, ileal digestibility, intestinal morphology, gastrointestinal pH, and serum biochemical traits in the broiler chickens. Can. J. Anim. Sci. (2024)
9. Y. Liu, C. Guo, C. Wang, Biochemical characterization of an organic solvent- and salt-tolerant xylanase and its application of arabinoxylan-oligosaccharides production from corn fiber gum. Int. J. Biol. Macromole. **280**, 136146 (2024)

10. F. Salzano, M. Aulitto, G. Fiorentino, D. Cannella, E. Peeters, D. Limauro, A novel endo-1,4-β-xylanase from Alicyclobacillus mali FL18: biochemical characterization and its synergistic action with β-xylosidase in hemicellulose deconstruction. Int. J. Biol. Macromole. **264**, 130550 (2024)
11. P.O. Magalhães, A.M.F. Milagres, Biochemical properties of a β-mannanase and a β-xylanase produced by Ceriporiopsis subvermispora during biopulping conditions. Int. Biodeterioration Biodegrad. **63**, 191–195 (2009)
12. A. Pereira de Almeida, I.P. Vargas, C.L. Marciano, F.F. Zanoelo, G.C. Giannesi, M.d.L.T.d. Moraes Polizeli, et al., Investigation of biochemical and biotechnological potential of a thermo-halo-alkali-tolerant endo-xylanase (GH11) from Humicola brevis var. thermoidea for lignocellulosic valorization of sugarcane biomass. Biocatal. Agric. Biotechnol. **44**, 102424 (2022)
13. H. Karaoglu, K.M.A. Ramadan, S.A. Al Hashedi, A. AlShoaibi, Z. Iqbal, R. Aydın et al., Selection, heterologous production, and functional characterization of a thermostable xylanase from anoxybacillus for dough and bread quality enhancement. Int. J. Biol. Macromole. **312**, 144000 (2025)
14. X. Liu, Q. Yan, Y. Xue, S. Wang, H. Yang, Z. Jiang, Biochemical characterization of a novel glycoside hydrolase family 11 xylanase from Chaetomium sp. suitable for bread making. Process Biochem. **117**, 1–9 (2022)
15. Z. Akpinar, H. Karaoglu, Characterization of a highly thermostable recombinant xylanase from Anoxybacillus ayderensis. Protein Express. Purif. **219**, 106478 (2024)
16. H. Wang, X. Qi, S. Gao, Y. Zhang, Y. An, Biochemical characterization of an engineered bifunctional xylanase/feruloyl esterase and its synergistic effects with cellulase on lignocellulose hydrolysis. Bioresource Technol. **355**, 127244 (2022)
17. H. Rahimian Gavaseraei, R. Hasanzadeh, M. Afsharnezhad, A. Foroutan Kalurazi, S.S. Shahangian, M.R. Aghamaali et al., Identification, heterologous expression and biochemical characterization of a novel cellulase-free xylanase B from the thermophilic bacterium Cohnella sp.A01. Process Biochem. **107**, 48–58 (2021)
18. Y. Zhang, X. Liu, M. Liu, L. Han, D. Zhao, H. Rao et al., Enzymatic modification of whole wheat dough gluten matrix development and bread quality by a novel wheat arabino-xylanase from Podospora comata with its properties and substrate specificity mechanism. Int. J. Biol. Macromole. **310**, 142860 (2025)
19. H. Yang, D. Sun, Y. Li, W. Zhao, Z. Jiang, Q. Yan et al., Development of a green, efficient and renewable biorefinery strategy for the production of xylooligosaccharides from corncobs based on a novel specific xylanase. Carbohyd. Polym. Technol. Appl., 100858 (2025)
20. P. Chutani, K.K. Sharma, Biochemical evaluation of xylanases from various filamentous fungi and their application for the deinking of ozone treated newspaper pulp. Carbohyd. Polym. **127**, 54–63 (2015)
21. X. Liu, Y. Liu, Z. Jiang, H. Liu, S. Yang, Q. Yan, Biochemical characterization of a novel xylanase from Paenibacillus barengoltzii and its application in xylooligosaccharides production from corncobs. Food Chem. **264**, 310–318 (2018)
22. M.E. MacDonald, N.G. Wells, B.A. Hassan, J.A. Dudley, K.J. Walters, D.M. Korzhnev et al., Effects of Xylanase A double mutation on substrate specificity and structural dynamics. J. Struct. Biol. **216**, 108082 (2024)
23. E.R. Kantrowitz, Allostery and cooperativity in Escherichia coli aspartate transcarbamoylase. Arch. Biochem. Biophys. **519**, 81–90 (2012)
24. S. Roy, T. Dutta, T.S. Sarkar, S. Ghosh, Novel xylanases from Simplicillium obclavatum MTCC 9604: comparative analysis of production, purification and characterization of enzyme from submerged and solid state fermentation. Springerplus **2**, 1–10 (2013)
25. M.S. Butt, M. Tahir-Nadeem, Z. Ahmad, M.T. Sultan, Xylanases and their applications in baking industry. Food Technol. Biotechnol. **46**, 22–31 (2008)
26. I.M. Hidayatullah, T. Setiadi, M.T.A.P. Kresnowati, R. Boopathy, Xylanase inhibition by the derivatives of lignocellulosic material. Bioresource Technol. **300**, 122740 (2020)

27. N. Tripathi, B. Goel, N. Bhardwaj, R.A. Vishwakarma, S.K. Jain, Exploring the potential of chemical inhibitors for targeting post-translational glycosylation of coronavirus (SARS-CoV-2). ACS Omega **7**, 27038–27051 (2022)
28. N. Bhardwaj, B. Kumar, P. Verma, A detailed overview of xylanases: an emerging biomolecule for current and future prospective. Bioresources Bioprocess. **6**, 40 (2019)

Chapter 5
Molecular Biology of Xylanase Genes

Abdul Sattar Jatoi, Bazil Owais Nasir, Saim Akram, and Owais Baig

Abstract Since xylanase genes have a crucial role in degradation of xylan, a main part of plant hemicellulose, the molecular biology of such genes has become a fundamental area of research. Xylanases, secreted by such a large number of microorganisms (fungi, bacteria, actinomycetes) play a key role in biomass degradation, which is important for use in biofuel production, the management of agricultural waste, as well as for many industrial applications. This chapter focuses on the genetic and biochemical pathways present for regulation, expression and evolution of xylanase genes and the molecular mechanisms for the utilization of xylan as a carbon source by microorganisms. In addition, the section discusses the genomic and metagenomic ways to search for new xylanase genes and their regulating elements using bioinformatics tools used to predict gene functions and gene clusters. This review presents a comprehensive overview of xylanase gene molecular biology that will enable further research and technological advancements in the future to take advantage of xylanase application in biotechnology and industrial processes.

Keywords Molecular biology · Xylanase gene

1 Introduction

Xylan is the most abundant hemicellulose present in the secondary cell walls of plants and is a major part of lignocellulosic biomass. The backbone is composed of a β-1,4 linked xylopyranosyl unit that may be substituted with arabinose, glucuronic acid, and acetyl group, according to the plant source. Degradation of xylan from biomass is efficient and of great importance in biotechnology including pulp and paper, biofuel, animal feed, and foods industries [1].

An example of enzymes that catalyse (via endo-hydrolysis of β-1,4-xylosidic linkages) xylanases (endo-1,4-β-xylanases) (EC 3.2.1.8). However, these enzymes

A. S. Jatoi (✉) · B. O. Nasir · S. Akram · O. Baig
Department of Chemical Engineering and Technology, Dawood University of Engineering & Technology, Karachi, Pakistan
e-mail: abdul.sattar@duet.edu.pk

A. Sattar Jatoi (ed.), *Xylanase from Microorganisms*,
https://doi.org/10.1007/978-981-95-0824-2_5

are produced by many microorganisms for example, bacteria, fungi, actinomycetes and some yeasts. They play a key role in sugar and other valuable product production from plant biomass. Particular producers of xylanase enzymes include bacteria like Bacillus and Streptomyces as well as fungi like Aspergillus, Trichoderma and Penicillium [2].

However, xylanases are far from biomass softening. Xylanases are used in the paper industry for bio-bleaching to lower use of chlorine based chemicals which solves the environmental pollution problem. Xylanases are used in the food industry to improve dough handling and improve the quality of bread. These enzymes in animal feed make plant cell walls more digestible. Furthermore, they play an important role in second generation biofuel production for economical and sustainable bioethanol production, since efficient saccharification of hemicellulose is required [3].

To optimize on production of xylanase genes and to tailor their properties for specific industrial application, it is important to understand the molecular biology of the xylanase genes. The xylanase genes have been cloned, expressed, and engineered in a variety of host organisms using molecular biology tools. Thus, such enzymes have become more efficient, thermostable, and pH stable for industrial processes. Also, investigation of the genetic regulation and expression of these genes reveals how the synthetic or repression of these genes can be promoted or repressed depending on different environmental and nutritional conditions [4].

This chapter is dedicated to present a general description of molecular biology of xylanase genes. However, we will discuss the classification, genetic organization of xylanases, their regulatory mechanisms, cloning and expression of xylanases, and techniques of molecular engineering. We will also discuss, respectively, insights from genomics and metagenomics, functional characterization of these enzymes and wide ranging biotechnological applications [5]. The chapter ends with some discussion on the future perspectives, challenges and emerging trends in the area.

In the context of rapid plant biomass deployment in a bioeconomy, xylanases are an essential component of enzyme toolkits needed for sustainable processing of plant biomass. The progress in genomics, synthetic biology and protein engineering remains on the path to the creation of highly efficient xylanases adapted to the requirements of different markets [6].

2 Classification of Xylanases

Xylanases are grouped mainly by sequence similarity, three dimensional structure and catalytic mechanism into a heterogeneous group of enzymes [1]. The most commonly used classification is based on the Carbohydrate-Active Enzymes (CAZy) database and defines the xylanases as strategic families of glycoside hydrolases. GH10 and GH11 are two most studied families but xylanases are present in GH5, GH8, GH30 and GH43 as well [2].

GH10 xylanases are generally high molecular weight (β/α)_8-barrel (TIM barrel) enzymes. They are broad substrate specific, able to cleave highly substituted xylans and hydrolyze shorter oligosaccharides. Since these enzymes are usually more tolerant to side chain decorations like arabinose or glucuronic acid, they have proved highly valuable in the degradation of complex hemicellulose from hardwoods and cereals [3].

GH11 xylanases differ, e.g. smaller, usually < 30 kDa, β-jelly roll structures rather than GH5 xylanases. As they are highly specific for unsubstituted parts of xylan, they are more efficient in cleaving linear xylan backbones. Being compact in size, possessing high catalytic activity and applicable for penetration of enzymes into fibrous materials, they (GH11 xylanases) are preferred for application in the paper and pulp industry [7].

The second part of the sentence describes the lesser studied GH5 and GH8 xylanases, which are less commonly studied, but are known to be involved in some specialized microbial systems. The structural and substrate specificity features of GH30 family xylanases are unique due to its capacity to characterize and hydrolyze glucuronoxylan linkages. Endo-xylanase activities are often supplemented by enzymes of the GH43 family that include arabinofuranosidases, which assist in removing side chains from the xylan backbone, and hence act synergistically with endo-xylanases [8].

Xylanases are also classified according to their mode of action other than sequence and structure.

- Almost endo-xylanases cleave β1,4 xylosidic bond of the xylan backbone and release the short xylo-oligosacchride.
- True exo-xylanases xylan units from the non-reducing ends of xylan chains but are rare exist and are usually known as β-xylosidases (EC 3.2.1.37).

Xylanases have a very large domain architecture that strongly affects their catalytic properties. Most microbial xylanases consist of modular structure that includes a catalytic domain (CD) with one or two carbohydrate binding modules (CBMs). Non catalytic CBMs function performing two main functions: binding to the xylan substrate and localizing the enzyme on the plant cell wall, thereby increasing the efficiency of the overall process. These domains are linked by linker regions that are often rich in proline, threonine and serine residues and give the protein flexibility [9].

In industrial applications enzyme function is affected by the presence or absence of CBMs. For instance, CBM containing xylanases are desirable where complex biomass hydrolysis, where CBM less variants are suggested in food processing for decreased binding of insoluble residues [10].

Xylanase genes analyzed phylogenetically from different microbial sources show a great diversity, indicating adaptation to several ecological niches. It is remarkable that xylanases produced by thermophilic and halophilic organisms are highly stable at high temperatures and salinities which make them attractive for industrial use under those truly harsh conditions [11].

Overall, classification of xylanases in glycoside hydrolase families, structural and catalytic features, reveals their functional heterogeneity. This element is crucial for choosing appropriate enzymes for various applications and guiding attempts at protein engineering focused on improving functional properties [12].

3 Genetic Organization of *Xylanase* Genes

Xylanase genes exhibit a wide range of genetic organization among different types of microbial species, and their expression is tightly regulated in response to the xylan or xylan derivatives in the environmental signals [13]. In prokaryotes, notably in the case of Bacillus subtilis, many xylanase genes are in operons with multiple functionally related genes transcribed as a single gene under control of one promoter. They usually contain not only the structural genes responsible for encoding xylanases, but also genes for auxiliary enzymes (e.g. β-xylosidases, arabinofuranosidases), the transporters and regulatory proteins [14].

Unlike in eukaryotic microorganisms (*Aspergillus niger, Trichoderma reesei*), where the xylanase genes are usually found as single copy genes with no clear genomic context, xylanase genes from filamentous fungi are usually present in multi-copy families containing several xylanase genes. Yet, they may still be subject to global transcription factor co regulation in response to environmental stimuli. The regulatory elements that recognize these inducers are found in the promoter regions of these genes. Among these are binding sites for transcription factors such as XlnR (in Aspergillus spp.), which activates the expression of a number of bite enzymes involved in degradation of hemicellulose [15].

The typical xylanase genes contain an open reading frame (ORF) with a catalytic domain, which is sometimes preceded by sequences for CBMs and linker sequences. The introns are mostly absent in many microbial xylanase genes (especially in prokaryotes), so that it can be directly transcribed into mRNA. Many fungal xylanases contain introns that must be spliced prior to translation in the mRNA. The transcription start site, TATA box, and enhancer or silencer elements upstream of the ORF play critical roles in gene regulation [16].

Expression of xylanase gene is usually regulated by carbon starvation or by the presence of complex polysaccharides, such as xylan. Carbon catabolite repression (CCR), as a common mechanism, inhibits the expression of xylanase gene expression when preferred carbon source is present, for example, glucose. In many fungi, CreA protein is the key mediator of CCR, and it binds to specific sites in promoter region and inhibits transcription [17].

The introduction of further complexity is of course epigenetic and post transcriptional regulation. In eukaryotic systems, histone modifications and chromatin remodeling can affect the accessibility of xylanase gene promoter. Furthermore, the 5′ and 3′ UTRs also affect mRNA stability, secondary structure and translational efficiency.

Many microbial genomes have been subjected to comparative genomic analyses and conserved gene clusters for xylan degradation pathway have been shown to be conserved among the microbial genomes [18]. In many clusters, sensor proteins, two components regulatory systems, ABC transporters, and multiple glycoside hydrolases are found. These clusters are organized so that xylan detection leads to a cascade of gene activation events in a coordinated response mechanism [19].

The design of artificial operons and expression cassettes have been realized using synthetic biology approaches for investigation and optimization of xylanase gene expression. Most of these constructs have strong, inducible promoters, ribosome binding sites and codon-optimized sequences to produce the protein in heterologous hosts like *Escherichia coli, Pichia pastoris* or *Saccharomyces cerevisiae* [20].

Genetic organization of xylanase genes is discussed, and concluded with the balance of structural gene components, regulatory sequences, and environmental responsiveness. Such genetic blueprints allow one to enhance enzyme production and control gene expression in industrial biotechnology applications [21].

4 Cloning and Expression of *Xylanase* Genes

Biotechnology techniques for xylanase gene cloning and heterologous expression are fundamental to producing high yield xylanase and can tailor its properties for industrial application [22, 23]. As these processes allow researchers to isolate xylanase genes from natural producers (e.g., fungi or bacteria) and to express them under controlled conditions in more amenable host organisms, they enable researchers to understand the function of certain genes and genes from different species [24].

4.1 Gene Cloning Techniques

The most commonly employed approach to cloning of xylanase genes is by way of genomic DNA or complementary DNA (cDNA) from xylanase producing microorganisms [25]. RT-PCR amplification of cDNA derived from xylan induced cultures is generally used to isolate xylanase gene for *Aspergillus species.* Gene discovery can be assisted by degenerate primers based on the conserved regions of known xylanase genes or genome mining tools [26].

After the gene fragments are amplified, they are then ligated into suitable plasmid vectors with the restriction enzymes and ligases. Alternately, the promoter elements, selectable markers, and fusion tags are often provided in these vectors for ease of purification or detection. The recombinant plasmids are introduced into host organisms by transformation (for example by electroporation or chemical methods), and then selection and screening for expression [27].

4.2 Expression Systems

Various heterologous expression systems are employed to produce xylanase enzymes efficiently. The most commonly used hosts include:

Name of microorganism	Summary
Prokaryotic systems (*Escherichia coli*)	These are widely used due to ease of manipulation, rapid growth, and well-characterized genetics. However, *E. coli* often fails to perform post-translational modifications such as glycosylation, which are critical for the activity and stability of fungal xylanases
Yeast systems (*Pichia pastoris*, *Saccharomyces cerevisiae*)	These are suitable for expressing eukaryotic xylanases due to their ability to perform glycosylation and secrete functional proteins into the medium. *P. pastoris* in particular has been successful for high-yield, methanol-induced expression of xylanases under the AOX1 promoter
Filamentous fungi (*Aspergillus niger*, *Trichoderma reesei*):	These are natural xylanase producers and can also serve as hosts for overexpression. Use of native or strong constitutive promoters (e.g., gpdA, xlnA, cbh1) enables efficient expression of xylanase genes. Moreover, these hosts facilitate secretion of active enzymes and correct post-translational processing

4.3 Expression of Xylanase Genes in Aspergillus Species

Extensive work has been done on studying xylanase genes, xlnA, xlnB and xynG2, in *Aspergillus niger, Aspergillus oryzae.* Typically, these genes are cloned under the control of inducible or constitutive fungal promoters in order to increase transcriptional levels. The xylose inducible xlnA promoter is controlled by the XlnR transcription factor which is activated in response to xylan or xylan derivatives [6].

Simple purification shortly after expression also may be accomplished by expression constructs that include a secretion signal sequence (e.g., glucoamylase or α amylase signal peptides). Protoplast mediated transformation is usually used for transformation of Aspergillus spp. and an integrant is rescued from the respective auxotrophic markers or resistant to antibiotic [28].

4.4 Regulatory Elements and Codon Optimization

To enhance expression in heterologous systems, codon optimization is frequently performed to match the host's codon usage bias [29]. Synthetic gene constructs also utilize engineered regulatory elements such as:

- **Strong Promoters**: e.g., AOX1 (methanol-inducible in *P. pastoris*), PGK1 (constitutive in yeast), or xlnA (xylan-inducible in *Aspergillus*).
- **Ribosome Binding Sites (RBS)**: Optimized RBS improve translational efficiency in prokaryotic hosts.
- **Terminator Sequences**: e.g., CYC1 terminator in yeast or trpC in fungi to ensure proper transcriptional termination.

4.5 Post-translational Modifications and Protein Folding

Since eukaryotic expression systems (such as yeasts and filamentous fungi) can do necessary posttranslational modifications, like N glycosylation and disulfide bond formation, they often complement the studies performed with *E. coli*. The stability and activity of some fungal xylanases requires these modifications. On the other hand, expression in *E. coli* tends to yield inclusion body formation that must be refolded in vitro [30, 31].

4.6 Fusion Tags and Purification

Since the protein is tagged with fusion tags (His tags, GST tags, FLAG tags ... etc.) for facilitated downstream purification at N or C terminal regions. In affinity chromatography based purification and immunodetection via immunoassays or Western blotting, these tags allow improvement [32].

4.7 Expression Challenges and Optimization Strategies

Some challenges in xylanase expression include proteolytic degradation, low yield, or misfolding. These can be addressed through:

- Co-expression of chaperones or foldases (e.g., PDI, BiP).
- Engineering protease-deficient host strains.
- Tuning culture conditions (pH, temperature, inducer concentration).
- Synthetic operons and gene stacking for co-expression of synergistic enzymes.

5 Regulation of Xylanase Gene Expression

Xylanase gene expression is regulated tightly with respect to environmental parameters, especially the availability of complex carbohydrates, xylan [33]. This regulation prevents wasting energy through the synthesis of xylanases for free. Numerous layers of regulatory mechanisms; including transcriptional, post-transcriptional, and epigenetic, are involved in controlling precisely the expression of xylanase genes in prokaryotic and eukaryotic systems [2].

In *Aspergillus niger* and *Trichoderma reesei*, the xylanolytic transcriptional activator XlnR (also referred to as Xyr1 in T. reesei) mediates transcriptional regulation, mostly. Binding of XlnR to conserved promoter motifs of xylanase genes and other hemicellulolytic enzymes promotes their expression upon exposure to xylan or its hydrolysis products (xylose and xylobiose). Deletion of xlnR in *Aspergillus* leads to a strong repression of xylan-degrading enzymes, in addition to which it is central for regulation [34].

However, repressive control of xylanase gene expression is usually carbon catabolite repression (CCR). CCR mechanisms repress transcription of xylanolytic genes in the presence of preferred carbon sources (glucose) in order to salvage carbon and energy via carbon efficient metabolism [12]. In filamentous fungi, this repression is mainly under control of the CreA protein, a Cys2His2 zinc finger transcription factor which represses transcriptional activation by binding to such sites (5′-SYGGRG-3′) in xylanase gene promoter regions [35].

Two component regulatory systems are used by prokaryotic organism Bacillus subtilis for the detection of xylan and induction of gene expression by the sensor kinases and the response regulators. Rapid transcriptional activation of operons encoding xylanases, β-xylosidases and associated transporters are made by these systems. There are also sigma factors (e.g., σ^X and σ^M in Bacillus) that can regulate transcription of xylanase genes in response to environmental and triggering signals.

Post-transcriptional regulation further fine-tunes xylanase gene expression. Therefore, in fungi untranslated regions (UTRs) of xylanase mRNAs are found to influence transcriptional stability and translation efficiency. Although small RNAs and RNA binding proteins also could participate in controlling mRNA turnover and ribosome accessibility, such roles are not yet proven. In bacteria, regulatory RNAs can inform translation initiation according to metabolite level.

- Epigentetic modifications (histone acetylation and methylation etc.) can also effect xylanase gene promoter chromatin structure and transcriptional machinery accessibility. It was found in Trichoderma that chromatin remodeling is associated with increases in cellulolytic and hemicellulolytic enzyme expression during lignocellulose degradation. Histone acetyl transferases (HATs) and deacetylases (HDACs) have been identified as targets for improving the enzyme production of industrial strains; promoter accessibility is modulated.

Currently, regulatory circuits for controlled xylanase expression are designed by synthetic biology approaches. Precise control on xylanase gene expression in engineered hosts such as E. coli, Pichia pastoris is provided by the inducible promoters, synthetic transcription factors and CRISPR based gene regulation systems. Indeed these strategic approaches are necessary for achieving the highest yields and enzyme properties that are tailored for industrial applications [36].

The regulation of transcription of xylanase genes is a complicated multilayer process involving transcriptional activators and repressors, environmental sensing systems, mRNA stability mechanisms and chromatin level controls in summary. These regulatory networks need to be understood in a comprehensive way for the production of xylanase and the full harnessing thereof in biotechnological applications [37].

6 Molecular Engineering and Mutagenesis

Consequently, molecular engineering and mutagenesis of xylanase genes have been developed into powerful strategies to enhance the properties of enzymes like thermostability, pH tolerance, substrate specificity and catalytic efficiency [38]. Modification of the xylanases is essential in order to fine tune these enzymes for use in a variety of industrial processes where harsh conditions exist, such as the pulp bleaching industry, feed for animals digestion other biofuel production. Themethods, routeshetics, and applications of rational design, directed evolution, and site specific mutagenesis of xylanase genes engineering are described in this section [39].

6.1 Rational Design of Xylanases

Rational design relies on the modification of some specific amino acid positions according to the known 3D structure and functional residues of the enzyme. Structural biology and bioinformatics are now enabling researchers to model xylanase enzymes and to find sites related to stability and activity. Scientists can design mutations that increase desirable traits by analyzing binding clefts, active sites and surface residues there [40].

For instance, if the thermostability can be improved by increasing the number of salt bridges, disulfide bonds, or hydrophobic interactions in specific regions, this will tremendously increase. Modulation of substrate specificity or reduction of product inhibition is achieved by substituting amino acids in the catalytic pocket. Several studies have demonstrated the successful introduction of such rationally designed mutations to generate enzymes that might work with higher temperature or broader pH ranges [7].

6.2 Directed Evolution

Natural selection in the laboratory means directed evolution generated large libraries of enzyme variants and screens for improved characteristics [41]. Generally, it involves random mutagenesis methods including error prone PCR, DNA shuffling, or saturation mutagenesis. The gene libraries are expressed in appropriate hosts and variants with improved activity or stability are selected by high throughput screening assays [42].

Unlike most methods, this method does not depend on knowledge of the enzyme structure, which is perfect when trying to explore new types of improvement [43]. Xylanase variants with improved thermal stability, resistant to proteolytic degradation and improved efficiency under industrial conditions have been produced by directed evolution. The synergistic enhancements can be obtained through successive rounds by combining beneficial mutations[44].

6.3 Site-Directed and Site-Saturation Mutagenesis

Site directed mutagenesis is the altering of specific nucleotides (of a DNA sequence) to change the amino acids of a particular protein [45]. This approach is commonly used to test for the involvement of a particular residue or residues in enzyme function and validate structural insights. An alternative of this method is site–saturation mutagenesis replacing a target amino acid with all potential alternatives producing a comprehensive mutational landscape [46].

This has helped in identifying residues that bind the substrate and participates in the catalytic activity, as well as stabilizing the fold [47]. An example is xylanases whose hydrolysis patterns are modified or resistance to inhibitor altered by alteration of conserved residues in the glycoside hydrolase family motifs. We have also demonstrated that mutations in the N- or C-terminal regions also can affect both the folding and secretion efficiency [48].

6.4 Fusion Proteins and Chimeric Xylanases

Chimeric enzymes made by using fusion protein technology is that catalytic domain is linked to carbohydrate binding modules (CBMs) or other functional domain. Substrate binding rate constant, synergistic activity with other enzymes or resistance to denaturation are these properties of engineered proteins [49].

Fused chimeric xylanases have been raised utilizing GH11 domains and CBMs from dissimilar sources to frame expanded energy effective xylanases with respect to degrading complex lignocellulosic substrates [50]. These fusion constructs are particularly valuable in biomass conversion insofar as enzyme synergy and targeting are an issue [51].

6.5 Engineering Secretion and Expression

An essential part of the molecular engineering is to improve the secretion and expression of xylanases [43]. Signal peptide engineering can be used to increase the efficiency with which recombinant xylanases are secreted by host cells, in fungal and yeast systems in particular. Moreover, there are mutations that lower aggregation or misfolding and increase total yields [52].

Another important tool is codon optimization, particularly where the xylanase genes are heterologous fungal or bacterial genes [14]. One can optimize gene expression (i.e. modifying promoter, ribosome binding sites and transcription terminators) modifying regulatory elements [53].

6.6 Computational Tools and Machine Learning

The advances in computational biology have so far been helpful in xylanase engineering [54]. The effect of mutation upon enzyme structure, and subsequently function is predicted from molecular dynamics simulations that can be used to test ideas about possible mutations, and docking studies, and stability prediction algorithms. Mutational datasets have patterns and they can predict useful substitution using machine learning models trained on these [55].

They used these tools to reduce experimental workload and also guide mutagenesis strategies better. For illustration, it is evident that artificial enzyme saturation mutagenesis with the aid of machine learning has been employed to identify mutations that improve the performance of the enzyme under industrial conditions [56].

6.7 Applications of Engineered Xylanases

Xylanases have cross applications in several industries. Pulp bleaching processes are in need of thermostable xylanases and pH tolerant variants are also useful in animal feed formulations that operate in the GI tract of an animal. Prebiotic xylo-oligosaccharides are produced with the help of xylanases having modified specificity [57].

Additionally, xylanases possessing multi-functional or bifunctional xylanase activity have been engineered with additional enzyme function, e.g. cellulase or β-xylosidase, activity. These enzymes provide synergistic degradation of lignocellulose and improve the economics for biorefineries [44].

7 Genomic and Metagenomic Insights

Through genomic and metagenomic approaches, xylanase gene diversity, evolution and ecological function have been profoundly increased [58]. Through the whole genomes sequencing of xylanase producing fungi and bacteria, xylanase gene clusters were mapped, glycoside hydrolases (GH) families (GH10 and GH11) including those families were characterized and xylanase transcriptional regulators XlnR and Xyr1 responsible for xylanase expression in response to environmental inducers like xylan and xylose were uncovered [59]. Typically, these genomes encode several xylanases with complementary functions for helping microbial adaptation to complex lignocellulosic substrates [60]. At the same time, metagenomics has provided a powerful means to discover enzymes based on access to the genes of uncultured microbes from many heterogeneous ecosystems like soils, marine sediments, compost and guts of animals. Metagenomics have been utilized both functionally and whole sequence, many xylanase genes have been identified with high throughput sequencing and bioinformatic tools such as HMMs and CAZy annotation [61]. In addition, metatranscriptomics and metaproteomics allow us to make inferences about active xylanase expression at the in situ level, how the microbial community coordinates biomass degradation at different environmental conditions. Horizontal gene transfer, domain shuffling, and duplication of the xylanase genes are found to occur in comparison genomics and phylogenetic analyses, thereby providing for xylanase gene functional diversity [62]. The structure is predicted by bioinformatics tools like MEGA, Phyre2, AlphaFold etc. and as far as functional prediction is concerned, network analysis could aid us to map metabolic pathways [63]. However, the challenges still remain, namely functional validation of predicted genes and low expression in heterologous hosts, however, integrating multi omics strategy and synthetic biology have a promise [64]. In total, genomic and metagenomic view expands the evolutionary complexity of xylanase systems, and simultaneously offers a collection of biocatalysts for industrial biotechnology and sustainable bioeconomy targets [65].

8 Functional Characterization

It is essential to understand the functional characterization of xylanases which will enable their knowledge of enzymatic behavior, substrate specificity, structural dynamics and their utility in industries [3]. The process is complete, including biochemical assays, kinetic profiling, structural analysis, substrate interaction studies and evaluation at different physicochemical conditions. First, since mass production of the enzyme is needed for analysis, recombinant expression of the xylanase genes in a suitable host like *Escherichia coli, Pichia pastoris* or *Bacillus subtilis* is done [66]. To study the protein in question in detailed in vitro studies, the protein can usually be purified by affinity chromatography [67]. Catalytic activity is generally assessed by enzyme assays done in reducing sugars quantification, including dinitrosalicylic acid (DNS) [68]. Optimal reaction conditions, pH tolerance and thermal stability are determined from these tests in a range of pH, temperature and substrate concentrations [69]. This provides the kinetic parameters (Km, Vmax) that for describing enzyme affinity and rate or comparing different xylanases or engineered variants [70]. The enzyme's selectivity is further shown in the substrate specificity tests using different natural and synthetic xylans such as beechwood xylan, oat spelt xylan, or arabinoxylan, which were used to hydrolyze diverse hemicelluloses. In addition, performing the product analysis via high performance liquid chromatography (HPLC), thin layer chromatography (TLC) or mass spectrometry (MS) allows for the identification of these hydrolysis products (i.e. xylose, xylobiose, longer xylo-oligosaccharides) thus aiding in the determination of the mode of action, either exo or endo oxidation [71]. The three-dimensional arrangement of catalytic residues and substrate binding sites is stalled by molecular study or X-ray crystallography, NMR spectroscopy, or cryo-electron microscopy [72]. Influence and comprehension of enzyme function and stability depends on such structural insight. Probing the roles of individual amino acids in catalysis, binding, stability or in the generation of improved industrial variants through mutation is most commonly performed using mutagenesis experiments, most commonly site directed or saturation based. Secondly, functional characterization includes the evaluation of enzyme synergism with other lignocellulolytic enzymes such as cellulases, ligninases and accessory debranching enzymes. The concerted action is important in this context in the biomass degradation application, where it is required for effective hydrolysis [73, 74]. The second part of functional screening relates to compatibility with industrial conditions: resistance to metal ions, detergents, and organic solvents, especially for applications in textile, pulp and paper, and detergent industries. In addition, atomistic techniques such as molecular dynamics simulations give an in silico view on enzyme stability, substrate binding dynamics and conformational changes during catalysis. Comparative studies of different xylanases from different glycoside hydrolase families allow a relationship between sequence diversity and functional properties, as needed to pick the best candidates for their particular tasks [75]. Testing enzymes on actual substrates such as wheat straw or corn stover or bagasse in pilot studies evaluates the applicability of enzymes in an industrial scale and determines how efficient the enzymes are, compatible the process is and how cost effective it is. This phase also takes into account environmental factor such as salinity, redox potential, and cofactor presence. In addition, proteomics and transcriptomics (combinations of omics) can be used to analyze xylanase expression under different growth and stress conditions for functional validation and identify the bioprocess optimization. Taken together, the functional characterization of xylanases is a multiaffiliated undertaking in which molecular biology combines with biochemistry and structural biology and industrial biotechnology. Based on it, xylanases can be tailored to address the requirements of different industrial applications with high robustness, thus being critical biocatalysts driving the bio-based economy.

9 Functional Metagenomics of Xylanase

It is essential to understand the functional characterization of xylanases which will enable their knowledge of enzymatic behavior, substrate specificity, structural dynamics and their utility in industries. The process is complete, including biochemical assays, kinetic profiling, structural analysis, substrate interaction studies and evaluation at different physicochemical conditions. First, since mass production of the enzyme is needed for analysis, recombinant expression of the xylanase genes in a suitable host like Escherichia coli, Pichia pastoris or Bacillus subtilis is done. To study the protein in question in detailed in vitro studies, the protein can usually be purified by affinity chromatography. Catalytic activity is generally assessed by enzyme assays done in reducing sugars quantification, including dinitrosalicylic acid (DNS). Optimal reaction conditions, pH tolerance and thermal stability are determined from these tests in a range of pH, temperature and substrate concentrations. This provides the kinetic parameters (Km, Vmax) that for describing enzyme affinity and rate or comparing different xylanases or engineered variants. The enzyme's selectivity is further shown in the substrate specificity tests using different natural and synthetic xylans such as beechwood xylan, oat spelt xylan, or arabinoxylan, which were used to hydrolyze diverse hemicelluloses. In addition, performing the product analysis via high performance liquid chromatography (HPLC), thin layer chromatography (TLC) or mass spectrometry (MS) allows for the identification of these hydrolysis products (i.e. xylose, xylobiose, longer xylo-oligosaccharides) thus aiding in the determination of the mode of action, either exo or endo oxidation [76]. The three-dimensional arrangement of catalytic residues and substrate binding sites is stalled by molecular study or X-ray crystallography, NMR spectroscopy, or cryo-electron microscopy [77]. Influence and comprehension of enzyme function and stability depends on such structural insight. Probing the roles of individual amino acids in catalysis, binding, stability or in the generation of improved industrial variants through mutation is most commonly performed using mutagenesis experiments, most commonly site directed or saturation based [78]. Secondly, functional characterization includes the evaluation of enzyme synergism with other lignocellulolytic enzymes such as cellulases, ligninases and accessory debranching enzymes. The concerted action is important in this context in the biomass degradation application, where it is required for effective hydrolysis [79]. The second part of functional screening relates to compatibility with industrial conditions: resistance to metal ions, detergents, and organic solvents, especially for applications in textile, pulp and paper, and detergent industries. In addition, atomistic techniques such as molecular dynamics simulations give an in silico view on enzyme stability, substrate binding dynamics and conformational changes during catalysis. Comparative studies of different xylanases from different glycoside hydrolase families allow a relationship between sequence diversity and functional properties, as needed to pick the best candidates for their particular tasks [80]. Testing enzymes on actual substrates such as wheat straw or corn stover or bagasse in pilot studies evaluates the applicability of enzymes in an industrial scale and determines how efficient the enzymes are, compatible the process is and how cost effective it is. This phase also takes into account environmental factor such as salinity, redox potential, and cofactor presence [81]. In addition, proteomics and transcriptomics (combinations of omics) can be used to analyze xylanase expression under different growth and stress conditions for functional validation and identify the bioprocess optimization [82]. Taken together, the functional characterization of xylanases is a multiaffiliated undertaking in which molecular biology combines biochemistry and structural biology and industrial biotechnology [27]. Based on it, xylanases can be tailored to address the requirements of different industrial applications with high robustness, thus being critical biocatalysts driving the bio-based economy [83].

10 Future Perspectives and Challenges

Knowledge of the enzymatic behavior (functional characterization) of xylanases in terms of substrate specificity, structural dynamics and possible industrial use is important for a better understanding of enzyme action. The process is complete, including biochemical assays, kinetic profiling, structural analysis, substrate interaction studies and evaluation at different physicochemical conditions. First, since mass production of the enzyme is needed for analysis, recombinant expression of the xylanase genes in a suitable host like Escherichia coli, Pichia pastoris or Bacillus subtilis is done. To study the protein in question in detailed in vitro studies, the protein can usually be purified by affinity chromatography. Catalytic activity is generally assessed by enzyme assays done in reducing sugars quantification, including dinitrosalicylic acid (DNS). Optimal reaction conditions, pH tolerance and thermal stability are determined from these tests in a range of pH, temperature and substrate concentrations. This provides the kinetic parameters (Km, Vmax) that for describing enzyme affinity and rate of comparing different xylanases or engineered variants. The enzyme's selectivity is further shown in the substrate specificity tests using different natural and synthetic xylans such as beechwood xylan, oat spelt xylan, or arabinoxylan, which were used to hydrolyze diverse hemicelluloses. In addition, performing the product analysis via high performance liquid chromatography (HPLC), thin layer chromatography (TLC) or mass spectrometry (MS) allows for the identification of these hydrolysis products (i.e. xylose, xylobiose, longer xylo-oligosaccharides) thus aiding in the determination of the mode of action, either exo or endo oxidation. Next, X-ray crystallography, NMR spectroscopy, or cryo electron microscopy are used to structural characterize catalytic residues and substrate binding sites and molecular modeling and docking studies are used to complement the 3D arrangement of catalytic residues and substrate binding sites. These structural insights are useful for explaining how the active site architecture affects enzyme function and stability. Probing the roles of individual amino acids in catalysis, binding, stability or in the generation of improved industrial variants through mutation is most commonly performed using mutagenesis experiments, most commonly site directed or saturation based. Secondly, functional characterization includes the evaluation of enzyme synergism with other lignocellulolytic enzymes such as cellulases, ligninases and accessory debranching enzymes. The concerted action is important in this context in the biomass degradation application, where it is required for effective hydrolysis. The second part of functional screening relates to compatibility with industrial conditions: resistance to metal ions, detergents, and organic solvents, especially for applications in textile, pulp and paper, and detergent industries. Additionally, in silico methods including molecular dynamics simulations offer atomistic insights on the enzyme stability, the substrate binding dynamics and conformational changes during catalysis. Comparative studies of different xylanases from different glycoside hydrolase families allow a relationship between sequence diversity and functional properties, as needed to pick the best candidates for their particular tasks. The industrial scale applicability is tested by running the enzymes on real world materials such as wheat straw, corn stover and bagasse in pilot studies to evaluate enzyme efficiency, process compatibility and cost effectiveness. This phase also takes into account environmental factor such as salinity, redox potential, and cofactor presence. In addition, proteomics and transcriptomics (combinations of omics) can be used to analyze xylanase expression under different growth and stress conditions for functional validation and identify the bioprocess optimization. The functional characterization of xylanases is a multiaffiliated undertaking in which molecular biology combines with biochemistry, structural biology, and industrial biotechnology. Based on it, xylanases can be tailored to address the requirements of different industrial applications with high robustness, thus being critical biocatalysts driving the bio-based economy.

11 Conclusion

Knowledge of the enzymatic behavior (functional characterization) of xylanases in terms of substrate specificity, structural dynamics and possible industrial use is important for a better understanding of enzyme action. The process is complete, including biochemical assays, kinetic profiling, structural analysis, substrate interaction studies and evaluation at different physicochemical conditions. First, since mass production of the enzyme is needed for analysis, recombinant expression of the xylanase genes in a suitable host like Escherichia coli, Pichia pastoris or Bacillus subtilis is done. To study the protein in question in detailed in vitro studies, the protein can usually be purified by affinity chromatography. Catalytic activity is generally assessed by enzyme assays done in reducing sugars quantification, including dinitrosalicylic acid (DNS). Optimal reaction conditions, pH tolerance and thermal stability are determined from these tests in a range of pH, temperature and substrate concentrations. This provides the kinetic parameters (Km, Vmax) that for describing enzyme affinity and rate or comparing different xylanases or engineered variants. The enzyme's selectivity is further shown in the substrate specificity tests using different natural and synthetic xylans such as beechwood xylan, oat spelt xylan, or arabinoxylan, which were used to hydrolyze diverse hemicelluloses. In addition, performing the product analysis via high performance liquid chromatography (HPLC), thin layer chromatography (TLC) or mass spectrometry (MS) allows for the identification of these hydrolysis products (i.e. xylose, xylobiose, longer xylo-oligosaccharides) thus aiding in the determination of the mode of action, either exo or endo oxidation. Next, X-ray crystallography, NMR spectroscopy, or cryo electron microscopy are used to structural characterize catalytic residues and substrate binding sites and molecular modeling and docking studies are used to complement the 3D arrangement of catalytic residues and substrate binding sites. These structural insights are useful for explaining how the active site architecture affects enzyme function and stability. Probing the roles of individual amino acids in catalysis, binding, stability or in the generation of improved industrial variants through mutation is most commonly performed using mutagenesis experiments, most commonly site directed or saturation based. Secondly, functional characterization includes the evaluation of enzyme synergism with other lignocellulolytic enzymes such as cellulases, ligninases and accessory debranching enzymes. The concerted action is important in this context in the biomass degradation application, where it is required for effective hydrolysis. The second part of functional screening relates to compatibility with industrial conditions: resistance to metal ions, detergents, and organic solvents, especially for applications in textile, pulp and paper, and detergent industries. Additionally, in silico methods including molecular dynamics simulations offer atomistic insights on the enzyme stability, the substrate binding dynamics and conformational changes during catalysis. Comparative studies of different xylanases from different glycoside hydrolase families allow a relationship between sequence diversity and functional properties, as needed to pick the best candidates for their particular tasks. The industrial scale applicability is tested by running the enzymes on real world materials such as wheat straw, corn stover and bagasse in pilot studies to evaluate enzyme efficiency, process compatibility and cost effectiveness. This phase also takes into account environmental factor such as salinity, redox potential, and cofactor presence. In addition, proteomics and transcriptomics (combinations of omics) can be used to analyze xylanase expression under different growth and stress conditions for functional validation and identify the bioprocess optimization. Taken together, the functional characterization of xylanases is a multiaffiliated undertaking in which molecular biology combines with biochemistry and structural biology and industrial biotechnology. Based on it, xylanases can be tailored to address the requirements of different industrial applications with high robustness, thus being critical biocatalysts driving the bio-based economy.

References

1. S.A. Nour, G.M. El-Sayed, H.A.A. Taie, M.T.H. Emam, A.F. El-Sayed, R.G. Salim, Safe production of Aspergillus terreus xylanase from Ricinus communis: gene identification, molecular docking, characterization, production of xylooligosaccharides, and its biological activities. J. Genetic Eng. Biotechnol. **20**, 121 (2022)
2. H. Niu, D. Zhu, J. Leng, Z. Wang, D. Liu, Y. Chen et al., Biofilm-based immobilized fermentation of engineered Komagataella phaffii for xylanase production. Biores. Technol. **418**, 131918 (2025)

3. H. Karaoglu, K.M.A. Ramadan, S.A. Al Hashedi, A. AlShoaibi, Z. Iqbal, R. Aydın et al., Selection, heterologous production, and functional characterization of a thermostable xylanase from anoxybacillus for dough and bread quality enhancement. Int. J. Biol. Macromolec. **312**, 144000 (2025)
4. Y. Du, P. Shi, H. Huang, X. Zhang, H. Luo, Y. Wang et al., Characterization of three novel thermophilic xylanases from Humicola insolens Y1 with application potentials in the brewing industry. Biores. Technol. **130**, 161–167 (2013)
5. L. Zhao, K. Meng, P. Shi, Y. Bai, H. Luo, H. Huang et al., A novel thermophilic xylanase from Achaetomium sp. Xz-8 with high catalytic efficiency and application potentials in the brewing and other industries. Process Biochem. **48**, 1879–1885 (2013)
6. F. Zheng, H. Zhang, J. Wang, J. Chen, H. Zhuang, A. Basit, Expression and characterization of a novel halophilic GH10 β-1,4-xylanase from Trichoderma asperellum ND-1 and its synergism with a commercial α-L-arabinofuranosidase on arabinoxylan degradation. Int. J. Biol. Macromol. **282**, 136885 (2024)
7. S. Ling, J. Xing, S. Li, L. Zhang, C. Shen, J. Hong et al., Enhancing the catalytic performance of xylanase XynASP through semi-rational design in the cord region to promote its application in juice clarification. Int. J. Biol. Macromol. **305**, 141138 (2025)
8. A. Saleem, S. Waris, T. Ahmed, R. Tabassum, Biochemical characterization and molecular docking of cloned xylanase gene from Bacillus subtilis RTS expressed in E. coli. Int. J. Biol. Macromol. **168**, 310–321 (2021)
9. X. Qi, X. Liu, Y. Zhang, J. Hao, H. Rao, D. Zhao et al., A novel wheat arabino-xylanase from Trichoderma gamsii: Study on biochemical characterization, substrate recognition mechanism and its application in wheat flour dough and Chinese bread. LWT **219**, 117550 (2025)
10. C.C.d.M. Capetti, O. Ontañon, L.E. Navas, E. Campos, R. Simister, A. Dowle et al., Sugarcane bagasse derived xylooligosaccharides produced by an arabinofuranosidase/xylobiohydrolase from Bifidobacterium longum in synergism with xylanases. Carbohyd. Polym. **339**, 122248 (2024)
11. Y. Zhang, X. Liu, M. Liu, L. Han, D. Zhao, H. Rao et al., Enzymatic modification of whole wheat dough gluten matrix development and bread quality by a novel wheat arabino-xylanase from Podospora comata with its properties and substrate specificity mechanism. Int. J. Biol. Macromol. **310**, 142860 (2025)
12. X. Yan Wang, Z. Li Ding, Z. Sheng Zhang, C. Bin Zhu, S. Yang, H. Fei, Application of two crucial hemicellulolytic enzymes in fish diet: Xylanase and mannanase. Anim. Feed Sci. Technol. **319**, 116178 (2025)
13. H. Askari, S. Soleimanian-Zad, M. Kadivar, S. Shahbazi, Creating a novel genetic diversity of Trichoderma afroharzianum by γ-radiation for xylanase-cellulase production. Heliyon **10**, e28349 (2024)
14. C. Liu, Y. Zhang, C. Ye, F. Zhao, Y. Chen, S. Han, Combined strategies for improving the heterologous expression of a novel xylanase from Fusarium oxysporum Fo47 in Pichia pastoris. Synth. Syst. Biotechnol. **9**, 426–435 (2024)
15. Z. Hu, Q. Liu, B. Ouyang, G. Wang, C. Wei, X. Zhao, Recent advances in genetic engineering to enhance plant-polysaccharide-degrading enzyme expression in Penicillium oxalicum: a brief review. Int. J. Biol. Macromol. **278**, 134775 (2024)
16. A. Kumar, R. Naraian, Chapter 6—Differential expression of the microbial β-1,4-Xylanase, and β-1,4-endoglucanase genes, in *New and Future Developments in Microbial Biotechnology and Bioengineering*, ed. by H.B. Singh, V.K. Gupta, S. Jogaiah (Elsevier, Amsterdam, 2019), pp. 95–111
17. Y. Abdulhadi, K. Kaur, A. Vyas, Chapter 3—Role of microbial xylanases in biorefinery platform and its impact on ecosystem services, in *Relationship Between Microbes and the Environment for Sustainable Ecosystem Services*, ed. by J. Samuel, A. Kumar, J. Singh (Elsevier, 2023), pp. 43–59
18. H. Tanaka, T. Nakamura, S. Hayashi, K. Ohta, Purification and properties of an extracellular endo-1,4-β-xylanase from Penicillium citrinum and characterization of the encoding gene. J. Biosci. Bioeng. **100**, 623–630 (2005)

19. A.P. MacCabe, M.T. Fernández-Espinar, L.H. de Graaff, J. Visser, D. Ramón, Identification, isolation and sequence of the Aspergillus nidulans xlnC gene encoding the 34-kDa xylanase. Gene **175**, 29–33 (1996)
20. M. Khanahmadi, I. Arezi, M.-S. Amiri, M. Miranzadeh, Bioprocessing of agro-industrial residues for optimization of xylanase production by solid- state fermentation in flask and tray bioreactor. Biocatal. Agric. Biotechnol. **13**, 272–282 (2018)
21. J. Sermsathanaswadi, S. Baramee, C. Tachaapaikoon, P. Pason, K. Ratanakhanokchai, A. Kosugi, The family 22 carbohydrate-binding module of bifunctional xylanase/β-glucanase Xyn10E from Paenibacillus curdlanolyticus B-6 has an important role in lignocellulose degradation. Enzyme Microb. Technol. **96**, 75–84 (2017)
22. P. Bajpai, Chapter 7—Molecular cloning and expression of xylanase gene, in *Microbial Xylanolytic Enzymes*, ed. by P. Bajpai (Academic Press, 2022), pp. 109–116
23. M. Jeya, S. Thiagarajan, J.-K. Lee, P. Gunasekaran, Cloning and expression of GH11 xylanase gene from Aspergillus fumigatus MKU1 in Pichia pastoris. J. Biosci. Bioeng. **108**, 24–29 (2009)
24. P. Dey, A. Roy, Cloning, characterization and expression of a gene encoding endo-1, 4- β-xylanase from the fungus Termitomyces clypeatus. Carbohyd. Res. **505**, 108333 (2021)
25. G.S. de Amo, C. Bezerra-Bussoli, R. Rodrigues da Silva, L.T. Kishi, H. Ferreira, E. Gomes et al., Heterologous expression of GH11 xylanase from Myceliophthora heterothallica F.2.1.4 in Pichia pastoris. Biocatal. Agric. Biotechnol. **61**, 103343 (2024)
26. D.M. Tran, T.U. Huynh, T.H. Nguyen, T.O. Do, A.D. Nguyen, Data on gene cloning, expression, purification, and characterization of the glycoside hydrolase family 11 from Bacillus velezensis. Data Brief **52**, 109834 (2024)
27. D. Sun, H. Qi, G. Dou, S. Mao, F. Lu, K. Tian et al., Ancestral sequence reconstruction of a robust β-1,4-xylanase and efficient expression in Bacillus subtilis. Int. J. Biol. Macromol. **282**, 137188 (2024)
28. X. Liu, Y. Zhang, X. Qi, D. Zhao, H. Rao, X. Zhao et al., Advances of microbial xylanases in the application of flour industries: a comprehensive review. Int. J. Biol. Macromol. **282**, 137205 (2024)
29. A. Dey, CRISPR/Cas genome editing to optimize pharmacologically active plant natural products. Pharmacol. Res. **164**, 105359 (2021)
30. Z.K. Bagewadi, G.H. Illanad, I.A. Shaikh, M.H. Mahnashi, S.S. Shettar et al., Molecular expression, purification and structural characterization of recombinant L-Glutaminase from Streptomyces roseolus. Int. J. Biol. Macromolec. **273**, 133142 (2024)
31. A. Yap, I. Glarcher, M. Misslinger, H. Haas, Characterization and engineering of the xylose-inducible xylP promoter for use in mold fungal species. Metab. Eng. Commun. **15**, e00214 (2022)
32. W.A. Raja, P. Çalık, Biochemical production from sustainable carbon sources by Komagataella phaffii. Biochem. Eng. J. **219**, 109702 (2025)
33. L.C.M. Neri, H. Guðmundsson, G. Meurrens, A. Robert, O.H. Fridjonsson, G.O. Hreggvidsson et al., Identification and characterization of endo-xylanases from families GH10 and GH11 sourced from marine thermal environments. Enzyme Microb. Technol. **187**, 110592 (2025)
34. C. Stefanello, Y.K. Dalmoro, H.V. Rios, M.S. Vieira, M.L. Moraes, O.F. Souza et al., A Bacillus subtilis xylanase improves nutrient digestibility, intestinal health and growth performance of broiler chickens undergoing an intestinal challenge. Poult. Sci. **104**, 104908 (2025)
35. Q. Shi, J. Ma, A.M. Abdel-Hamid, Y. Li, P. Zhong, D. Wang et al., Mining of latent feruloyl esterase resources in rumen and insight into dual-functional feruloyl esterase-xylanase from Pecoramyces ruminantium F1. Biores. Technol. **418**, 131854 (2025)
36. Ritika, S. Pant, A. Prakash, P. R. Vundavilli, K. C. Khadanga, A. Kuila et al., Concomitant inhibitor-tolerant cellulase and xylanase production towards sustainable bioethanol production by Zasmidiumcellare CBS 146.36. Fuel **375**, 132593 (2024)
37. J. Wang, C. Bao, H. Cao, F. Huang, Y. Liu, Y. Cao, Multi-copy expression of a protease-resistant xylanase with high xylan degradation ability and its application in broilers fed wheat-based diets. Int. J. Biol. Macromol. **257**, 128633 (2024)

38. J.P. Acevedo, M.T. Reetz, J.A. Asenjo, L.P. Parra, One-step combined focused epPCR and saturation mutagenesis for thermostability evolution of a new cold-active xylanase. Enzyme Microb. Technol. **100**, 60–70 (2017)
39. H. He, J. Liu, Y. Wang, Y. Chen, K. Lu, M. Liu, Site-directed mutagenesis of family GH10 Aspergillus fumigatus xylanase A and the interaction with Oryza sativa xylanase inhibitor protein. Biocatal. Agric. Biotechnol. **54**, 102920 (2023)
40. Z.-Y. Bai, S. You, F. Zhang, Z.-W. Dong, Y.-F. Zhao, H.-J. Wen et al., Efficient fermentable sugar production from mulberry branch based on a rational design of GH10 xylanase with improved thermal stability. Renew. Energy **206**, 566–573 (2023)
41. H. Xing, G. Zou, C. Liu, S. Chai, X. Yan, X. Li et al., Improving the thermostability of a GH11 xylanase by directed evolution and rational design guided by B-factor analysis. Enzyme Microb. Technol. **143**, 109720 (2021)
42. C. Du, H. Li, M. Zhang, X. Wang, L. Xing, L. Wan et al., Improving the thermostability of ulvan lyase from polysaccharide lyase family 25 based on multiple computational rational design strategies. Int. J. Biol. Macromol. **302**, 140468 (2025)
43. L.F. Ribeiro, G.P. Furtado, M.R. Lourenzoni, R.J. Ward, Design, construction and characterization of laccase-xylanase chimeras by insertional fusion, in *Methods in Enzymology* (Academic Press, 2025)
44. S.R. Lee, D.R. Khan, J.Y. Park, S. Kaczmarek, E.J. Choi, Novel thermal stability enhanced xylanase improves the performance and digestibility parameters in broilers. Poult. Sci. **103**, 104447 (2024)
45. Z. Lai, C. Zhou, X. Ma, Y. Xue, Y. Ma, Enzymatic characterization of a novel thermostable and alkaline tolerant GH10 xylanase and activity improvement by multiple rational mutagenesis strategies. Int. J. Biol. Macromol. **170**, 164–177 (2021)
46. Q. Wu, C. Zhang, W. Dong, H. Lu, Y. Yang, W. Li et al., Enhanced thermostability of xylanase XynA via computationally designed assembly of multiple N-terminal disulfide bridges. Process Biochem. **138**, 67–78 (2024)
47. S. Sharma, P. Verma, K. Agrawal, Harnessing the potential of fungal xylanases: An insight into its application and technological advancements. Ind. Crops Prod. **222**, 119967 (2024)
48. M. Liu, J. Li, A.U. Rehman, S. Luo, Y. Wang, H. Wei et al., Sensitivity of family GH11 Bacillus amyloliquefaciens xylanase A (BaxA) and the T33I mutant to Oryza sativa xylanase inhibitor protein (OsXIP): An experimental and computational study. Enzyme Microb. Technol. **156**, 109998 (2022)
49. A. Purohit, G. Singh, S.K. Yadav, Chimeric bi-functional enzyme possessing xylanase and deacetylase activity for hydrolysis of agro-biomass rich in acetylated xylan. Colloids Surf. B **204**, 111832 (2021)
50. A. Arsalan, Y. Ravikumar, X. Tang, Z. Cao, M. Zhao, W. Sun et al., Chimeric enzymes in the pulp and paper making industry: current developments. Biotechnol. Adv. **79**, 108530 (2025)
51. S. Shahid, S. Batool, A. Khaliq, S. Ahmad, H. Batool, M. Sajjad et al., Improved catalytic efficiency of chimeric xylanase 10B from Thermotoga petrophila RKU1 and its synergy with cellulases. Enzyme Microb. Technol. **166**, 110213 (2023)
52. E. Abedi, F. Fatemi, Y. Sefidbakht, S.E.R. Siadat, Development and characterization of a thermostable GH11/GH10 xylan degrading chimeric enzyme. Enzyme Microb. Technol. **149**, 109854 (2021)
53. Y. Lai, J. Wang, N. Xie, G. Liu, D.C. Lacap-Bugler, Identification of a novel forkhead transcription factor MtFKH1 for cellulase and xylanase gene expression in Myceliophthora thermophila (ATCC 42464). Microbiol. Res. **294**, 128097 (2025)
54. L. Yang, R. Luo, Z. Tong, Y. Song, G. Zhou, Z. Peng et al., Achieving efficient co-expression of endo- β-1,4-xylanase and α-arabinofuranosidase in Trichoderma reesei and application in the production of arabino-xylo-oligosaccharides. Int. J. Biol. Macromol. **306**, 141599 (2025)
55. H. Yang, D. Sun, Y. Li, W. Zhao, Z. Jiang, Q. Yan et al., Development of a green, efficient and renewable biorefinery strategy for the production of xylooligosaccharides from corncobs based on a novel specific xylanase. Carbohyd. Polym. Technol. Appl. **10**, 100858 (2025)

56. Y. Li, C. Li, S. Muhammad Aqeel, Y. Wang, Q. Zhang, J. Ma et al., Enhanced expression of xylanase in Aspergillus niger enabling a two-step enzymatic pathway for extracting β-glucan from oat bran. Bioresource Technol. **377**, 128962 (2023)
57. C.K. Kim, H.S. Choi, S.J. Lee, J.H. Lee, J.H. Lee, H.Y. Yoo et al., Production of xylanase from a novel engineered Pichia pastoris and application to enzymatic hydrolysis process for biorefinery. Process Biochem. **65**, 130–135 (2018)
58. Y. Liu, H. Chen, Y. Zhang, C. Liu, L. Song, Metagenomics-resolved genomics provide novel ecological insights into resistome community coalescence of wastewater in river environment. Water Res. **267**, 122473 (2024)
59. K. Junge, K. Cameron, B. Nunn, Chapter 12—Diversity of psychrophilic bacteria in sea and glacier ice environments—insights through genomics, metagenomics, and proteomics approaches, in *Microbial Diversity in the Genomic Era*, ed. by S. Das and H.R. Dash (Academic Press, 2019), pp. 197–216
60. E. Zhang, S. Breselge, N. Carlino, N. Segata, M.J. Claesson, P.D. Cotter, A genomics-based investigation of acetic acid bacteria across a global fermented food metagenomics dataset. iScience **28**, 112139 (2025)
61. K.H. Sim, J. Ho, J.Q. Lim, S.H. Chan, A. Li, K.R. Chng et al., A metagenomics-based workflow for the detection and genomic characterization of GBS in raw freshwater fish. Microbiol. Spectrum **12** (2024)
62. T.A. Pellegrinetti, G.G.T.N. Monteiro, L.N. Lemos, R.A.C.d. Santos, A.G. Barros, L.W. Mendes, PGPg_finder: a comprehensive and user-friendly pipeline for identifying plant growth-promoting genes in genomic and metagenomic data. Rhizosphere. **30**, 100905 (2024)
63. A.K. Rout, P.S. Tripathy, S. Dixit, D.U. Behera, B. Behera, B.K. Das et al., Metagenomics analysis of sediments of river Ganga, India for bacterial diversity, functional genomics, antibiotic resistant genes and virulence factors. Curr. Res. Biotechnol. **7**, 100187 (2024)
64. C.C. Saak, E.C. Pierce, C.B. Dinh, D. Portik, R. Hall, M. Ashby et al., Longitudinal, multi-platform metagenomics yields a high-quality genomic catalog and guides an in vitro model for cheese communities. mSystems. **8** (2023)
65. S. Zeng, S. Wang, D. Mu, Metagenomics for IgA-coated gut microbiota: from taxonomy to function. Trends Microbiol. (2025)
66. D. Talens-Perales, E. Jiménez-Ortega, P. Sánchez-Torres, J. Sanz-Aparicio, J. Polaina, Phylogenetic, functional and structural characterization of a GH10 xylanase active at extreme conditions of temperature and alkalinity. Comput. Struct. Biotechnol. J. **19**, 2676–2686 (2021)
67. S.F. Ullah, A.A. Souza, P.R.V. Hamann, A.R.P. Ticona, G.M. Oliveira, J.A.R.G. Barbosa et al., Structural and functional characterisation of xylanase purified from Penicillium chrysogenum produced in response to raw agricultural waste. Int. J. Biol. Macromol. **127**, 385–395 (2019)
68. Y. Huang, P.K. Busk, L. Lange, Cellulose and hemicellulose-degrading enzymes in Fusarium commune transcriptome and functional characterization of three identified xylanases. Enzyme Microb. Technol. **73–74**, 9–19 (2015)
69. D.Y. Kim, S.H. Lee, M.J. Lee, H.-Y. Cho, J.S. Lee, Y.H. Rhee et al., Genetic and functional characterization of a novel GH10 endo-β- 1,4-xylanase with a ricin-type β-trefoil domain-like domain from Luteimicrobium xylanilyticum HY-24. Int. J. Biol. Macromol. **106**, 620–628 (2018)
70. Y. Nakamichi, T. Fouquet, S. Ito, M. Watanabe, A. Matsushika, H. Inoue, Structural and functional characterization of a bifunctional GH30-7 xylanase B from the filamentous fungus Talaromyces cellulolyticus. J. Biol. Chem. **294**, 4065–4078 (2019)
71. M. Gaffney, S. Carberry, S. Doyle, R. Murphy, Purification and characterisation of a xylanase from Thermomyces lanuginosus and its functional expression by Pichia pastoris. Enzyme Microb. Technol. **45**, 348–354 (2009)
72. D. Driss, J.G. Berrin, N. Juge, F. Bhiri, R. Ghorbel, S.E. Chaabouni, Functional characterization of Penicillium occitanis Pol6 and Penicillium funiculosum GH11 xylanases. Protein Expr. Purif. **90**, 195–201 (2013)
73. H. Jun, Y. Bing, Z. Keying, C. Daiwen, Functional characterization of a recombinant thermostable xylanase from Pichia pastoris: a hybrid enzyme being suitable for xylooligosaccharides production. Biochem. Eng. J. **48**, 87–92 (2009)

74. Y. Song, Y.G. Lee, I.S. Choi, K.H. Lee, E.J. Cho, H.-J. Bae, Heterologous expression of endo-1,4-β-xylanase A from Schizophyllum commune in Pichia pastoris and functional characterization of the recombinant enzyme. Enzyme Microb. Technol. **52**, 170–176 (2013)
75. T. Collins, M.-A. Meuwis, I. Stals, M. Claeyssens, G. Feller, C. Gerday, A novel Family 8 Xylanase, functional and physicochemical characterization*. J. Biol. Chem. **277**, 35133–35139 (2002)
76. I. Zghab, K.u. Rehman, M. Alissa, S.A. Alghamdi, A. Alghamdi, M.A. Alshehri et al., Xylanase functionalized hydrothermal synthesis of novel XLNs/Pt-$BiVO_4$ nanocomposite: ultra efficient nanocatalyst with enhanced photocatalytic and biocidal applications. Inorg. Chem. Commun. **172**, 113586 (2025)
77. A. Rezayat, A. Kakanejadifard, A. Taheri-Kafrani, F. Trotta, Xylanase immobilized on β-cyclodextrin functionalized magnetic nanoparticles: Implications for antioxidant component extraction and fruit juice quality improvement. J. Mol. Liq. **387**, 122660 (2023)
78. S. Kharazmi, A. Taheri-Kafrani, Bi-enzymatic nanobiocatalyst based on immobilization of xylanase and pectinase onto functionalized magnetic nanoparticles for efficient fruit juice clarification. LWT **183**, 114914 (2023)
79. D. Alagöz, N.E. Varan, A. Toprak, D. Yildirim, S.S. Tukel, R. Fernandez-Lafuente, Immobilization of xylanase on differently functionalized silica gel supports for orange juice clarification. Process Biochem. **113**, 270–280 (2022)
80. H. Wang, X. Qi, S. Gao, G. Kan, L. Damdindorj, Y. An et al., Characterization of a novel multifunctional β-glucosidase/xylanase/feruloyl esterase and its effects on improving the quality of Longjing tea. Food Chem. **453**, 139637 (2024)
81. K. Sharma, I.L. Antunes, V. Rajulapati, A. Goyal, Low-resolution SAXS and comparative modeling based structure analysis of endo-β-1,4-xylanase a family 10 glycoside hydrolase from Pseudopedobacter saltans comb. nov. Int. J. Biol. Macromol. **112**, 1104–1114 (2018)
82. S. Wu, N. Zhang, Q. Wan, Disulfide bonds enhance thermal stability and thumb region drives activity of the glycoside hydrolase 11 xylanase rMxylcd. J. Struct. Biol. **217**, 108209 (2025)
83. S. Singh, J. Ahmed, P.V. Gavande, C.M.G.A. Fontes, A. Goyal, Structural and functional insights into the glycoside hydrolase family 30 xylanase of the rumen bacterium Ruminococcus flavefaciens. J. Mol. Struct. **1272**, 134155 (2023)

Chapter 6
Immobilization and Stabilization of Xylanase

Amama Tariq, Abdul Sattar Jatoi, Talha Rana, Muhammad Farhan, and Abbas Ali

Abstract The hydrolytic enzyme xylanase is a chemically relevant enzyme that has broad industrial applications including pulp and paper, food and feed, textile, and biofuels industry. However, the bioactivity and stability of the enzyme have historically been affected by some factors, e. g. its ability to be reused, its operational durability in harsh processing conditions, and its use as a catalyst under very severe conditions. Immobilization techniques are useful as possible approaches to enhance the thermal and operational stability of xylanase and allow repeated use of the enzyme, as well as maintain degradation efficiency over long periods of time. This review (or study) reviews various immobilization techniques, including adsorption, covalent binding, entrapment, and encapsulation, as well as their ability to modify the biochemical properties of the enzyme. The bioactivity of the enzyme with respect to supportive material effects, such as natural polymers, synthetic matrices, and nanomaterials, are investigated critically. Finally, the latest progress in the methods of stabilization of xylanase is discussed, including genetic engineering, cross-linking, and additives for building resistance to denaturation.

Keywords Immobilization · Stabilization · Xylanase

1 Introduction

Xylanase (endo-1,4-β-xylanase, E.C.3.2.1.8A) is a member of hemicelluloses, acting in synergy with other hydrolases such as β-xylosidase, α-arabinofuranosidase, and acetyl Xylan esterase to depolymerize hemicellulose structure. The enzyme specifically hydrolyses the β-1,4-glycosidic bonds of Xylan backbone, **i.e.,** of β-D-Xylo pyranose units. This hydrolysis leads to the formation of xylo-oligosaccharides, which are further broken down by β-xylosidases to release xylose [13].

A. Tariq · A. S. Jatoi (✉) · T. Rana · M. Farhan · A. Ali
Department of Chemical Engineering, Dawood University of Engineering and Technology Karachi, Karachi, Pakistan
e-mail: abdul.sattar@duet.edu.pk

A. Sattar Jatoi (ed.), *Xylanase from Microorganisms*,
https://doi.org/10.1007/978-981-95-0824-2_6

Xylanases are produced by numerous microorganisms such as fungi and bacteria. The extremophile nature of fungi, in particular filamentous fungi, has been well researched in relation to xylanase production because of their potential to secrete large amounts of enzyme extracellularly, and they can therefore be used for industrial large-scale production. In contrast, bacterial xylanases, especially those isolated from extremophilic bacteria, have been the focus of great attention for industrial applications owing to their high activities. These bacteria are commonly well suited for biotechnological applications, as they can produce xylanases under neutral to alkaline pH, thereby decreasing the number of downstream processing steps [11]. Xylanases are found also in marine animals, in crustaceans, snails, green algae, protozoans and they synthesize such enzymes. Marine-derived xylanases exhibit remarkable characteristics, such as hyper thermostability, cold adaptation, and the ability to withstand high salt concentrations, which make them particularly valuable for industrial processes under extreme conditions [11]. However, there is controversy in the selection of the best microbial producer for xylanase production, in the past filamentous fungi have been preferred because of their ability to produce high amounts of xylanase in the growth medium. However, bacterial xylanases have also emerged as a formidable competitor, especially because bacterial production is often more economical and may also be easily upscaled to an industrial production scale. By contrast to fungi, bacteria do not concatenate cellulases with xylanases, simplifying the work downstream [11].

Both solid-state fermentation (SSF) and submerged fermentation (SMF) are used for xylanase production and have their advantages. SSF has also been proven to yield higher productivity and lower costs, so, it has become more popular. However, SMF is easier to operate, generates a more concise and straightforward downstream processing such as enzyme extraction which can be very expensive and labor-intensive in SSF [14]. Selection of the starting materials is one of the crucial problems for xylanase production, and especially the lignocellulosic plant biomass (e.g., wheat bran or rice husk) commercially available and economically feasible (Fig. 1). However, commercial Xylan still is very expensive, which is a limit for large-scale Xylan production. For these reasons, alternative molecular strategies have been employed, including gene overexpression of xylanases encoding genes in microorganism hosts such as *E. coli*, which results in better enzyme productivity and activity of the recombinant enzymes [14].

In summary, xylanase enzymes are key biocatalysts in various industrial processes, with a wide range of applications in biorefineries, food processing, textile industries, and beyond. Continuous development in microbial production, fermentation strategies and genetic engineering further optimizes the productivity of xylanase for scaling with the goal of being a sustainable industrial enzyme.

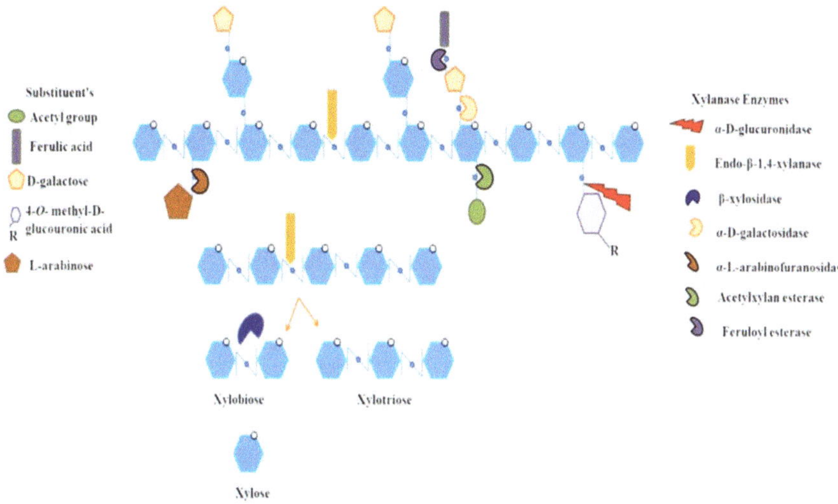

Fig. 1 Structural representation of xylan with targeted glycosidic bonds for hydrolysis by *Xylanolytic* enzymes [3]

1.1 Industrial Applications of Xylanase Enzymes

Xylanase enzymes, which break Xylan, a major plant polysaccharide of hemicellulose, have been widely used in many fields because of their xylanolytic activity. Xylanases can also serve as eco-friendly substitutes for strong chemical bleaching agents to prevent environmental pollution. When hemicellulose is degraded by xylanases, the pulp brightness can be enhanced, and the use of chlorine-based chemicals can be minimized [11]. Xylanases can enhance paper quality through enhanced brightness, strength, and smoothness. They are also able to break down hemicellulose for enhanced fiber bonding and enhanced paper properties [11]. Xylanases in animal feed can hydrolyze complex polysaccharides (i.e., arabinoxylans) into simpler sugars, which are more digestible. This can enhance nutrient absorption and animal growth [6]. Xylanases can be used to improve bread quality through a reduction in loaf volume, softening and shelf life. When hemicellulose is fragmented in wheat flour xylanases the rheology and fermentation of dough can be improved [6]. In the fruit juice industry, xylanases can be used for clarification of fruit juices due to their ability to hydrolyze the polysaccharides responsible for turbidity. This leads to cleaner and more attractive juice products [6]. Xylanases can enhance beer quality by stimulating the release of fermentable sugars from malt. This may result in increased alcohol concentration and improved flavor [6]. Xylanases are capable of decomposing hemicellulose, a key component of lignocellulosic biomass, into simple sugars (Fig. 2). These sugars can subsequently be fermented to obtain bioethanol or other biofuels [14]. Xylanases can be employed to neutralize gums and pectin's from natural fibers such as flax and jute, preparing them for fabric production [14].

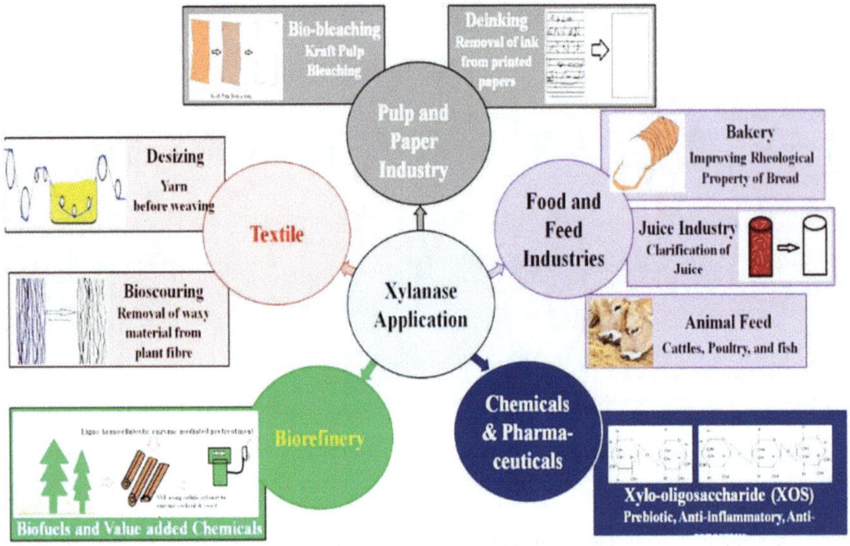

Fig. 2 Xylanase eco-friendly enzyme applications across industries [3]

Once specific uses of xylanases are understood, it is possible for various industries to exploit the potential for greener, more efficient technologies.

1.2 Need for Immobilization

Enzymes are biocatalysts and catalyze biochemical and chemical reactions with significant importance in the regulation of life processes in living organisms. For centuries, natural materials derived from biological sources have been utilized to catalyze chemical reactions, such as the production of acrylamide through nitrile hydratases [9].

Recent advancements in biocatalysis have enabled the synthesis of complex molecules, with enzymes proving particularly effective due to their ability to direct group-controlling selectivity. Furthermore, the activity and/or selectivity of enzymes can be engineered to operate both in organic and aqueous solvents without any inactivation or loss of selectivity. Enzymes are more and more exploited in different fields, including food, pharmaceutics and biotech, utilizing them for production of dairy, vegetable juices, beer and fermented starch, providing an environmentally and human friendly alternative to chemical catalysts [7].

Nevertheless, enzymes have also some challenges, such as low stability, low selectivity and low activity. The potency of their use is dramatically sensitive to their three-dimensional shape, which confers chemo-, regio-, and stereoselectivity to allow the synthesis of complex and chiral ligands. Although enzyme technology can be promising, for practical applications enzymatic technologies are limited by problems of high cost, short-term stability, and poor reusability. To mitigate these drawbacks, enzyme immobilization has become an important approach. This process involves fixing enzymes onto solid support materials, enhancing their stability, reusability, and catalytic efficiency. Immobilization enhances enzyme characteristics, such as heat resistance, pH resistance, and functional stability, which are very important in industrial applications. Moreover, immobilized enzymes can be repeatedly used prior to irreversible denaturation which enhances dramatically their economic success in high-volume processes [7].

Immobilization also offers some practical advantages like making the separation of the enzymes from the reaction products easier, lowering the contamination and increasing the product purity. This is especially useful, for example, in food and pharmaceuticals industries, for maintaining the quality of the product. In addition, the immobilization of enzyme can help overcome the autocatalysis issue that makes industrial reactions difficult by changing the reaction condition and destroying the enzymes. For instance, immobilization of urease on amino resin beads improved the resistance to denaturation and suppressed the pH changes during the reaction of urease-catalyzed urea hydrolysis reaction and thus improved the reaction reproducibility. In addition, immobilization maintains enzymes against enzyme inactivation triggered by protein unfolding or subunit release, and thus provides structural rigidity and stability to multimeric [8].

In summary, enzyme immobilization addresses key challenges in enzyme use, such as instability, inefficiency, and limited reusability, making it an indispensable technique for improving enzyme performance in a variety of industrial and commercial applications.

1.3 Brief Overview on Functionalized Nanoparticles of Immobilization of Enzyme

Nanoparticles have become a unique platform for immobilization of enzymes, with the following advantages over conventional protocols. Because of their specific characteristics as having high surface area, adjustable pore size, and increased stability, they are excellent candidates toward the development of very efficient and stable biocatalysts. Due to their large surface area, the nanoparticles have higher enzyme loading, and the resulting catalytic activity is stronger. Due to the large available surface area, mass transfer from/to substrates and products to/from the enzyme active sites are optimized, eliminating diffusional bottlenecks.

Enzymatic immobilization on nanoparticles can greatly increase the thermostability of enzymes by enabling them to be used at physiological temperature. Nanoparticles protect enzymes from denaturation at extreme pH. Enhanced resistance to proteolytic degradation: Immobilization can protect enzymes from the action of protease enzymes extending their lifetime.

Immobilized enzymes can be conveniently isolated from the reaction mixture and can be repeatedly used with payback costs.

Immobilization can enhance the long-term stability of enzymes by decreasing the frequency of replacement [10]. The nanoparticle pore size may be adjusted to maximize the enzyme's access and the product transport in and out of the particle. The surface of nanoparticles can be functionalized with target groups to improve enzyme binding, orientation and activity [10]. Hydrolysis of complex carbohydrates, proteins, and lipids to generate sugars, peptides, and fatty acids. Synthesis of sweeteners, flavors, and nutraceuticals. Transform low-value substrates to high-value products, for example, lactose to glucose and galactose.

Detection of contaminants, allergens, and other quality parameters. Real-time monitoring of food quality and freshness[1].

Despite the great promise of nanoparticles, there are challenges to be tackled by cost, toxicity, and regulatory issues. To address these limitations, researchers have been investigating the application of biocompatible degradation nanoparticles as well as the development of more effective and cost-effective production techniques.

2 Properties and Characteristics of Xylanase

2.1 Structure and Function of Xylanase

Xylanase enzymes are generally from glycoside hydrolase families 10 or 1. They all possess a conserved architecture, containing a catalytic domain responsible for Xylan hydrolysis and a carbohydrate-binding module (CBM) that can mediate substrate binding [3].

When xylanase is immobilized, β-1,4-glycosidic bond hydrolysis activity in Xylan is maintained. Immobilization has a stabilizing effect and, therefore, allows it to be used in demanding environments, namely, at elevated temperatures and for long periods of time. This renders immobilized xylanase intensively attractive in industrial processes including biofuel production, food processing, and paper bleaching [3].

2.2 Substrate Specifics and Kinetic Parameters

If a xylanase molecule is immobilized, its substrate specificity and kinetic parameters are not the same as its free molecule. Here are some specifics:

Xylanase is the major substrate of Xylan, which is a major constituent in plant cell walls. Xylo oligosaccharides are immobilized xylanase can also catalyze Xylo oligosaccharide transformation into simple sugars [5].

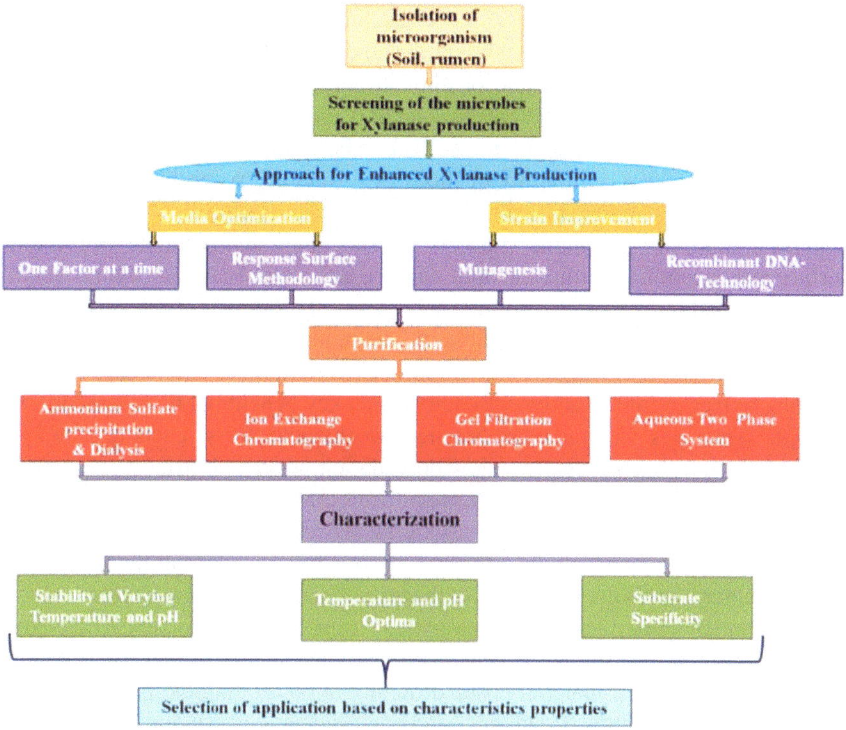

Fig. 3 Schematic overview of xylanase production and purification process [3]

Michaelis Constant (Km) value can vary depending on immobilization technique and support material. [e.g., Km values for immobilized xylanase in a perovskite composite, nanocomposite or single support vary from 0.639 to 5.430 mg/mL].

Maximum Velocity (Vmax) parameter also changes with immobilization. [e.g., values for immobilized xylanase can vary from 0.167 to 0.226 μmol/min/mL] [2].

Immobilized xylanase is frequently reported with elevated thermostability, which is retained with higher activity at high temperatures. The immobilized xylanase can be recycled several times retaining a significant portion of its activity in multiple cycles [4] (Fig. 3).

2.3 Existing Challenges and Limitations

Both immobilization and stabilization can play a major role in improving the industrial application of xylanase, but there are still problems and limitations to be faced. Shortcoming: Immobilization and stabilization can cause xylanase to lose its enzymatic activity during or after immobilization. This problem is most likely due to conformational changes of the enzyme's structure when it reacts with the immobilization matrix (mainly in covalent bonding and cross-linking) and can either restrict substrate access to the active site or alter the enzyme's three-dimensional structure which adversely affects its catalytic activity. Another limitation is the support material. While many supporting materials can be synthesized in both organic and inorganic forms, there is difficulty in choosing a suitable carrier that would provide optimal binding, stability, biocompatibility and cost optimization. Most support materials, especially inorganic supports such as silica and metal oxides, require surface modification to further enhance enzyme affinity and can increase production complexity and cost. As well, physical and chemical properties of the support (such as porosity, particle size and hydrophilicity) are required to be controlled to promote uniform distribution of enzymes and minimize restriction of mass transfer.

Immobilized xylanase is typically touted as a key benefit but its reusability over multiple cycles has been mixed in practice. Mechanical stress, leaching of enzyme from poorly bound supports, and slow degradation under industrial conditions can reduce enzyme activity over time. Immobilization also limits the flexibility of the enzyme as well as ability to adapt to changing process conditions, reducing performance in applications where pH, temperature, or substrate concentrations may fluctuate. Another issue is the scalability of the immobilization techniques. Methods that work at the laboratory scale e. g. entrapment using sol-gels or nanoparticle-based immobilization systems may not be economically feasible or technically feasible at large scale due to their material costs, difficulty in controlling the process or difficulties in mass production. Regulatory and environmental considerations may occur due to the non-biodegradable and potentially toxic materials used in certain immobilization systems. Finally, the lack of insight into the interactions between xylanase and immobilization matrices (particularly at the molecular level) delays the rational design of efficient immobilization strategies. The focus must therefore remain on the research on the structural–functional interfaces, the interfaces between enzymes and surfaces, and novel hybrid materials, in order to overcome these limitations and realize the full potential of xylanase immobilization in industrial biotechnology.

3 Immobilization Techniques of Xylanase Enzyme

Immobilization of the xylanase enzymes are performed by bonding enzymes on a solid support or matrix, to enhance the stability, reusability, and operational effectiveness in industrial scale applications (Fig. 4). Various methods are used for this, such as physical adsorption, covalent ligation, entrapment, cross-linking, which provide each their own set of benefits and drawbacks.

The present technique is based on Weeks's interactions (van der Waals force, hydrogen bonds, and electrostatic force) for immobilization of an enzyme on the support material. Despite their simplicity and low cost, physical adsorption is also known to be less stable because the enzyme can be desorbed due to environmental changes such as pH or temperature [11].

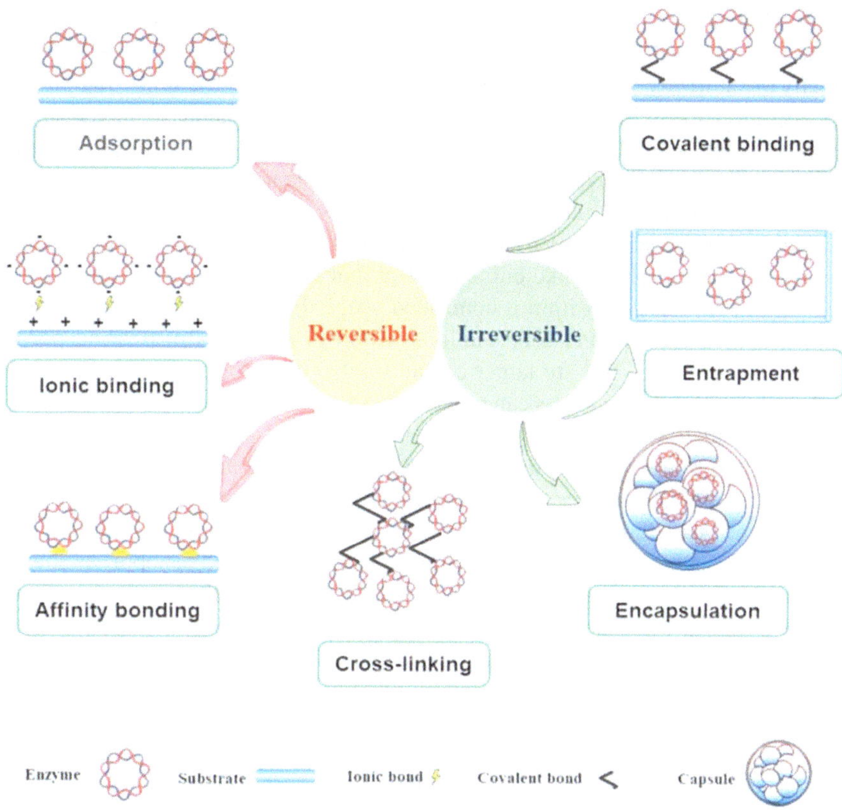

Fig. 4 Various methods of immobilization [7]

Enzymes are covalently anchored to the support matrix by chemical linkers, and strong and covalent immobilization is achieved. E.g., recombinant β-xylanase from Thermology anthophilia was immobilized onto silica-coated Fe_3O_4 magnetic nanoparticles by carbodiimide activation. This method improves both enzyme stability and reusability and has the further advantage of easy magnetic separation for the industrial process [12]. In this method, enzymes are locally immobilized in a matrix but do not adhere to it lying on each other, to the study of their interactions with each other and their environment, whether explicitly (for instance, a drug binding to an enzyme) or implicitly (for instance, a drug interacting with an enzyme that can alter the environment of a target protein) or even by connecting them indirectly (such as an enzyme in one step of a metabolic pathway facilitating the transformation of a compound that could have been the product of a related reaction in another step of the same pathway). Magnetic and starch nanoparticles are used as effective carriers with high enzyme loading and the ability to resist environmental influences. Especially for enzymes and microenvironment control entrapment are very useful, as depicted in the immobilization of different enzymes on graphene and starch nanoparticles for activity enhancement and recycling [1]. Cross-linking is the process of covalent bonds being established between enzyme molecules, either directly or indirectly via a bifunctional reagent. This method has improved thermal and operational stability, such that xylanase is robust to extreme industrial conditions. Specifically, applications of β-cyclodextrin-functionalized magnetic nanoparticles have been discussed for the diacylation of xylanase in the industry, which can enhance its reusability and performance [14]. The encapsulation immobilization technique entails entrapping various biomolecules in different polymeric matrices. Encapsulation is markedly like entrapment in that both the enzymes and the cells are free in solutions yet within a controlled space. Encapsulation aims to immobilize sensitive enzyme and cell solutions within tiny vesicles possessing porous membranes. Enzymes that are larger cannot freely enter or exit from the capsules, while small substrates/products can move across the semipermeable membrane with ease. Encapsulation keeps biological systems embedded in an extremely thin film to prevent the biocatalysts from contacting the interfering environment that could reduce their efficiency. Thus, encapsulation leads to prolonged activity of biocatalysts. Many materials (such as cellulose nitrate and nylon) can be used to build microcapsules. Ionotropic gelation of alginates and silica-based nanoporous sol–gel glasses has also proven effective for enzyme encapsulation. This method is distinguished by the fact that the encapsulation of enzyme could be very easily performed. Developments in materials science have contributed to synthesized encapsulates with improved properties, such as increased morphological stability, designed physicochemical permeability, and diminished enzyme leakage. There is also a medium for co-immobilization, whereby the enzymes are immobilized in any combination recommended for a specific use. Nonetheless, some limitations share a roundabout connection with this technique. For instance, the diffusion problem is quite limiting in quick accumulation of the products from the reaction that determines the rupture of membranes [7].

These methods, individually or in combination, have furthered the industrial utilization of xylanase to the point of economic and ecological solutions for biomass depolymerization, biofuel synthesis, and other applications.

4 Support Material for Xylanase Enzyme

Likewise used as matrices to construct recoverable biocatalytic systems with. improved enzyme durability in reaction environments. They are plentiful in nature and readily accessible, Solution with outstanding biocompatibility and does not require any further advanced processing. Purification, which renders them relatively inexpensive. Additionally, the existence of numerous functional, Functional groups (i.e., –OH, COOH, C=O, –SH, $-NH_2$) on the mineral surfaces are used to generate the. Even covalent bonds between the enzyme and support material facilitate tailoring of the minerals. Introduction of additional functional groups increases the area of adhesion and hydrophobicity. although the most frequently Lipases, α-amylases, tyrosinases, and glucose oxidases are immobilized. Enzymes fixed on minerals are primarily utilized in environmental engineering for the treatment of waste and wastewater, as well as. In biosensors to increase the linear range and detection threshold. For instance.

4.1 Inorganic Supports for Enzyme Immobilization

Inorganic supports are an important component of enzyme immobilization because of their high mechanical strength, chemical stability, and resistance to microbial degradation. Materials such as silica, alumina, controlled pore glass, zeolites, and various metal oxides provide high surface area, pore size control, and surface functionalization mechanisms that allow for covalent binding, adsorption, or entrapment of enzymes (Table 1). These supports are particularly suitable for industrial applications requiring high temperatures or extremely low pH values where organic carriers may degrade. Inorganic matrices can improve the thermal and operational stability of enzymes, enhance their reusability, and minimize leaching of their substrate to extend the enzyme catalytic lifetime. Significant advances in nanostructured inorganic materials and hybrid composites contribute to efficient enzyme immobilization systems which bring increased activity, specificity, and process efficiencies. These supports create a stable and inert environment for the immobilization of enzymes and thus are applicable for many different biocatalytic purposes [12].

4.2 Organic Support for Enzyme Immobilization

4.2.1 Natural Polymer

Natural polymers such as collagen, chitosan, carrageenan, alginate, cellulose, starch, and agarose are widely used as support matrices for enzyme immobilization due to their ability to form inert gels, their chemical structures that allow easy activation, and their capacity to bind proteins and enzymes reversibly or irreversibly (Table 2). These polymers are cheap, designable in large scale, and retain high thermal/mechanical stability upon being cross-linked with bifunctional reagents. Similarly, organic substrates, biopolymers and synthetic polymers, are also favored for enzyme immobilization because of their biocompatibility, good binding interaction with biomolecules and self-assembly capability of covalent bond with enzymes. Compared with inorganic matrices, where cross-linking reagents are necessary and biocompatibility and geometrical tunability are constrained, organic matrices bring improved catalytic reactivity and wider range of application possibilities [12].

Table 1 Inorganic Material, Their Properties, and Immobilized Enzymes

S.no.	Inorganic material	Properties	Enzyme immobilized
1	Zeolite (molecular sieves)	Large specific surface area of the zeolite substrate resulted in high enzyme loading	Glucose oxidase (GOx), α-chymotrypsin, lysozyme
2	Ceramics	Both macro- and micropores efficient in lowering diffusion rate and increasing specific surface area	Candida antarctica lipase
3	Celite	Inexpensive, low polarity, large adhesion area, chemically inert, resistance against pH, temperature, urea, detergents, and organic solvents	Lipase, polyphenol oxidase, β-galactosidase
4	Silica	Nanosized structures, high surface area, ordered arrangement, high stability to chemical and mechanical forces, high mechanical strength	Lignin peroxidase, horseradish peroxidase, α-amylase
5	Glass	Highly viscous liquid	α-Amylase, nitrite reductase
6	Activated carbon	Large contact sites, high surface area	Acid protease, acidic lipase
7	Charcoal	Excellent adsorbent, minimum fine particulate matter release	Papain, amyl glucosidase

4.2.2 Synthetic Polymer

Synthetic polymers, which are manufactured by careful selection of monomers, are extensively employed as support matrices for enzyme immobilization owing to the adjustable chemical and mechanical properties. The monomer choice imprints upon the polymer the degree of its solubility, porosity, stability and hydrophobic/hydrophilic behavior, to enable a specific interaction of the polymer with enzymes by adsorption or covalent immobilization. Functional groups, such as hydroxyl, carboxyl, amine and epoxy, can be exploited for surface functionalizing and enhancing enzyme binding. Synthetic polymers provide degree of freedom on matrix-enzyme spacers, improving the enzyme deformability, thermal properties and the reusability thereof. Polystyrene, PVC, polyurethane, and UV-activated polyethylene glycol are examples of suitable polymers for immobilizing specific enzymes, such as α-amylase, lipase, and glucose oxidase (and hence obtaining improved biocatalytic activity, thermal or oxidative stress resistance, etc.) [10].

Table 2 Organic support system, their properties, and immobilized enzymes

S.no.	Natural polymers as support	Properties	Enzyme immobilized
1	Alginate	Used as xanthan—alginate beads, alginate—polyacrylamide gels, and calcium alginate beads can be reused, improves the stability of enzymes	
2	Chitosan and chitin	Used in combination with alginate less leaching effect, more reliable for enzyme trapping, in form of beads can entrap more enzymes	D-Hydantoinase
3	Collagen	Retain significant activity after many cycles of reuse	Tannase, catalase
4	Carragen (linear sulfated polysaccharide)	Improve stability, as pseudoplastic in nature it helps it to thin under shear stress and recover its viscosity once stress is removed cheap, durable, better entrapment	Lipase, α-galactosidase
5	Gelatin (hydrocolloid material, high in amino acid)	High adsorption ability, promote loading efficiency	
6	Cellulose	Longer storage capacity, better formability, and flexibility	Fungi laccase, penicillin G acylase, glucoamylase, α amylase, tyrosinase, lipase, and β-galactosidase
7	Starch (homopolysaccharide)	Stable	Bitter gourd peroxidase
8	Pectin (heteropolysaccharide)	Used along with glycerol as plasticizer to reduce brittleness of support pectin–chitin and pectin–calcium alginate have enhanced thermal and denaturant resistance and catalytic properties	Papain
9	Sepharose	Porous, easy adsorption, retain catalytic properties at extremes of pH, temperature, high salt concentration	Amylase, glucoamylase

4.3 Composite Support for Enzyme Immobilization

So-called hybrid and composite materials combine the advantages of inorganic and organic matrices to produce biocatalytic systems with improved properties. These include the provision of a suitable environment for biomolecules; stabilization of enzyme-support interactions; and improved mechanical stability and resistance.

The properties exhibited by hybrids and composites could not be traced to those existing on any of the individual components. Hybrid supports stabilize enzyme-support interactions and protect them from conformational changes. Biocatalytic systems are indeed reinvented through hybrid supports. Composite supports will conduct from one organic across the board to another, a wide variety of enzymes, irrespective of catalytic class. For hybrids, organic-organic, inorganic-inorganic, or organic–inorganic precursors should aim at an equilibrium.

4.3.1 Organic-Organic Hybrids

Organic-organic hybrids are prepared by mixing synthetic and natural polymers to obtain materials with improved properties for enzyme immobilization. Organic-organic hybrids have improved properties, including pH and thermal resistance, mechanical stability, and biocompatibility. The hydrophobic/hydrophilic character of the matrix can be controlled to enhance enzyme-support interactions and catalytic activity. Organics and organic hybrids come in forms including particles, fibers, tubes, beads, membranes or sheets in varied forms. All classes of catalytic compatible with adsorption covalent immobilization encapsulation. This system ensures that hybrids can be suited for immobilization of the entire catalytic class for application. Reusability/stability—stable and recyclable catalytic systems via biocatalytic for retention of the catalytic property [15].

4.3.2 Organic-Inorganic Hybrids

Organic–inorganic hybrids combine materials of organic and inorganic origin to create support for enzyme immobilization. Organic-inorganic hybrids are good in stability and mechanical resistance due to their inorganic precursors. The organic components of these hybrids offer many functional moieties for interacting with biocatalysts. Organic-inorganic hybrids can be used for several enzyme immobilization methods including adsorption, covalent binding, and encapsulation. These hybrids have numerous practical applications due to stability, functionality, and versatility. Organic-inorganic hybrids enhance enzyme activity and stability to be applied in various biocatalytic applications [15].

4.3.3 Inorganic-Inorganic Hybrids

Inorganic-inorganic hybrids involve combining inorganic materials to prepare supports for enzyme immobilization. Inorganic-inorganic hybrids exhibit good pH and thermal stability, mechanical resistance, and chemical inertness. These hybrids have various functional groups, such as hydroxyl, carbonyl, and carboxyl groups, for enzyme immobilization. Presence of functional groups in inorganic-inorganic hybrids, which could make their surfaces easy for functionalization and modification.

This category of hybrid can be employed using many immobilization techniques such as adsorption, covalent binding, or entrapment. Unique properties make inorganic-inorganic hybrids interesting for further exploitation in various fields related to biosensors, biocatalysis, etc. [15].

5 Discussion

Xylanase enzymes are versatile biocatalysts that are applied in the pulp and paper, animal feed, food, biofuel and textile industries as they catalyze the depolymerization of hemicellulose to smaller sugars. Most of their microbial production, especially using fungi and bacteria, has been improved using fermentation strategies and each source provides various benefits in enzyme production and business mode. Weaknesses for instance enzyme stability, activity and reusability have been overcome by immobilization processes, which incorporate enzyme adsorption, covalent attachment, entrapment and cross linking. Functionalized nanoparticles, as characterized by large surface area, controlled pore size and improved thermal and pH stability have been found to be excellent support carriers for immobilized enzymes which has taken functionalization of nanoparticles to a higher level of industrial applications. Nevertheless, some problems have not been solved until now, for example, cost and difficulties in mass transfer; however, new immobilization methods should be investigated.

6 Conclusion

Optimization of xylanase enzymes and the enhancement of immobilization procedures give hope to eco-efficient methodologies for industries. Immobilization not only improves enzymes stability and reusability, but also allows the enzymes to be used in extreme conditions, which is very essential in industries such as biofuel and food processing. Functionalization of nanoparticles has extended the applications of enzymes to the next level of enhancement in terms of performance and reusability. Though some drawbacks are still present, further investigation on Microbial production and Immobilization methods thus offers a window towards the actualization of a broader utilization of the xylanase enzymes in more environmentally friendly and efficiency improved industrial applications. These efforts ethos the enzyme as the key driver for green technology innovations.

References

1. D. Agyei, B.K. Shanbhag, T.O. Akanbi, L. He,. Enzyme immobilization and engineering for food applications, in *Improving and Tailoring Enzymes for Food Quality and Functionality* (Elsevier, 2024), pp. 259–293. https://doi.org/10.1016/b978-0-443-15437-9.00001-x
2. B. Bakar, G. Dik, A. Ulu, B. Ateş, Immobilization of xylanase into zeolitic imidazolate framework-67 (ZIF-67) and manganese-doped ZIF-67 (Mn/ZIF-67): a comparison study. Top. Catal. **67**(9–12), 698–713 (2024). https://doi.org/10.1007/s11244-023-01898-1
3. N. Bhardwaj, B. Kumar, P. Verma, A detailed overview of xylanases: an emerging biomolecule for current and future prospective, in *Bioresources and Bioprocessing*, vol. 6, Issue 1 (Springer Science and Business Media Deutschland GmbH, 2019). https://doi.org/10.1186/s40643-019-0276-2
4. covalent-immobilization-of-xylanase-purified-from-aspergillus-niger-dfr-5-on-the-surface-of-glutar. (n.d.)
5. E.A. Karam, N.A. Elattal, F.H. Motawe, S.S. Hanafy, *Production, Immobilization and Evaluation of Xylanase's Efficacy from Streptomyces Sp. NRC-EAK-H26 Grown on Lignocellulosic Waste* (n.d.). https://doi.org/10.36632/csi/2024.13.3.36
6. B.K. Kashyap, M.K. Solanki, D.V Kamboj, A.K. Pandey, Waste to energy: prospects and applications, in *Waste to Energy: Prospects and Applications* (Springer, Singapore, 2021). https://doi.org/10.1007/978-981-33-4347-4
7. Y.R. Maghraby, R.M. El-Shabasy, A.H. Ibrahim, H.M.E.S. Azzazy, Enzyme immobilization technologies and industrial applications, in *ACS Omega*, vol. 8, Issue 6 (American Chemical Society, 2023), pp. 5184–5196. https://doi.org/10.1021/acsomega.2c07560
8. R.A. Meryam Sardar, Enzyme immobilization: an overview on nanoparticles as immobilization matrix. Biochem. Anal. Biochem. **04**(02) (2015). https://doi.org/10.4172/2161-1009.1000178
9. H.H. Nguyen, M. Kim, An overview of techniques in enzyme immobilization. Appl. Sci. Convergence Technol. **26**(6), 157–163 (2017). https://doi.org/10.5757/asct.2017.26.6.157
10. I. Onoka, J.J. Makangara, Micro- and nanocarriers for immobilization of enzymes, in *Marine Biopolymers: Processing, Functionality and Applications* (Elsevier, 2024), pp. 545–572. https://doi.org/10.1016/B978-0-443-15606-9.00019-X
11. F.I. Qeshmi, A. Homaei, P. Fernandes, R. Hemmati, B.W. Dijkstra, K. Khajeh, Xylanases from marine microorganisms: a brief overview on scope, sources, features and potential applications, in *Biochimica et Biophysica Acta—Proteins and Proteomics*, vol. 1868, Issue 2 (Elsevier B.V., 2020) https://doi.org/10.1016/j.bbapap.2019.140312
12. V.L. Sirisha, A. Jain, A. Jain, Enzyme immobilization: an overview on methods, support material, and applications of immobilized enzymes, in *Advances in Food and Nutrition Research*, vol. 79 (Academic Press Inc., 2016), pp. 179–211. https://doi.org/10.1016/bs.afnr.2016.07.004
13. M. Srivastava, N. Srivastava, R. Singh, *Clean Energy Production Technologies Series Editors: Neha Srivastava P. K. Mishra Bioenergy Research: Commercial Opportunities & Challenges* (n.d.). http://www.springer.com/series/16486
14. D. Tyagi, D. Sharma, Production and industrial applications of xylanase: a review, in *International Journal of Scientific Research & Engineering Trends*, vol. 7 (n.d.)
15. J. Zdarta, A.S. Meyer, T. Jesionowski, M. Pinelo, A general overview of support materials for enzyme immobilization: characteristics, properties, practical utility, in *Catalysts*, vol. 8, Issue 2. MDPI (2018). https://doi.org/10.3390/catal8020092

Chapter 7
The Role of Xylanase in Agriculture and Animal Feed

Shabana Memon, Babar Ali Shaikh, Hamid Ali Memon, Muhammad Ali Memon, Sadaf Memon, and Tayyaba Memon

Abstract Xylanase plays a crucial role in enhancing nutritional values in agriculture and animal feeds. This enzyme has hydrolytic activities, cleave β-1, 4, a vital component of plant cell wall polysaccharide Xylan. It breaks down Xylan to xylo-oligosaccharides and releases sugars and nutrients used for digestibility in animals and poultry. and is being applied in a wide range of areas, including baking food, animal feed, beverages, oil, solvents, chemicals, paper, and industrial biofuels. It reduces viscosity in the digestive tract and helps feed efficiency. Xylanase provides anti-nutritional effects on various non-starch polysaccharides, enhancing a balanced digestive system in animals. In most of the feeds of animals and industrial products such as chemicals, fuels, textiles, bleaching of cellulose pulp, solvents, alcohols, oils, sugar, and organic acids xylanase has been added for several decades. Significantly, xylose from biomass is being used in bio-based commercial products. Hence, the use of xylanases would greatly enhance the overall economic materials and help on sustainability of agriculture through processing lignocellulosic for the generation of liquid fuels and chemicals. Consequently, biowastes such as sugarcane and corn substrates can be efficiently valorized for biofuel production, and used an alternative transportation fuel from renewable energy sources, so that to deprive the petroleum-derived fuel challenges.

S. Memon (✉) · S. Memon · T. Memon
Department of Plant Breeding and Genetics, Sindh Agriculture University Tandojam, Tando Jam, Pakistan
e-mail: smemon@sau.edu.pk

S. Memon
e-mail: smemon@sau.edu.pk

B. A. Shaikh
Agriculture Research Sindh, Tando Jam, Pakistan

H. A. Memon
Livestock and Fisheries Department, Government of Sindh, Karachi, Pakistan

M. A. Memon
Ziauddin University, Karachi, Pakistan

A. Sattar Jatoi (ed.), *Xylanase from Microorganisms*,
https://doi.org/10.1007/978-981-95-0824-2_7

Keywords Xylanses · Bio-products · Wheat · Hemicellulose · Organic acid · Fuel and animal feed

1 Introduction

Xylanase is a hydrolase that breaks down hemicellulose, a component of the cell wall, into fermentable sugars in plants. It is the second significant polysaccharide found in plant cells [15]. Xylanase is a xylan-degrading enzyme, acting with beta-1, 4-linkages in Xylan and generates hydrolysis of hemicellulose to pentose sugars. Xylanases belong to a prominent group of carbohydrases supporting degradation and have prominent activities in xylan degradation [29]. Their are many applications of xylanases used for extraction in starch, clarification of wine and juices, kraft pulps bleaching, thickening of food, and improving dough textures. Xylanses have been produced from various lignocellulosic residues such as rice husk, soya bean hulls, wheat brans, and sugarcane bagasse [14]. These materials may provide a feasible amount of xylanase production which can be significant economically. This enzyme is used in agriculture to add nutritional value to animal and agricultural waste degradation [25, 19]. Agriculture wastes including sorghum straw, orange peels, rice bran, grape pomace contains SSF which may be used for production of xylanase [24]. Moreover, fermentable sugars from agricultural waste can be second-generation ethanol [2].

The significance of this enzyme has increased in the past few years as diverse biotechnological applications used in industries for food preparation, paper, pulp, and animal feed production. Some applications include fruit juice clarification, bioconversion of lignocellulosic agricultural residues to fuels and chemicals, bread making as used in dough conditioners improving dough rheology and other quality baked products [27]. Many microorganisms such as *Psychrophilic haloplanktis*, *C. Adeliae* and *C. Albidus* produce cold-active xylanases [31, 10]. Xylanses not only improve the dough stability and strength of wheat but also improve the volume and texture of the bread [7]. This is because of the efficient activity of *psychrophillic xylanses* converted into insoluble hemicellulose in the dough into soluble sugars before baking, making the dough fluffy and firm with a voluminous loaf of bread having improved softness and elasticity [28]. Microorganisms such as fungi, bacteria, algae, and yeast are profitable for increasing the enzyme through different biotechnological techniques. Agro-waste materials are not being utilized in a valuable form. Most of the pulp wood leftovers are discarded during processing and degradation. Globally enzyme market has been raised to create 75% of the synthesized enzyme used in detergents (37%), followed by textiles (11%), starch (11%), baking (8%), and animal feed (6%). Significant attention is required to improve the performance and are less cost-effective in the market. These enzymes are being used in fermentation methods and have a great advantage in recombinant DNA technology/gene cloning [16, 12]. Other enzymes are being used in the market such as celluloses and pectinases currently account 20% of the global enzymatic market.

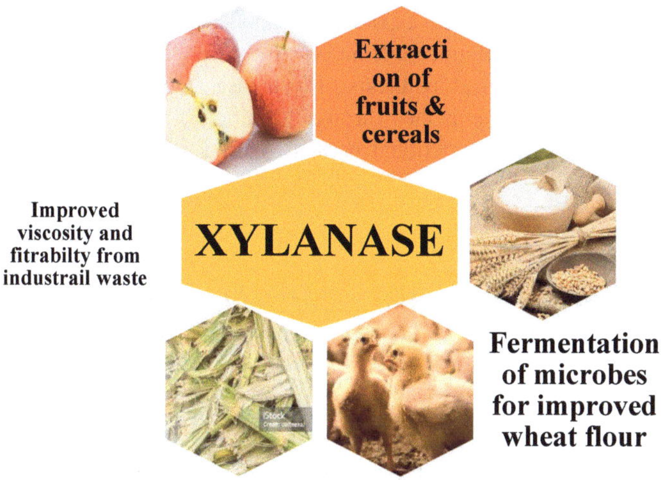

Fig. 1 Efficient methods of extraction of fruits and cereals used in various, pulp, and dough making. Paper industry and animal digestion

Significant attention is needed to improve performance and guarantee cost-effective manufacturing. The goal should be primitive producing xylanase when cultivated on various carbon and nitrogen sources. Natural resources such as mushrooms may be used for synthesis of enzymes. Hence, such strategies are required to improve the quantity and quality of xylanases through various microorganisms for Agro-bioproducts for animals and poultry feed (Fig. 1).

The natural resources around the surroundings can be used as bio-reservoirs containing xylan. Hemicellulose is a natural polymer, having polysaccharides obtained from lignocellulose biomass and vegetable gums, which is used for feed and biofuel production [1]. Many microorganisms are involved in fermentation and degradation for the production of xylanases [34]. For xylan hydrolysis many microbial sources including fungal, algea and bacterial strains have been detected in the production of enzymatic activities such as *Penicillium digitatum, Bacillus amyloliquefaciens, Rhizoctonia solani* exhibiting significant results from lignocellulosic biomass biodegradation [33]. The major producers of xylanase enzymes are fungal microbes as compared to bacteria, which are efficient at a pH range of 4–6 and have 50 °C below the temperature [4]. Thus, fungi have higher xylanase activity than bacteria, producing higher yields than bacteria [1]. Hence, microorganisms play a vital role in the production of xylanases, ultimately supporting agriculture sustainability [30].

Despite the usage of natural cellulosic resources globally cellulose-containing waste products and raw materials have been tremendously unidentified for their beneficiary uses [18]. However, many industries are using carbon sources microbes for producing organic acids, cellular proteins, enzymes, and prebiotic oligosaccharides [32]. Moreover, fermented sugars from agricultural waste is being utilized for making ethanol [8, 23]. Consequently, biotechnological applications have been successful in producing inexpensive enzymes, especially xylanase synthesis by purifying xylan. It has been detected that lignocellulosic wastes are environmentally friendly and are cheaper for the industrial synthesis of xylanase. Mostly, agricultural leftover usage produces microbial enzymes through significant biotechnological methods [30]. Xylanases have multifunctional uses in food, animal feed, fruit juices, and pulping.

1.1 Application of Xylanases in Agriculture, Food, and Animal Feed

Xylan has a diverse range of mechanisms, substrates, and chemical properties that are useful for various food and animal products. It consists of xylose linkages in lignocellulose which vary in physical and chemical properties [26]. This metabolizes a particular enzyme which is xylanases. Several xylanases have different functions which occur due to post-translation changes such as distinct glycosylation and proteolysis. Additionally, gene expression in abundance s caused by a variety of microbes. A single gene showing pleiotropy effects may create different xylanases, which may have a tremendous effect on crop species. It has been reported that genes have been encoded for producing xylanases, cloned in homogenous and heterogeneous hosts [6].

Mostly xylanases are being used in the digestibility of ruminant animal feed, extraction, and clarification of juices, pulp bleaching, paper industry, oils, cosmetics, and biofuel. The lignocellulosic material and agricultural waste such as rice husk, corn, bean pods, maize cobs, sugarcane bagasse, coffee pulp, wheat straw, and hard and soft woods contain hemicellulose and lignin [28]. Many manufacturing industries, forestry, and agriculture produce a huge amount of lignocellulose waste, unfortunately, most of it is wasted and only a few amount is used in making different products [11].

Currently, it has been discovered that cellulose-free xylanases create an environmentally friendly methodology for the paper and pulp industry [6]. Microorganisms play a vital role in regulating biosynthesis through various positional isomers. Lignocellulosic materials can generate liquid fuels and chemicals. Hence, making valuable petroleum free products improves the economy. For the near future, understanding the significance of xylanase activity and its mechanism is essential for improved or alternative applications of xylanases (Table 1).

1.2 Bioproducts Utilization for Xylanases Production

The main constituents of lignocellulose are cellulose, hemicellulose and lignin, which are metabolized by cellulase xylanase, and laccase, respectively [11]. From this

Table 1 Multifunctional application of xylanases enzyme from agricultural crop species, waste products, and other materials

Various crops	Genus/species/stuff	Application/uses	References
Fruits (oranges, peaches, pineapple, apples)	Streptomyces	Extraction of juices	[35]
Insoluble sugars	Arabinoxylan	Baking of bread	[3, 14]
Wheat and cereals	*Triticum aestivum* L	Animal feed	
Chemical bleach	$NaClO_2$	Paper industry	[16]
Biomass	Xylo-oligosaccharides	Pharmaceutical and cosmetics	[15]
Bio-based commercial products	Xylose	Industrial products: oils, solvents	[24]
Hardwoods and softwoods	Angiosperms and gymnosperms	Food, drinks and beverages	[6]
Fermentable sugars	β-xylosidases	Essential oils and fragrances	[9]
Wheat, barley, rice husk	*Pleurotus ostreatus* and *Pleurotus eryngii*	Agro-industrial byproducts	[19]
Industrial waste (lignocellulose waste) and sugarcane bagasse, soyabean pods, wood flakes	*Citronella bagasse (Cymbopogon nardus)*	Cultivation of mushrooms	[24]

lignocellulose sugars are utilized in industry. Some researchers reported that potato dextrose agar supplements separated with 1% carboxyl methyl cellulose (CMC), 1% xylan, and 0.1% of 2,2′-azino-bis (3-ethylbenzthiazoline-6-sulphonic acid), along with the white rot fungi have the ability to synthesis the activity of the enzymes [19]. Similarly, *Pleurotus ostreatus* and *Lentinus squarrosulus* possesed the greater average relative cellulase and laccase activity (1.07 ± 0.073 and 1.43 ± 0.286), respectively.

Pleurotus ostreatus grown on rice straw substrate obtained maximum growth and output with a total yield of 399.70 g per kg of substrate [9]. Wood flakes and sugarcane bagasse obtained the lowest growth at 13.54 g per kg of the substrate. Rice straw has been found the highest biological efficiency (39.40 g) and wood flakes obtained the lowest (1.35 g), respectively. A species, *Citronella bagasse* (*Cymbopogon nardus*), has played a significant role on biological efficiency and yield. Lignocellulosic waste has been used for cultivation of mushrooms, providing an economical and sustainable approach in agriculture [16].

1.3 Extraction of Fruit and Vegetable Juices

The juice and wine industries are the main contributors to the enzyme market. Fruit and vegetable juices must be produced using the extraction, clarifying, and stabilisation processes. By dissolving xylans into separate components that can be utilised to create flavours or sweeteners like xylitol, endo-1,4-β-xylanase helps increase the amount of juice that can be extracted from fruits and vegetables. It also helps with the maceration process [22]. Furthermore, it improves the fruit juice's filterability by reducing its thickness. Additionally, endo-1, 4-β-xylanase improves the extraction of other fermentable sugars from barley, which is advantageous for the manufacture of beer and the conversion of discarded barley into animal feed. Additionally, endo-1, 4-β-xylanase contributes to the reduction of brewing liquid viscosity, which in turn improves filterability.

1.4 Bioconversion of Lignocellulose in Biofuels

The best sources of nitrogen and carbon are ammonium sulphate and maize cob, respectively. In *A. niger*, both wild-type and UV-mutated, the addition of Mn^{2+} enhanced the synthesis of xylanase, but Cu^{2+} had the reverse effect. Ammonium sulphate precipitation was used to partially purify the xylanases from the wild-type and UV-mutated strains. 45 °C and pH 3.0 are optimal for the activity of the partly purified xylanase. At 45 °C and pH 3.0, the xylanases exhibited the highest stability. Compared to sawdust as a substrate, maize cob increased the enzyme activity higher. The maximal xylanase activity of the UV-mutated *A. niger* strain was 9.3641 U/mL, higher than that of the wild-type strain, which was 4.124 U/mL. Using sawdust and corncob as xylanase manufacturing substrates reduces waste and production expenses, which benefits the environment and the economy [33].

The usage of agricultural waste for making enzymes is essential to meet the goals for using biowaste and evoking future agriculture production. Under optimal conditions, maize cob and wheat bran waste are the best carbon sources for xylanase production, followed by pigeon pea pod waste. *Lentinus polychrous* synthesis xylanases and cellulase activity at around 35 °C. According to Windari et al. [32], the temperature during incubation has a significant impact on the formation of cellulose and xylanase. From 4.0 to 7.0, the xylanase showed stability over a broad pH range. At pH values between 4.0 and 5.0, the cellulase maintained its stability with 94% activity. Whereas xylanase activity increases up to 60 °C, cellulase activity peaks between 30 and 60 °C. Both enzymes work well in the pH range of 4.0–6.0. L. polychrous Lev.LP-PT-1's xylanase and cellulase's have high heat resistance which is useful for industrial biotechnological applications [8]. Among carbon sources, wheat bran, a cheap agricultural waste, produced the most xylanase (469.45 IU/mL), while soybean meal was the most efficient source of nitrogen (478.5 IU/mL). The optimal pH and temperature for enzyme activity are 6.4 and 55 °C, respectively. Xylanase is stable at 85 °C and pH 9.0. The best carbon sources for xylanase synthesis were found in oat spelt Xylan and wheat bran.

1.5 The Paper and Pulp Industries

Many of the microorganisms are essential for breakdown of plant nutrients to thrive plant resources. Xylanase, so called endo-1,4-β-xylanase or β-1,4-xylan xylanohydrolase, is an enzyme that breaks down hemicellulose in plant cell walls by converting beta-1,4-xylan into xylose. Fungi, bacteria, algae, snails, insects, seed except mammals are nor responsible for producing this enzyme. Most of the biowastes contains a huge amount of xylanase and is used in commercial applications such as the digestibility of silage and whitening wood pulp for the utilization of paper making [23]. Lignocellulose sugars and edible mushrooms have been found for processing xylanase activity [4]. The maximum stability between Ph 4.0 and 10.0 was obtained through various enzymatic activity tests. Maximum xylanase activity was seen at 60 °C. Many metal compounds are also responsible for synthesis of xylanase as $ZnCl_2$ and has been used for industry at various pH levels and temperatures. The use of inexpensive agriculture waste used as a carbon source may help to reduce the costs through fermentation [3]. Euclyptus karft pulp treated with enzymes are more effective substitute for chlorine and bleaching. *Bacillus licheniformis* 77-2's enzyme output was considerably enhanced, and the substrate was the only factor influencing the production rate. When birch wood was utilised as a substrate, the enzyme's activity peaked at 46.47 μ/mg of activity units, whilst *Bacillus licheniformis* 77-2 cultivated on maize straw generated 42.27 μ/mg of activity units [35].

1.6 The Pharmaceutical, Food and Feed Industries

The enzyme have a primarily importance in the pharmaceutical and feed industries, improving digestibility of feed by breaking down the complex plant cell wall, especially in animal feed. Moreover, xylanase is being used in the pharmaceutical industry to produce prebiotics (xylo-oligosaccharides) beneficial for gut health and immune system [20]. The animals access to allow more nutrients by breaking down hemicellulose in plants-based feed. This may provide a better feed conversion ration and raised growth rates. Many animals are fed high fiber diets, xylanase can reduce the viscosity of intestinal components and prevent digestive issues. The ingredients added in the feed are cereals for instance wheat, rice, corn, barley and soyabean mean, they all enhance digestion in the animals. Many researchers are investigating xylanase activity for immune function and gut health, potentially leading to dietary supplements or therapeutic agents.

Since thirty years xylanase was first introduced in broiler diets in Europe [5]. However, many animal dietary by-products were made through enzyme technology,

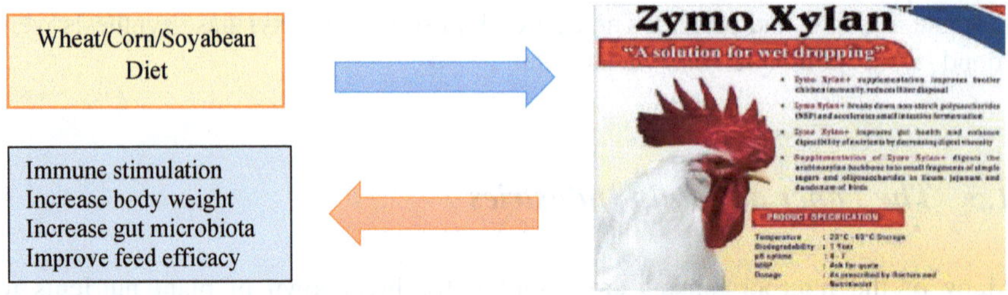

Fig. 2 Dietary byproducts supplements of ruminant animal and poultry feed

which still need to be improved as for commercial purpose. Advance feed xylanase technology has to be revolutionized for broiler nutrition. The need for improvement is due to gut viscosity has been a problem in broiler chickens when fed by wheat based diets. Many dietary products have been made by corn/soyabean, beneficial for lower disease risks from reduce level of anti-nutritional factors and no effects of viscosity in the intestine. It was reported by Kljak et al. [13] that adding Endo-1, 4-β-xylanase, and cellulase to forages produces higher-quality silage and improves ruminants' ability to digest plant cell walls. Cows and other ruminant animals have higher levels of helpful sugars for digestion as a result of receiving endo-1, 4-β-xylanase therapy. It is also known that endo-1,4-β-xylanase produces compounds that many rumen microbes use as food (Fig. 2).

1.7 Agro-residues Used for Fungal Xylanase Production

In the past decades, xylanases have been found a significant need in the market, due to its potential value on byproducts for feed of poultry and animals. Nowadays, insufficient amount of enzymes are being practiced in the market. To avoid the need of affordable enzymes various biotechnological application can help to produce higher amount of xyanases. This enzyme have been produced through various techniques of microbial processes. Microbes produce xylanase from agro-residues and reduce the enzyme costs and increase the productivity. Agro-industrial waste are used in fermentation systems, which promote fungal growth and development of the product can be made through selection of agro-industrial waste. Microorganisms are rich carbon sources, and identifies sustainable substrates enhancing the synthesis of xylanases [20]. Oat spelt xylan, has high substituted xylan increasing the synthesis of xylanases. Crude xylanase, which contains very little cellulose, affects the bio bleaching process carried out by *Bacillus licheniformis.*

1.8 Next Future Level of Xylanase Improving Feed Application

It has been seem that most of the bioproducts containing xylanase are valuable, but still needs to be improved for gut viscosity and entire disease effects. Birds have small intestine using nutrients undigested from various feed stuff. The importance lies on feed which are beneficial for diet and have less cost formulation [17].

Globally, the broiler diets are made up of corn and soyabean formulation, containing higher amount of arabinoxylan. Thus, reduce gut viscosity and less weight in laying hens and turkeys.

Commercially xylanses breakdown insoluble NSPs which have substrate activity. Byproducts derived from cereals contain more complex fiber components, processing the raw material. The bird or animal interaction with feed relies with the gastrintestinal tract (GIT), which varies within the species and its age [23]. In the market, it has been observed that the feed containing xylanase is not that much high as that of phytase. Hence, future strategies is required to improve the xylanase feed application. The feed should have the following capabilities.

1. The ability to break down soluble and insoluble arabinoxylan in a wide range of feed products.
2. Optimal Ph for preferred section of gastrointestinal tract.
3. No inhibitory activities in the presence of xyalanase.
4. Comprehensive feed processing thermostability.

Before GH 11 family of xylanases was commonly used in feed, which were not suitable for the diets. They reduced soluble NSPsin wheat-based diets, lowering the viscosity level in the gut of broilers and less effective having arabinoxylan complex [26]. GH11 xylanases hinder the branches or side chains on arabinose backbones. Hence, they are highly specific and low branching wheat backbone. Significantly, broader range of feedstuffs having potential should be incorporated into the complete diet. Therefore, GH10 produce higher levels of xyalanses with cheaper ingredients, also having significant value proportion of further reducing feed costs (Fig. 3).

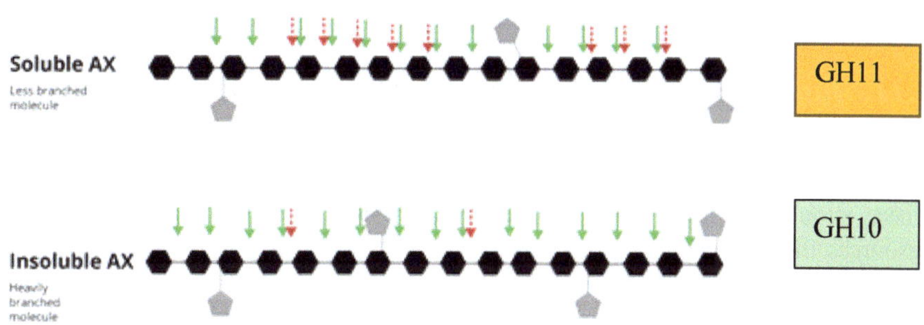

Fig. 3 GH10 xylanase effective against GH11 for feed stuffs and GIT improvement

1.9 Other Applications of Xylanase Enzyme in Food Industry

Despite of the various used of xyalanse enzyme in byproducts of animal feed and agriculture, this enzyme is being used in increasing the bread quality and texture [30], animal feed from nutritional supplement, biofuel production from plant biomass, textile processing from hard woods of hemicellulose and removing plant fibers from raw materials, extraction of oils from seeds of plants processing and starch extraction from plants and even production of beverages such as coffee by enhancing extraction efficiency. Essential for break down of hemicellulose (xylan) has a significant role [9].

Efficient degradation of plant biomass are challenging that requires hydrolysis of lignocellulosic biomass. The biomass can be degraded by combination of multiple enzymes that hydrolyze complex polysaccharides into fermentable sugars under industrial level. These enzymes are active with a wide range of temperatures and Ph [33]. Hence, microorganisms exploring thermophilic extreme environment using through biotechnological methods helps in biomass degradation [24]. Nowadays, xylanases are being produced from microorganisms for commercial use, especially industries are interested for biofuel production, chemical and pharmaceuticals industry, wood pulp bio-bleaching, papermaking, and manufacture of food and beverages and animal nutrition.

1.10 Genesis of Xylanases

Xylanses are enzymes mostly found in microorganisms but they are also produced by marine, alsae, protozoans, snails, insects, seeds, plants and other natural sources [21]. The nature have tremendous species that breaks down hemicellulose and have thermostable enzymes for industrial purposes. The microorganisms survive in the extreme environment and biological sources including bacteria, fungi and yeasts are the highly recommended organisms for xyanase production through natural processes [13].

1.11 Innovative Future Strategies for Sustainable Agriculture

There should be some future strategies for evaluating and processing natural resources like wood gums, plant seeds, corn and rice husk, wheat-based products, applications for the synthesis of Xylan (hemicellulose). Biotechnical techniques and molecular biology can interpreted the genes interacted with the linkages for breakdown of hemicellulose. High viscosity and low cost animal and poultry feed need to be produced for sustainable agriculture. Hence, future strategies are required for innovative research on xylanase synthesis through natural resources and Agri-product wastes.

References

1. A. Burlacu, C.P. Cornea, F. Israel-Roming, *Screening of Xylanase Producing Microorganisms* (2016)
2. A.K. Patel, C.D. Dong, C.W. Chen, A. Pandey, R.R. Singhania, Production, purification, and application of microbial enzymes, in *Biotechnology of Microbial Enzymes* (Academic Press, 2023), pp. 25–57
3. Z. Ahmad, M.S. Butt, A. Ahmed, M. Riaz, S.M. Sabir, U. Farooq, F.U. Rehman, Effect of Aspergillus niger xylanase on dough characteristics and bread quality attributes. J. Food Sci. Technol. **51**, 2445–2453 (2014)
4. N. Akhtar, D. Goyal, A. Goyal, Characterization of microwave-alkali-acid pre-treated rice straw for optimization of ethanol production via simultaneous saccharification and fermentation (SSF). Energy Convers. Manage. **141**, 133–144 (2017)
5. A. Arczewska-Wlosek, S. Swiatkiewicz, D. Bederska-Lojewska, S. Orczewska-Dudek, W. Szczurek, D. Boros et al., The efficiency of xylanase in broiler chickens fed with increasing dietary levels of rye. Animals **9**(2), 46 (2019)
6. P. Bajpai, *Developments and Applications of Enzymes from Thermophilic Microorganisms* (Elsevier, 2023)
7. J. Both, B. Biduski, M. Gómez, T.E. Bertolin, M.T. Friedrich, L.C. Gutkoski, Micronized whole wheat flour and xylanase application: dough properties and bread quality. J. Food Sci. Technol. **58**, 3902–3912 (2021)
8. W. Braide, I.A. Kanu, U.S. Oranusi, S.A. Adeleye, Production of bioethanol from agricultural waste. J. Fundam. Appl. Sci. **8**(2), 372–386 (2016)
9. F. Carmen, C. Frances, L. Barthe, Trends on valorization of pitaya fruit biomass through value-added and green extraction technology–A critical review of advancements and processes. Trends Food Sci. Technol. **138**, 339–354 (2023)
10. D. Kour, K.L. Rana, T. Kaur, B. Singh, V.S. Chauhan, A. Kumar, et al., Extremophiles for hydrolytic enzymes productions: biodiversity and potential biotechnological applications. Bioprocess. Biomol. Prod. 321–372 (2019)
11. A. Dukare, K. Sharma, S. Kautkar, J. Dhakane-Lad, R. Yadav, V. Nadanathangam, S. Saxena, Microbial xylanase aided biobleaching effect on multiple components of lignocelluloses biomass based pulp and paper: a review. Nord. Pulp Pap. Res. J. **38**(3), 459–480 (2023)
12. S.K. Gupta, P. Shukla, Sophisticated cloning, fermentation, and purification technologies for an enhanced therapeutic protein production: a review. Front. Pharmacol. **8**, 419 (2017)
13. K. Kljak, M. Belović, M. Duvnjak, V. Jurišić, M. Radosavljević, Applications of Sourdough in animal feed. Sourdough Innov. 391–434 (2023)
14. M. Liu, R. Xiao, X. Li, Y. Zhao, J. Huang, A comprehensive review of recombinant technology in the food industry: Exploring expression systems, application, and future challenges. Compr. Rev. Food Sci. Food Saf. **24**(2), e70078 (2025)
15. M. Kuddus, P.W. Ramteke (eds.), *Value-Addition in Agri-food Industry Waste Through Enzyme Technology* (Elsevier, 2023)
16. M. Akram, F. Jabeen, M. Daniyal, R. Zainab, U. ul Haq, C.O. Adetunji, A.B. Ogbo, Genetic engineering of novel products of health significance: Recombinant DNA technology. *Functional Foods and Nutraceuticals: Bioactive Components, Formulations and Innovations* (2020), pp. 595–611
17. J. Ma, P. Zhong, Y. Li, Z. Sun, X. Sun, M. Aung et al., Hydrogenosome, pairing anaerobic fungi and H2-utilizing microorganisms based on metabolic ties to facilitate biomass utilization. J. Fungi **8**(4), 338 (2022)

18. Ma, R. *Screen Printed Moisture Sensor on Barrier Coated SBS Board: The Characterizations of the Hemicellulose-Based Biofilms and Their Applications for Smart Packaging* (2018)
19. Z. Mengqi, A. Shi, M. Ajmal, L. Ye, M. Awais, Comprehensive review on agricultural waste utilization and high-temperature fermentation and composting. Biomass Convers. Biorefinery **13**(7), 5445–5468 (2023)
20. M. Muhammad, A. Basit, K. Ali, W.J. Li, L. Li, H.I. Mohamed, Endophytic fungi as potential bio-control agents of soil-borne pathogen. J. Crop Health **76**(3), 617–636 (2024)
21. K.A. Nguyen, S. Wikee, S. Lumyong, Brief review: lignocellulolytic enzymes from polypores for efficient utilization of biomass. Mycosphere **9**(6), 1073–1088 (2018)
22. G. Ozyilmaz, E. Gunay, Clarification of apple, grape and pear juices by co-immobilized amylase, pectinase and cellulase. Food Chem. **398**, 133900 (2023)
23. M.P. Perini, V. Pedrinelli, P.H. Marchi, L.B.F. Henríquez, R.V.A. Zafalon, T.H.A. Vendramini et al., Potential effects of prebiotics on gastrointestinal and immunological modulation in the feeding of healthy dogs: a review. Fermentation **9**(7), 693 (2023)
24. B.K. Puja, S. Mallick, T. Dey, S. Chanda, S. Ghosh, Xylooligosaccharide recovery from sugarcane bagasse using β-xylosidase-less xylanase, BsXln1, produced by *Bacillus stercoris* DWS1: characterization, antioxidant potential and influence on probiotics growth under anaerobic conditions. Int. J. Biol. Macromol. **285**, 138307 (2025)
25. S. Bharathiraja, J. Suriya, M. Krishnan, P. Manivasagan, S.K. Kim, Production of enzymes from agricultural wastes and their potential industrial applications, in *Advances in Food and Nutrition Research*, vol. 80. (Academic Press, 2017), pp. 125–148
26. S.M. Kamande, G.I. Omwenga, M.P. Ngugi, Production of cellulases by *Xylaria* sp. and *Nemania* sp. using lignocellulose substrates for bioethanol production from maize cobs. Heliyon **10**(17) (2024)
27. V. Saharan, S. Tushir, J. Singh, N. Kumar, D. Chhabra, R.K. Kapoor, Application of MOGA-ANN tool for the production of cellulase and xylanase using de-oiled rice bran (DORB) for bioethanol production. Biomass Convers. Bioref. **14**(11), 11987–11999 (2024)
28. F. Sarmiento, R. Peralta, J.M. Blamey, Cold and hot extremozymes: industrial relevance and current trends. Front. Bioeng. Biotechnol. **3**, 148 (2015)
29. G. Singh, S. Kumar, S. Afreen, A. Bhalla, J. Khurana, S. Chandel et al., Laccase mediated delignification of wasted and non-food agricultural biomass: recent developments and challenges. Int. J. Biol. Macromol. **235**, 123840 (2023)
30. T. Li, R. Wang, B. Hua, L. Cao, Q. Zhang, Y. Zhai, et al., Improving the thermal stability of GH11 xylanase XynASP through cord region engineering. J. Agric. Food Chem. (2025)
31. U.S. Annapure, N. Pratisha, Psychrozymes: a novel and promising resource for industrial applications, in *Microbial Extremozymes* (Academic Press, 2022), pp. 185–195
32. T. Windari, Y. Yurliasni, E. Mariana, D.A. Marpaung, R.R. Jinnani, Comparative analysis of the physicochemical and microbiological properties of *Lactobacillus plantarum* and *Lactobacillus acidophilus* fermented milk with prebiotic Kepok Banana (*Musa acuminata*×balbisiana). Asian J. Environ. Res. **2**(1), 48–60 (2025)
33. S. You, Z. Zha, J. Li, W. Zhang, Z. Bai, Y. Hu et al., Improvement of XYL10C_Δ N catalytic performance through loop engineering for lignocellulosic biomass utilization in feed and fuel industries. Biotechnol. Biofuels **14**, 1–14 (2021)
34. Z. Zhang, J. Xing, X. Li, X. Lu, G. Liu, Y. Qu, J. Zhao, Review of research progress on the production of cellulase from filamentous fungi. Int. J. Biol. Macromol. 134539 (2024)
35. G. Zhang, F. Wang, S. Deng, G. Chen, H. Liu, H. Zhang, A pH-stable alkaline pectate lyase produced by the newly identified strain *Bacillus altitudinis* CAS-WZS-08. J. King Saud Univ. Sci. **35**(4), 102649 (2023)

Chapter 8
Xylanase in Waste Treatment and Environmental Applications

Zubair Hashmi, Abdul Sattar Jatoi, Abdul Qayoom, Ibrahim Maina Idriss, Muhammad Tanzeel Usman, Syed Hassan Abbas, Syed Osama Ali, Faheem Kareem, Rieza Zulrian Aldio, and Hajah Nur Aqidah Binti Haji Donglah

Abstract Xylanase, an enzyme that degrades xylan, a major component of lignocellulosic biomass, has emerged as a powerful tool in waste treatment and environmental remediation. This chapter provides a comprehensive overview of xylanase, including its definition, types (endo- and exo-xylanases), and enzymatic mechanisms. The various microbial sources of xylanase such as fungi, bacteria and yeasts will be discussed with respect to the factors influencing its activity. The methods of xylanase production, purification and characterization will be reviewed along with their relative importance for industrial application in maximizing enzyme efficiency. This chapter indicates the various uses of xylanase in agricultural, pulp and paper and food-processing waste treatment and its possible role as an environmental pollutant reducer in terms of chemicals and organic waste. The environmental benefits

Z. Hashmi (✉) · A. S. Jatoi · A. Qayoom · M. T. Usman
Department of Chemical Engineering, Dawood University of Engineering and Technology, Karachi, Pakistan
e-mail: zubair.hashmi@duet.edu.pk

Z. Hashmi · R. Z. Aldio · H. N. A. B. H. Donglah
Microalgae and Bioremediation Lab, Faculty of Integrated Technology, Universiti Brunei Darussalam, Gadong, Brunei Darussalam

I. M. Idriss
University of Maiduguri, Maiduguri, Nigeria

S. H. Abbas
Department of Chemical Engineering and Applied Chemistry, University of Toronto, Toronto, Canada

S. O. Ali
Department of Chemical Engineering, University of NewBrunswick, Fredericton, Canada

F. Kareem
Biosensors and Nanobiotechnology Laboratory, Integrated Science Building, Faculty of Science, Universiti Brunei Darussalam, Gadong, Brunei Darussalam

R. Z. Aldio
Department of Mechanical Engineering, Universitas Islam Riau, Pekanbaru, Riau, Indonesia

A. Sattar Jatoi (ed.), *Xylanase from Microorganisms*,
https://doi.org/10.1007/978-981-95-0824-2_8

of xylanase include soil remediation, pesticide degradation, and removal of heavy metals and spilling oil. Case studies are there to show practical use of xylanase in wastewater treatment plants and show how it is effective in the real world. If these potential applications are exciting, they will throw up problems such as enzyme stability, cost of production and substrate specificity. The chapter ends with perspectives on future innovations and directions, including enzyme engineering, hybrid enzyme systems and integration of xylanase to circular economy models. Xylanase is undoubtedly emerging as a suitable solution to sustainable waste management and environmental rehabilitation.

1 Introduction

The worldwide environmental crisis has seen increasing generation of waste; depletion of resources and pollution have entered the scene in the pursuit of more sustainable and effective options for waste management technology. In particular, the focus now has narrowed onto the treatment of lignocellulosic waste, which is widespread in agricultural residues, forestry by-products, and industrial waste streams [1, 2]. Among many biotechnological approaches explored, the enzyme xylanase seems to have emerged as a better solution owing to its specificity, environment-friendly character, and operational efficiency under mild conditions [3, 4].

Xylan hydrolysis- that's what Xylanase does best as it comprises the entire catalysis for hemicellulose, the natural constituent of 35 percent dry weight for many plants [5]. Xylan is one of the largest consisting biopolymers in nature as its production annually is estimated at 1–2 billion tons worldwide from agricultural residues, forestry wastes, and food industry by-products [6, 7]. The lignocellulosic materials in question, however, are not easily degraded by chemical and physical methods; therefore, their accumulation causes waste that is injurious to the environment [8]. Xylanase refers to the enzyme that brings a biological conversion process between xylan and the product of the xylan reaction-the split simple sugar xylose; hence, it turns useless complex residues into important by-products valuable in biomass valorization [9].

Recently, a tendency has grown into applying xylanase in bioremediation and the treatment of wastes. It has gained considerable attention as a result of recently generated publicity regarding its uses in sustainable and environmentally friendly technologies. By 2030, the global market of waste management is projected to be valued at $3.1 trillion, climbing on a compound annual growth rate of 5.5% from an estimated value of $2.08 trillion in 2021 (Fortune Business Insights, 2023) [8, 10, 11]. Waste management, pollution, and resource conservation are becoming ever more urgent issues of concern in many parts of the world today, and hence the tenets of enzyme-mediated treatment have gained a higher level of importance [12, 13].

These paper and pulping industries contribute significantly to incurring industrial waste in the environment as well as causing water and air pollution. The production of paper and paperboard is estimated to reach 200 million tons globally per year and results in over 40 billion cubic meters of wastewater discharges as it has been mentioned that the industry produces paper at a scale of about 200:1-flooding the total with wastewater [14]. The effluent resulting from various processes in paper mills contains a heavy load of organic matter which includes lignin and hemi-cellulose, both of which are highly resistant to conventional chemical treatments [15]. Xylanase offers a promising alternative by breaking down these recalcitrant compounds. Studies show that the use of xylanase can reduce the chemical oxygen demand (COD) of paper mill effluent by up to 40–50% [1, 2, 16], It was dramatically improving the quality of water and making it feasible to reclaim or reuse waste water in the paper industry [9, 16].

On par, agricultural waste has evolved into a growing concern. Coupled with this, the Food and Agriculture Organization (FAO) avers that about 1.3 billion tons of food waste are produced worldwide every year. Such acts contribute greatly to environmental degradation [17, 18].

For instance, in the agricultural area, crop residuals are usually abundant as corn stover, rice husks, and wheat straw, but are frequently burnt, causing air pollution and soil degradation. These include decomposition of xylanases, which liquefy these residues into simpler sugars, end products that can then be used or converted to biofuels, animal feed, or compost. Processing crop residues such as corn stover, rice husks, and wheat straw into some sugar, by which they can be used as raw materials in producing biofuels and animal feed or in compost production is beneficial in this sector. All of them are abundant in agriculture but most times end up burnt [19, 20]. For example, corn stover alone, which is produced at an estimated rate of 300 million tons per year in the U.S., is an excellent source of lignocellulose for xylanase treatment [21]. In order to reduce environmental implications, the effective utilization of these wastes using enzymes will generate the useful by products.

In the food processing sector, a large amount of organic waste is generated, the predominant component of which is typically xylan. More than 1.3 billion tons of waste are generated annually by food and beverage industries across the world (Waste and Resources Action Program 2022) [22, 23]. Using an enzyme xylanase, the sugar polysaccharides would be further converted into simpler sugars during the treatment of food processing effluent. More importantly, this sugar could further participate in bioenergy production or be used as raw materials for industrial processes. The xylanases are useful in the brewing industry for the breakdown of residual barley polysaccharides and increase wort clarification efficiency, thereby saving about 10–15% of water input [24].

Xynalase has applications that extend from waste treatment to environmental remediation. A global problem is the toxicity of soils and waters because of pesticides and petrochemicals. The world pesticide market was approximately $76.4 billion in 2022, and an estimated 2.5 million metric tons of pesticides are used globally annually [25]. Several chemicals, including glyphosate and chlorpyrifos identified, are persistent environmental pollutants and cut across the terrestrial and aquatic ecosystems, including human health [26]. This information provides some hope for the detoxification of these toxic compounds by xylanases. Published studies reveal that xylanases can enhance pesticide degradation, which is accelerated by 40% compared to the natural soil environment, thus aiding in bioremediation [27, 28].

In the oil-spill industry, xylanase holds great promise for bioremediation. According to the International Tanker Owners Pollution Federation (ITOPF), an estimated 1 million tons of oil are spilled into the oceans worldwide every year, devastating the marine life ecosystem, including flora and fauna [29]. In recent studies, xylanase acted with other microorganisms in the degradation of the petroleum hydrocarbons, thus mitigating the effects of oil spills in the environment [30, 31]. The significance of such enzymes in encouraging biodegradation in nature owing to the progressive upward trend in oil-degrading organisms thereby accelerating cleanup processes at contaminated sites is thereby brought out. Xylanases provide entrance into some of the critical environmental challenges this world is facing today. They can degrade lignocellulosic wastes and decontaminate hazardous pollutants, thus possessing the main potential for sustainable waste management and environmental remediation [4, 5]. There are various industrial applications of xylanases ranging from treatment of agricultural residues and industrial effluents to detoxification of spills. The potential of xylanase in the circular economy and therefore, in mitigating environmental effects is becoming a bigger issue as industries are moving towards sustainability. In this chapter, we will deal with the mechanism, current applications, bottlenecks, and future avenues for xylanase in waste treatment and environmental applications, providing an overview of how xylanase can contribute to a brighter, sustainable future.

2 Fundamentals of Xylanase

Xylanase within the enzymes constitutes the major functional portion in the degradation of Xylan, a major polysaccharide component of plant cell walls, especially in the hemicellulose fraction. A good knowledge of the basic mechanisms involved, types of xylanase, and sources of xylanase is important in evaluating its possible applications in diverse environmental and industrial processes. This chapter deals with basics of xylanase, such as its enzymatic action of xylanase, classification, sources, and factors affecting xylanase activity.

2.1 Definition and Function of Xylanase

Xylanase is an enzyme belonging to the glycoside hydrolase category and is particularly specified by EC code 3.2.1.8. It hydrolyzes an intricate polysaccharide, Xylan. This highly composite polysaccharide is headings acquired, consisting of xylose units linked by β-1,4-glycosidic links; such chemical structure makes xylan one of the abundant biopolymers available on earth and comprise thick portions of plant biomass. Such polymer is found mostly in the cell walls of grass, hardwoods, agricultural residues, and any plant material. Xylan is mostly linked with lignin-cum-cellulose and is highly unapproachable to most degrading organisms [32–34].

The main function of xylanase is to break down this complex structure into xylose, a simple sugar that can further be processed by different enzymes for several applications, including biofuel production, animal feed, and fermentation processes [35]. Hydrolysis of Xylan yields xylo-oligosaccharides and xylose, which can either be utilized by microorganisms or converted into value-added products [29]. This enzyme activity is vital in the efficient deconstruction of plant biomass and also for improving the accessibility of lignocellulosic materials to microbial degradation, which is important from the perspective of waste treatment and bioremediation [31, 33].

2.2 *Types of Xylanases*

Generally, xylanases are divided according to action mechanism and xylans that are attacked. With respect to these aspects, xylanases are generally classified into two classes.

2.2.1 Endo-xylanases (EC 3.2.1.8)

These enzymes hydrolyze the β-1,4-glycosidic bonds in xylan to generate xylooligosaccharides of different lengths. Endo-xylanases find the most application in industries ranging from pulp and paper to biofuels and sewage treatments [36, 37].

2.2.2 Exo-xylanases

The xylan-degrading enzymes sequentially cleave the xylan chains from both ends and liberate xylo-oligosaccharides of decreasing size, some of which may be of such low molecular weight that they have no significant effect on the kauri wood cellulose or lignin. Jointly, the enzymes may be more effective with endo-xylanases in achieving exhaustive xylan degradation [38, 39].

2.2.3 β-xylosidases (EC 3.2.1.37)

In other instances, xylo-oligosaccharidases finish hydrolyzing the terminal xylose units from xylo-oligosaccharides, converting the interstitial substrates further into free xylose [40, 41]. It is of importance to highlight that all three kinds of xylanases-endoxylanases, exoxylanases, and xylo-hydrolases-are often found to cooperate in

Table 1 The parametric conditions and survival characteristics of various xylanases [33, 34, 36, 38, 39]

Characteristic	Details
Enzyme type	Endo-xylanases, Exo-xylanases, β-xylosidases
Source organisms	*Trichoderma reesei, Thermomyces lanuginosus, Bacillus subtilis, Aspergillus niger*
Optimal pH	4.5–6.0
Optimal temperature	50–70 °C
Substrate specificity	Hemicellulose, xylan, lignocellulosic materials
Industrial applications	Pulp and paper, textiles, agriculture, biofuel production
Advantages	Biodegradable, environmentally friendly, cost-effective

nature and on an industrial scale toward efficient and effective degradation of xylan. Xylanases can, depending also on the source, be classified according to optimum temperature and optimum pH as well as substrate specificity, which has been known to vary significantly. This diversity allows xylanases to be tailored for specific applications in both laboratory and industrial settings. Table 1 presents detailed information on the parametric conditions and survival characteristics of various xylanases.

2.3 *Enzymatic Mechanism of Xylanase*

The enzymatic mechanism of xylanase typically follows a two-step process. First, the enzyme binds to the xylan substrate through specific interactions between the enzyme's active site and the xylan chain [42, 43]. This binding brings the xylan molecule into proximity to the enzyme's catalytic site. Next, the enzyme catalyzes the hydrolysis of the β-1,4-glycosidic bond between two xylose units [38]. The bond cleavage results in the formation of shorter xylo-oligosaccharides, which are released from the active site [37]. The product can be hydrolyzed or utilized in various biochemical processes by microorganisms [39, 40, 42]. Different kinds of xylanase enzymes operate with different mechanisms of inversion or retaining depending on the enzyme itself and the configuration of the active site. Acidic residues present within the active site stabilize the transition state during the hydrolysis reaction. The other factor that can affect both the efficiency as well as the specificity of xylanase is the structure of the xylan substrate along with the amino acid sequence of the xylanase and existence of any cofactors or inhibitors [33, 35, 40].

2.4 *Sources of Xylanase*

These enzymes break down plant cell wall components to allow the organisms to access the nutrient reserves contained in plant fibers. They are produced naturally by a variety of organisms, mainly microorganisms such as bacteria, fungi, and yeast [32–34, 36]. The principal sources of xylanases include.

2.4.1 Fungi

Among fungi, *Trichoderma reesei*, *Aspergillus niger*, and Penicillium spp. produce xylanases in vast amounts. Due to its high yield, thermotolerance, and broad substrate specificity, fungal xylanases are generally preferred for industrial applications [44–46].

2.4.2 Bacteria

In addition to fungi, bacterial classes like Bacillus subtilis and Clostridium produce xylanases. Of late, they, too, have generated a lot of interest because of their stability in high temperatures and alkaline conditions making them eligible for possible large-scale exploitation in industries such as textile manufacturing and biofuel production [47, 48].

2.4.3 Yeasts

Certain yeasts, such as Saccharomyces cerevisiae, produce xylanase as an end product through different metabolic processes. Yeasts are, however, easier to grow and handle but are less applied at an industrial scale in the production of enzymes in comparison to fungi or bacteria [12, 49].

Apart from the naturally occurring sources, there has been a growing trend of using genetically engineered organisms in the production of recombinant xylanase enzymes. The advancements in biotechnology have included the production of xylanase in Escherichia coli, Saccharomyces cerevisiae, and a number of other organisms, thus making such methods of enzyme production cost-effective and scalable.

2.5 *Factors Affecting Xylanase Activity*

Efficiency and stability are consequently determined by several controlling factors acting on xylanases in the natural and industrial fields. Temperature stands out as one definite parameter affecting these enzymes since each of them has a particular range with optimal activity being defined for any given source. The peak activity of fungal xylanases lies typically between 40 and 60 °C [42], for instance, xylanases from *Trichoderma reesei*, exhibit maximum activity at 50 °C for 4 h and retain 85% of their activity at this temperature [44]. In contrast, bacterial xylanases from Thermotoga maritima exhibit a higher degree of thermostability with optimal activity between 70 and 80 °C. Such enzymes retain 90% activity even after 10 h at 75 °C and thus may be applied in all processes involving high temperature [46]. Thermostability becomes highly important in scenarios like biofuel production, which will involve long operational periods at higher temperature ranges.

Another parameter well known to affect xylanase action is the reaction pH. Most fungal xylanases are reported to have maximum activity somewhere between slightly acidic to neutral pH values (4.0–7.0). For example, xylanase from Aspergillus niger achieves maximum activity at pH 5.5, retaining 95% activity within the pH range of 4.5–6.5 [50]. Bacterial xylanases, like those from *Bacillus subtilis*, exhibit optimal activity in alkaline conditions, peaking at pH 8.0 and maintaining over 80% activity between pH 7.5 and 9.0 [51]. Stability across a wide pH range is particularly important in processes such as paper bleaching, where pH fluctuations occur.

Substrate concentration also influences xylanase activity. At moderate xylan concentrations (10–30 g/L), xylanase activity increases proportionally with substrate availability. However, concentrations exceeding 50 g/L lead to substrate inhibition, reducing enzyme efficiency by up to 40% due to product accumulation [52]. For instance, in systems with xylan concentrations of 60 g/L, xylanase activity dropped by 35% compared to optimal conditions at 30 g/L. This issue can be mitigated by enzyme immobilization techniques, which enhance substrate turnover rates and prevent localized substrate saturation, improving efficiency by up to 50% [53].

Furthermore, the control of co-factors and inhibitors becomes essential for the modulation of xylanase activity. Divalent metal ions, like Ca^{2+} and Mg^{2+}, enhance enzyme activity by stabilizing its structure; for example, Ca^{2+} increases xylanase activity by 25% in *Thermomyces lanuginosus* xylanases [54]. While high concentrations of organic solvents (20% acetone or ethanol) can decrease xylanase activity by 40–60% [55]. In the study with 30% ethanol, the activity dropped by 55% and therefore reaction conditions needed to be optimized for industrial use [30]. The abovementioned parameters, "temperature, pH, substrate concentration, cofactors, and inhibitors," need to be controlled to optimize the performance and stability of xylanase, especially in challenging industrial arenas like biofuel production, food processing, and paper manufacturing.

3 Production and Characterization of Xylanase

Production and characterization of xylanase greatly implicate its use in industrial and environmental needs; the means of production, purification, and characterization ultimately determine xylanase efficiency, cost, and functionality [52, 53]. Certainly, xylanase production has been accomplished through fermentation or expression systems, each with its distinct advantages. On the one hand, xylanase production via microbial fermentation, especially solid-state fermentation (SSF), is relatively cheap; on the other hand, recombinant expression permits high yield and dedicated design of the enzyme [42, 46, 49]. Optimization of these methods will guarantee higher activity of xy1anase, thus providing more diverse applications.

3.1 *Microbial Sources of Xylanase*

Xylanase has been produced by fungi, bacteria, and yeasts; however, other fungal species that have been studied in the production of xilanases include *Trichoderma reesei* and *Aspergillus niger* [3, 56], which reported more than 500 IU/g yields on their substrates, like wheat bran [57]. When subjected to solid-state fermentation (SSF). When selected temperatures range between 40 and 60 °C, and the appropriate pH levels are maintained from 4 to 7, their activity and production of xylanase are optimum [45]. *T. reesei* xylanase, for instance, maintains 90% of its activity after 6 h at 50 °C [43]. Numerous bacteria, such as *B. subtilis* and *Clostridium thermocellum*, exhibit the property of producing highly active xylanases with the ability to withstand extreme conditions [39, 58]. *For B. subtilis alone, submerged fermentation (SmF) can generate approximately 700 IU/mL of xylanase that shows 85% activity at 80 °C and pH 9.0 for 8 h* [59]. In contrast, yeasts like *Saccharomyces cerevisiae* produce lower yields (10–100 IU/mL), but their genetic malleability makes them valuable for recombinant enzyme production, particularly in bioethanol production systems [60]. Figure 1 illustrates the microbial sources and production methodologies utilized for xylanase synthesis.

3.2 *Xylanase Production Methods*

Xylanase production relies on microbial fermentation and recombinant expression. Solid-state fermentation (SSF), using substrates like sugarcane bagasse or rice husks, provides yields ranging from 400 to 600 IU/g, depending on substrate composition and moisture content. SSF is cost-effective, with a 30–50% reduction in production costs compared to SmF, making it suitable for developing countries [46, 61]. Submerged fermentation (SmF), preferred for large-scale production, offers tighter control of environmental conditions, achieving yields up to 800 IU/mL, particularly with bacterial strains like *Bacillus licheniformis* [52, 61]. SmF also supports higher enzyme recovery rates, with up to 90% purification efficiency using ultrafiltration [37].

Recombinant expression systems allow for high-yield and high-purity production. *Escherichia coli* and *Pichia pastoris* are widely used due to their rapid growth and ease of genetic manipulation [62]. Recombinant xylanase expressed in *E. coli*

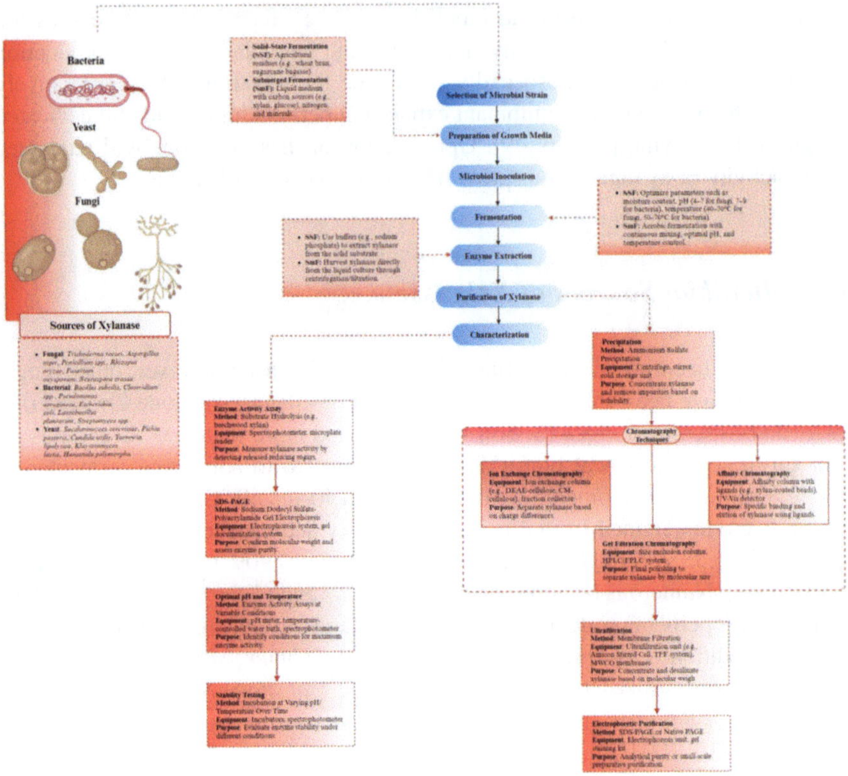

Fig. 1 A detailed workflow of xylanase production, from microbial source selection (fungi, bacteria, yeast) to characterization, highlighting fermentation, enzyme extraction, purification (precipitation, chromatography, ultrafiltration), and characterization techniques (activity assays, SDS-PAGE, stability testing)

achieves yields exceeding 1000 IU/mL, retaining 95% activity at 70 °C after 12 h [63]. In another study, recombinant GH11 xylanase (Mtxyn11C) from *Myceliophthora thermophile* was optimized in *Pichia pastoris* using N-terminal signal peptide engineering, achieving 2503 U/mL in a 7.5 L fermentor. Mtxyn11C demonstrated superior hydrolytic efficiency, converting birchwood xylan into xylose with 40% higher efficiency than commercial xylanases, highlighting its potential for sustainable biomass utilization and biofuel production [64]. Figure 1 depicted workflow of xylanase production, from microbial source selection (fungi, bacteria, yeast) to characterization, highlighting fermentation, enzyme extraction, purification (precipitation, chromatography, ultrafiltration), and characterization techniques (activity assays, SDS-PAGE, stability testing).

Table 2 Comparison of methods for fermentation and recombinant expression regarding their advantages and disadvantages, as well as their optimal conditions for xylanase production through these methods

Method	Advantages	Disadvantages	Yield	Cost	References
Solid-state fermentation	Low cost, uses agricultural waste	Low enzyme yield requires optimized conditions	Low to medium	Low to moderate	[65]
Submerged fermentation	High enzyme yield, better control over conditions	Higher cost of raw materials requires more energy	High	Moderate to high	[66]
Recombinant expression	High purity, easier scale-up	Expensive, requires genetic modification of hosts	High	High	[64]
Cell-free systems	No contamination risk, flexible	Expensive, needs efficient immobilization	Medium to high	High	[67]

3.3 Other Key Metrics

Substrate-to-product conversion rates in SSF range from 60–80%, depending on the lignocellulosic material used, while SmF achieves 70–85% under optimized conditions [53]. Industrial-scale recombinant systems reduce production costs by 20–30% compared to microbial fermentation, as enzyme recovery and purification processes are more streamlined [46]. Immobilized xylanase systems further enhance efficiency, with up to a 50% increase in activity retention after repeated uses, compared to free enzymes that lose 30–40% activity after three cycles [54]. These advancements highlight the versatility and efficiency of xylanase production methods. SSF remains the most economical approach, while recombinant systems provide unmatched precision and high yields, catering to the diverse needs of industrial applications. Table 2 encapsulated the pros and cons of different method of fermentation.

3.4 Purification and Characterization Techniques

Xylanase purification and characterization should determine its possible further applications in industrial and biotechnological processes. Xylanase is purified after production using a combination of techniques to give the required purity and activity [66, 68]. Ammonium sulfate precipitation is among the first techniques, wherein a salt is used to precipitate proteins selectively [69]. In this procedure, xylanase becomes concentrated, while impurities are removed, followed by dialysis for removing any excess salt. Usually, advanced chromatographic techniques are employed subsequently for the further purification of xylanase, such as ion exchange chromatography [50]. Classically, through this technique, the proteins are separated based on charge to form services of great purity of xylanase [65]. An advanced degree of purity may, finally, be ensured in affinity chromatography employing appropriately designed ligands or antibodies to specifically tag xylanase for applications where high-quality specification is required [70].

In the purification and characterization of xylanase, various biochemical analyses are performed to delve into the properties of the enzyme and its enzymatic actions. The enzyme assays look at the ability of the xylanases to hydrolyze particular substrates, i.e., beechwood xylan or oat spelt xylan [64], With specific assays, they yield information on the efficacy and possible applications of the enzyme. Sodium dodecyl sulfate–polyacrylamide gel electrophoresis (SDS-PAGE) is one of the methods commonly used to establish the molecular weight and purity of the enzyme. Clear bands on the gel indicate that purification has been successful and that no contaminants are present [71]. Conditions for optimal activity including temperature and pH are assessed by a range of controlled assays that have shown xylanase is optimally active within a temperature range of 40–60 degrees and pH of 5.0–7.0 depending on the source or mode of production [72]. All of these ensure that the enzyme is produced within the desired criteria for industrial applications. All these strategies combined improved the chances for producing a high-purity end-xylanase with multiple applications, including biofuel formulations, food processing, and paper brightening, and set the stage for a scale-up and sustainable industrial process.

Figure 2 shows all steps of purification and characterization of xylanase using various analytical methods. The chromatogram (a) depicts the elution profile of xylanase fractions during purification at controlled conditions, flow rate of 0.5 mL/min with sodium phosphate buffer, 50 mm (pH 7.0). It can be observed that the specific zones of elution clearly separated High Molecular Weight Xylanase (HMWX) and Low Molecular Weight Xylanase (LMWX) fractions, while the gray regions show all positive fractions and the black areas indicate the fractions with the highest activity. These fractions were further studied by SDS-PAGE to show purification based on the unique bands corresponding to the various molecular weights. Panels (b) and (c) show the enzymatic activity of the HMWX and LMWX fractions, respectively, concerning the elution volumes. Activity peaks of HMWX were found to be between 73 and 90 mL, while peaks of LMWX activity were between 193 and 210 mL, thus confirming separation dance in chromatographic processing. Panel (d) states that SDS-PAGE analyzes purified xylanase from *F. proliferatum* NRRL 26517. It should be noted that the analysis demonstrates a distinct prominent single band. This band matches the expected molecular weight compared to the protein standards validated the purity and specificity of the enzyme. Panel (e) illustrates the study focused on the characterization of XYL13, its purity and molecular weight confirmed through SDS-PAGE. Panels (f–h) refer to analysis of recombinant xylanases expressed in *E. coli* Rosetta (DE3). The distinct bands in panel (g) highlight soluble and insoluble fractions of protein. Further, panel (h) confirms the successful purification of recombinant xylanase with a sharp band around 27.98 Kda. This result, together with those of activity, corroborates the purification process and implies that highly purified xylanase can be produced for downstream applications. The integrated chromatographic-electrophoretic approach thus provides a clean, effective, and reproducible xylanase purification and characterization method for use in industry and biotechnology.

3.5 Factors Affecting Xylanase Activity

Xylanase activity is influenced by various factors, including temperature, pH, substrate concentration, and the presence of inhibitors or activators, all of which are critical for optimizing enzyme performance in industrial applications. Temperature plays a pivotal role in xylanase efficiency, with fungal xylanases such as those from *Trichoderma reesei* and *Aspergillus niger* exhibiting optimal activity at moderate ranges of 50–60 °C and 40–50 °C [68, 76], respectively, and bacterial xylanases like those from *Thermomyces lanuginosus* being thermostable up to 70 °C, making them suitable for high-temperature processes like biofuel production and wastewater treatment. Indeed, xylanase from the fungus *Thermomyces lanuginosus* is reported to retain about 90% activity at 60 °C, while the xylanase from *Trichoderma reesei* is active at 55 °C for more than three hours [6, 48, 51].

Stability under alternative pH environments has also been considered an important criterion: fungal xylanases are typically optimized at pH 4.5–6.0, as seen with *Trichoderma reesei* [77], while the bacterial counterparts, like those from *Bacillus subtilis*, are optimized at 6.0–7.5 and so are used more often in alkaline processes like textile processing [78] and bioremediation. Another important factor is substrate concentration, which has an influence on enzymatic activity such that xylanases follow Michaelis–Menten kinetics. The xylanase of *Thermomyces lanuginosus* presents Km values of 4.2 mg/mL and a Vmax of 175 U/mL at a concentration of 10 mg/mL of xylan, indicating a substrate with high affinity [52]. The presence of salts in the assay promoted xylanase of *Aspergillus niger* activity in a 25% increase, while EDTA acted as an inhibitor by chelating essential cofactors [72, 79]. These collectively underline the importance of adjusting operational conditions according to enzyme sources to achieve maximum efficacy across a broad range of industrial applications from biofuel generation to wastewater treatment. Optimal conditions for xylanase production for different microorganisms are summarised in Table 3.

4 Applications of Xylanase in Waste Treatment

This kind of xylanase is an enzyme that mainly breaks down xylan-a hemicellulose polysaccharides-and is increasingly finding diverse applications in bioconversion of many types of industrial waste [5, 54]. From agricultural residues to wastewaters

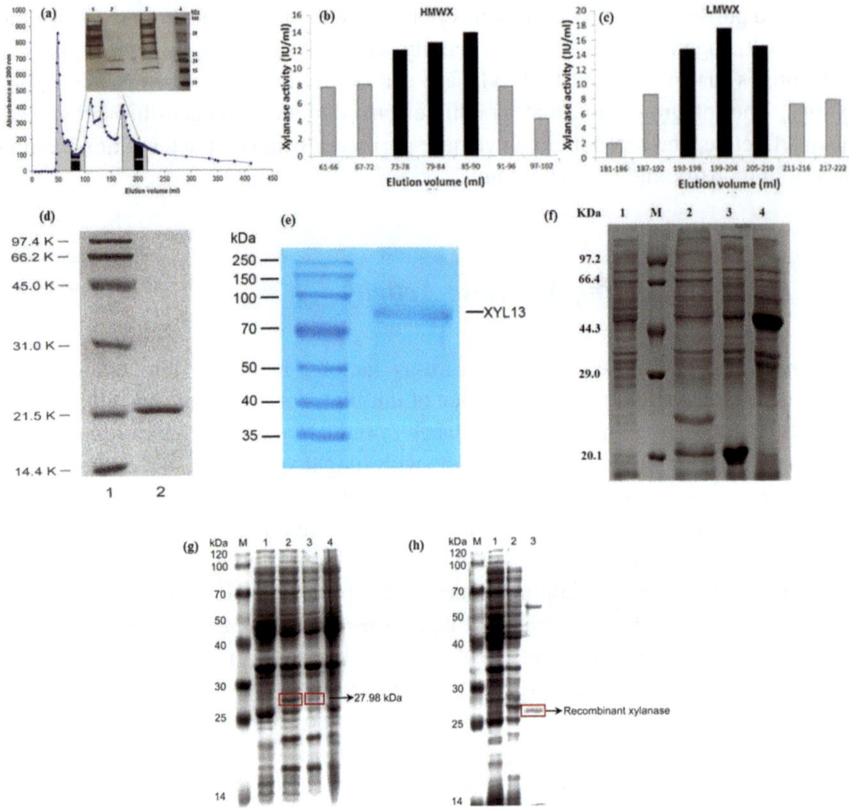

Fig. 2 **a** Chromatogram showing the elution profile of xylanase fractions under purification conditions (0.5 mL/min flow rate, 50 mM pH 7.0 sodium phosphate buffer) with corresponding SDS-PAGE analysis. SDS-PAGE lanes: 1—HMWX fractions (73–90 mL), 2—LMWX fractions (193–210 mL), 3—crude enzyme preparation, 4—marker protein. Grey areas indicate xylanase-positive fractions, while black areas highlight fractions with the highest activity. **b** HMWX activity and **c** LMWX activity profiles, with activity peaks aligning with chromatogram fractions. **d** SDS-PAGE analysis of purified xylanase from F. proliferatum NRRL 26517, with a single distinct band (~ 20 μg protein) confirming purification, analyzed against molecular weight standards (phosphorylase B—97.4 kDa, bovine serum albumin—66.2 kDa, ovalbumin—45 kDa, carbonic anhydrase—31 kDa, trypsin inhibitor—21.5 kDa, and lysozyme—14.4 kDa). **e** SDS-PAGE analysis of XYL13 showing purity and molecular weight. **f** SDS-PAGE of GH10 and GH11 xylanases, highlighting molecular differences. **g** SDS-PAGE analysis of recombinant xylanase expressed in *E. coli* Rosetta (DE3), displaying total proteins (Lane 1), recombinant protein expression (Lane 2), soluble protein (Lane 3), and inclusion bodies (Lane 4). **h** SDS-PAGE of purified recombinant xylanase with a sharp band at 27.98 kDa confirming successful expression and purification. Data adapted from respective studies [71–75]

Table 3 Summary of optimal temperature, pH, and substrate concentration for different sources of xylanase

Microorganism	Optimal pH	Optimal temperature (°C)	Activity	Applications	References
Trichoderma reesei	4.5–6.0	50–60	High activity at 55 °C	Pulp and paper, biofuel production	[80]
Thermomyces lanuginosus	4.5–5.5	55–70	Thermostable, stable at 60 °C	Wastewater treatment, agriculture	[81]
Bacillus subtilis	6.0–7.5	30–45	Moderate activity	Textile industry, bioremediation	[82]
Aspergillus niger	4.0–5.5	40–50	High activity at 45 °C	Food industry, paper processing	[83]
Penicillium oxalicum	5.0	50	High activity	Animal feed, biofuel production	[51]
Streptomyces sp.	6.5	55	Thermostable	Pulp bleaching, bioethanol production	[84]
Bacillus pumilus	7.0	50	Alkaline xylanase	Paper bleaching, textile processing	[85]
Aspergillus oryzae	5.0	50	High activity	Bread making, food industry	[86]
Paenibacillus campinasensis	7.0	60	Thermostable	Animal feed, biofuel production	[87]
Bacillus licheniformis	6.0	50	Alkaline xylanase	Pulp bleaching, textile industry	[88]
Aspergillus terreus	5.5	55	High activity	Biofuel production, animal feed	[89]

from pulp and paper, textiles, and food processing industries, xylanase is fast gaining ground in being recognized as potently suitable for sustainable waste management [15, 46]. This will cover diverse applications of this enzyme in different forms of treatment, ranging from industrial to agri-waste and its role in degrading complex lignocellulosic materials to enhance biodegradation processes of waste.

4.1 Xylanase in the Treatment of Agricultural Waste

Agricultural waste, especially crop residues, for example, from wheat, corn, rice, sugarcanes, and other similar crops, comes mainly from crop residues. Such residues are kept for a longer duration, which seriously raises an environmental issue, as they essentially contain lignocellulosic compounds and more importantly, xylan, which is hardly degraded in nature. Xylan supports the degradation of more complex compounds to simpler sugars, which are further needed for biofuels production and animal feed or composting [71, 73–75].

The xylanase enzymes catalyzed reaction of agricultural residues will denature them, reducing their material output to a large extent while at the same time making them easily digestible-an ideal recycled raw material. In corn stover, roughly about 25–30% xylan content in dry weight [90]. These xylanase treatment diminished hemicellulose and cellulose consistence of the material treated, thus improving its digestibility at a great increase to its value in biofuel production [21, 70]. Across wheat straw, an increase in anaerobic digestion biogas production of 15–20% by applying xylanase to the straw when compared to that achieved on untreated straw is possible [91]. As in most cases, biomass burning causes air pollution with reduced greenhouse gas emissions. In this sense, xylanase will deliver far more sustainable agricultural practices when incorporated in the bioenergy production process or any other useful product. The very illustration manifests the effect of xylanase degradation upon the lignocellulosic components (cellulose, hemicellulose, and lignin) in agricultural residues. Table 4 illustrates the various substrate xylanase treatment and effect (Fig. 3).

4.2 Xylanase in Pulp and Paper Industry Waste Treatment

The pulp and paper industry generates large volumes of wastewater that contain high levels of organic matter, particularly xylan from wood pulp [99]. These effluents often have high chemical oxygen demand (COD), biological oxygen demand (BOD), and total suspended solids (TSS), making them difficult to treat using conventional methods [15, 99]. Xylanase plays an important role in improving the biodegradability of these effluents by breaking down xylan and other complex polysaccharides in the waste, reducing COD and BOD levels [15, 100].

In another study, 40 GH10 xylanase genes from pulp and paper wastewater microbiota were expressed in *E. coli*, with PW-xyl9 showing optimal activity at pH 7 and 60 °C (470.1 U/mg protein) and PW-xyl37 at pH 7 and 55 °C (113.7 U/mg protein). These findings highlight wastewater microbiota as a valuable source of industrially relevant xylanases [55]. The enzyme treatment helped to accelerate the breakdown of hemicellulose and other organic pollutants, making the wastewater more amenable to biological treatment [101]. Xylanase treatment also helps reduce the amount of

Table 4 Impact of xylanase treatment on the biodegradation of lignocellulosic substrates like sugarcane bagasse, corn stover, and rice husks

Substrate	Xylanase treatment conditions	Biodegradation improvement	Reference
Sugarcane bagasse	30 °C, pH 5.0, 24-h enzymatic action	Enhanced hemicellulose degradation by 45%, increased sugar release for ethanol production by 38%	[92]
Corn Stover	40 °C, pH 5.5, 18-h pretreatment	Improved xylose yield by 50%, reduced lignin interference in downstream processes	[93]
Rice husks	35 °C, pH 6.0, 20-h incubation	Increased cellulose accessibility by 60%, boosting fermentable sugar recovery by 42%	[94]
Wheat straw	37–50 °C with addition of a non-ionic surfactant, Tween 20	A synergistic effect between cellulases, FAE and xylanase (10 U/g of cellulases, 3 U/g of xylanase and 10 U/g of FAE). The yield of enzymatic hydrolysis (36.6%)	[95]
Oil palm empty fruit bunch fiber (OPEFB)	Recombinant xylanase (rXyn) Pre-treatment of OPEFB with optimized reaction conditions	Xylo triose (0.361 mg/mL) and traces of xylose (0.044 mg/mL)	[96]
Oil palm empty fruit bunch fiber (OPEFB)	mXyn-CLEA-BSA (mutated xylanase immobilized). Pre-treatment of OPEFB with optimized reaction conditions	Enhanced xylanase activity (137.08% recovery)	[96]
Wheat straw	Immobilized β-xylanase on magnetic nanoparticles (MNPs), Low Temperature Alkali (LTA) Pretreatment: 6% NaOH, 1/10 S/L, 30 °C, 72 h	20.61% saccharification, 15% residual activity after 11th cycle	[97]
Meranti wood sawdust (MWS)	Immobilized Xylanase (entrapped and covalent binding), Chlorite delignification for hemicellulosic Xylan extraction	XOS production: 0.36 mg/mL after 60 h, 70% of initial XOS production after second cycle	[98]

chlorine bleach required in the pulp bleaching process, thereby lowering the generation of toxic chlorinated by-products [102]. The potential for xylanase to enhance effluent treatment is significant, as paper mills are some of the largest industrial consumers of water and contributors to wastewater pollution. Table 5 encapsulated the capability of xylanase functionality to eradicate the COD and BOD of different industrial waste.

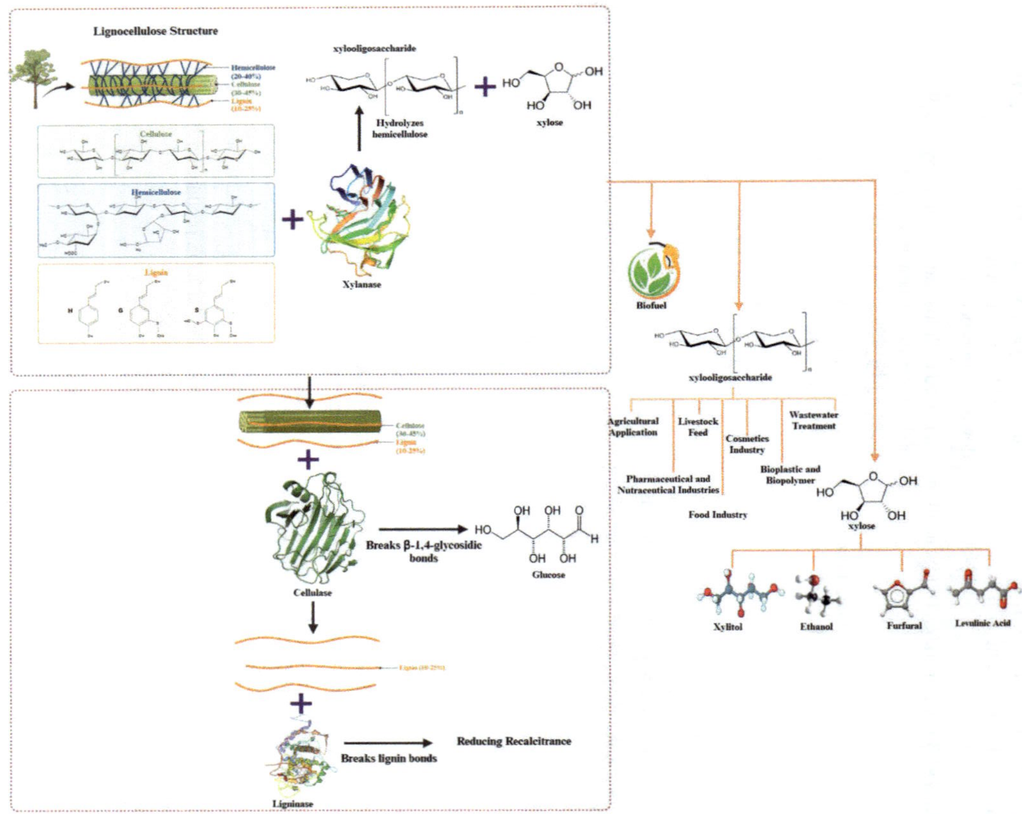

Fig. 3 Effect of xylanase treatment on the degradation of lignocellulosic components (cellulose, hemicellulose, and lignin) in agricultural residues

Table 5 Impact of xylanase on the reduction of COD, BOD, and TSS in pulp and paper mill wastewater

Industry/ application	Treatment process	Xylanase function	Reduction in pollutants	Efficiency gains	References
Agriculture (crop waste)	Degradation of agricultural residues like straw and husks	Degrades hemicellulose and Xylan	50–60% reduction in COD, 30–40% in BOD	25–30% faster degradation	[103]
Pulp and paper	Pulping and bleaching wood fibers	Break down hemicellulose to reduce chemical use	30–40% reduction in chemical consumption	20% improvement in energy use	[104]
Food processing	Wastewater treatment from food industry	Reduces organic matter and COD	25–30% reduction in COD	20% decrease in chemical flocculants	[105]

4.3 *Xylanase in Food Processing Waste Management*

Food processing industries generate substantial waste, including fruit and vegetable peels, starch-rich by-products, and plant-based residues that are rich in xylan [106]. Xylanase can be used to degrade these wastes, improving the recovery of valuable components such as fermentable sugars or organic acids and contributing to waste reduction and resource recovery [107, 108].

In a case study, 15 fungal strains were evaluated for their ability to produce endoglucanase and xylanase from food waste using solid-state fermentation (SSF). *Aspergillus niger* exhibited the highest enzyme activities, reaching 17.37 U/g ds for endoglucanase and 189.24 U/g ds for xylanase, with optimal production at 77.67% moisture content, 0.5 mL inoculum, and 25 °C for 6 days [109]. In a recent study, banana peel, guava bagasse, orange bagasse, and restaurant waste were treated with different alkaline solutions to solubilize xylan and produce xylooligosaccharides (XOS). The highest solubilization (90.70%) was achieved with restaurant waste using hydrogen peroxide, while sodium hydroxide was most effective for guava bagasse and banana peel. XOS production reached 59.86% with guava bagasse and 54.14% with banana peel [110]. Similar applications of xylanase in the fruit and vegetable processing industries have led to higher yields of sugars and organic acids, which can be utilized in food, beverage, and biofuel production. Figure 4 shows the diagram showing the treatment of food waste with xylanase for sugar extraction and subsequent bioethanol production.

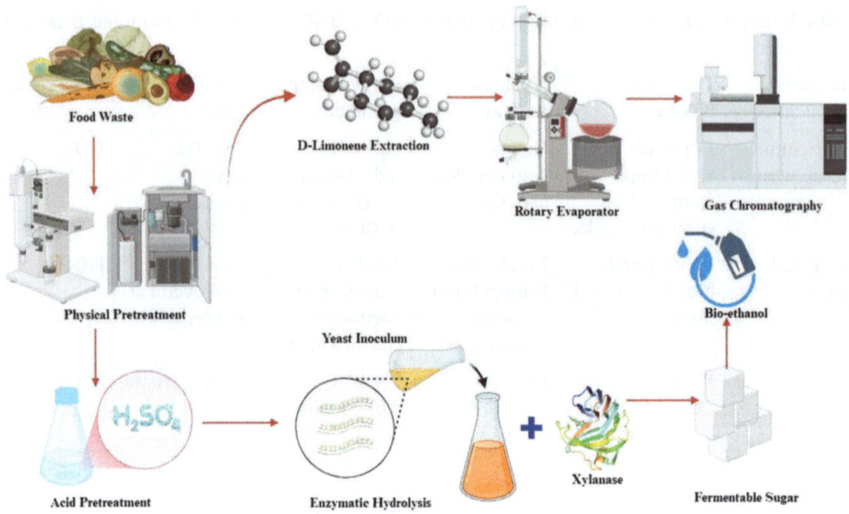

Fig. 4 Process diagram showing the treatment of food waste with xylanase for sugar extraction and subsequent bioethanol production

5 Environmental Benefits of Xylanase in Waste Treatment

Xylanase is rapidly gaining recognition as an environmentally sustainable tool for waste treatment. The enzyme's ability to degrade complex organic materials, particularly xylan, leads to several key environmental benefits, including reductions in chemical usage, enhanced biodegradation, and improvements in waste-to-energy conversion processes [100, 104, 110]. Additionally, xylanase plays a role in the breakdown of toxic compounds and supports the cleanup of oil spills and heavy metals. This section discusses these benefits in greater detail, supported by quantitative data and specific applications in various environmental contexts.

5.1 Reduction of Chemical Usage in Waste Treatment

Industrial wastewater treatment, especially in industries like pulp and paper, textiles, and food processing, traditionally relies on harsh chemicals such as chlorine, sodium hydroxide, and chlorine dioxide [111]. These chemicals contribute significantly to environmental pollution and escalate treatment costs [26, 97, 111]. Xylanases depolymerize xylan and polysaccharides in wastewater: thus, exhibit maximum viability for treatment [42], Optimistically, it brings down COD and BOD levels of effluents. A study found that treatment of pulp and paper mill effluents with xylanase led to a reduction in COD by 60% and in BOD by 55% as opposed to controls that were untreated [112]. The Bigger Pollution Drop helps the mills reduce the use of chlorinated bleach by 10–40% and thus reduces chances of on-site contamination with the toxic by-products such as dioxins and furans [113]. On the other hand, in textile sectors, xylanases could decrease COD by 35–50% in effluents laden with complex polysaccharides and dyes [114]. This ability to alternate or reduce the use of harmful chemicals gives xylanases a prime position in the sustainable future of wastewater treatment technologies. Xylanases also partially degrade some toxic substances, like phenolic compounds and synthetic dyes [54, 79, 113].

5.2 Xylanase in Soil Remediation and Pesticide Degradation

Organic pollutants such as hydrocarbons, agricultural residues, and pesticides are a significant reason for concern in the environment [115, 116]. These substances are very important as xylanases help in biodegradation by the breakdown of lignocellulosic materials to elevate microbial activity [70, 101]. Pseudoallescheria sp. 18A was a major isolate involved in the isolation of hydrocarbon-degrading fungi from clay soil. It was able to utilize aliphatic hydrocarbons (AH) while producing xylanase. During the remediation experiment, Pseudoallescheria significantly stimulated the degradation of AH when used in conjunction with a substrate mixture of wheat straw and poplar wood chips, with an overall reduction of 79.7%; detoxification reduced the mortality of *Folsomia candida* from 90 to 24%. Similar results were observed in the non-inoculated control soil, suggesting that xylanase production by the fungus enhanced both hydrocarbon degradation and lignocellulosic breakdown [117].

5.3 Xylanase in Environmental Remediation: Oil Spill Cleanup and Heavy Metal Removal

Xylanase has demonstrated significant potential in environmental remediation, particularly in oil spill cleanup and heavy metal removal [118]. Traditional oil spill cleanup methods, relying on mechanical or chemical processes, can be costly and harmful [119]. Xylanase offers an eco-friendly alternative by enhancing the biodegradation of petroleum hydrocarbons [120]. In different studies, it has been reported that xylanase treatment of oil-contaminated soil increased the rate of hydrocarbon biodegradation [112, 119, 121]. The enzyme facilitated microbial access to oil droplets, accelerating the overall breakdown. When combined with bioaugmentation, xylanase-treated areas exhibited about 50% reduction in total petroleum hydrocarbons (TPH) [122].

In addition to oil spill remediation, xylanase also plays a role in the removal of heavy metals like cadmium (Cd), lead (Pb), and arsenic (As) from contaminated soils [123]. While xylanase does not directly bind to heavy metals, it enhances microbial activity, which aids in the sequestration of these pollutants [124]. The bioaccumulation of cadmium (Cd) in plants during the biochar amendment process was likely influenced by the activity of xylanases. Xylanases are enzymes that break down xylan, a major component of plant cell walls, and their activity could enhance the availability of metal-binding sites in plant tissues, facilitating Cd^{+2} uptake and accumulation. This enzymatic activity may also influence the biochar-soil–plant interactions, potentially aiding in the sequestration and uptake of Cd^{+2} by the plant [125]. The significance of xylanase in enhancing bioremediation processes for organic pollutants and heavy metals stands to benefit cleaner soils through healthier ecosystems. Figure 5 shows the Impact of Pesticides on Soil and its exposure to human adversity along with the potential degradation of Pesticides via xylanases (fungi, bacteria) and in panel (c) Impact of Oil spill on ecosystem as well human, along in panel its treatment method with advantages and disadvantages.

6 Case Studies

It has been used in a variety of industries, and indeed, its commercial application as a sustainable and effective waste treatment and valorization manager has been proven. This section discusses exhaustive cases on xylanase operation in waste treatment plants, pilot valorization trials, and industry large-scale application-specific performance metrics, experimental results, and field data.6.1 Xylanase in Wastewater Treatment Plants.

WWTP stands for Wastewater Treatment Plant in general. How it works is treating wastewater with many different types of pollutants such as organic compounds, dyes, residual chemicals, etc. The addition of xylanase would enhance the biodegradation of complex polysaccharides such as xylan, lignin, and lignocellulosic materials that are usually resistant to the normal treatment procedure.

In a notable case study by Cuiyi Liang et al., cellulase and hemicellulase activities, critical bottlenecks in lignocellulosic biorefinery processes, were enhanced by the inclusion of β-glucosidase (BGL) and xylanase (XYL) in an enzyme cocktail. The study demonstrated that higher BGL and XYL activities could be achieved using wastewater and activated sludge as an induction medium, mixed with 5% rice straw by Hypocrea sp. W63. Analysis of ionic content revealed that a relatively low dose of sludge boosted enzyme production. Remarkably, a 1:10 ratio of sludge to wastewater, rich in nutrient elements, led to a 3.4-fold increase in BGL and a 3.7-fold increase in XYL production compared to using freshwater. This research highlights the potential of reusing substrates from domestic wastewater treatment and agricultural wastewater treatment residues and agricultural, residues, offering a sustainable approach to offering enzyme a sustainable approach production in biorefinery practices [126].

In another study on xylanase enzymes, 40 GH10 family xylanase genes were cloned and expressed from a pulp and paper wastewater treatment (PPWT) microbiota. Among them, two enzymes, PW-xyl9 and PW-xyl37, exhibited high xylanase activity on beechwood substrate. These enzymes had optimal conditions at pH 7 and

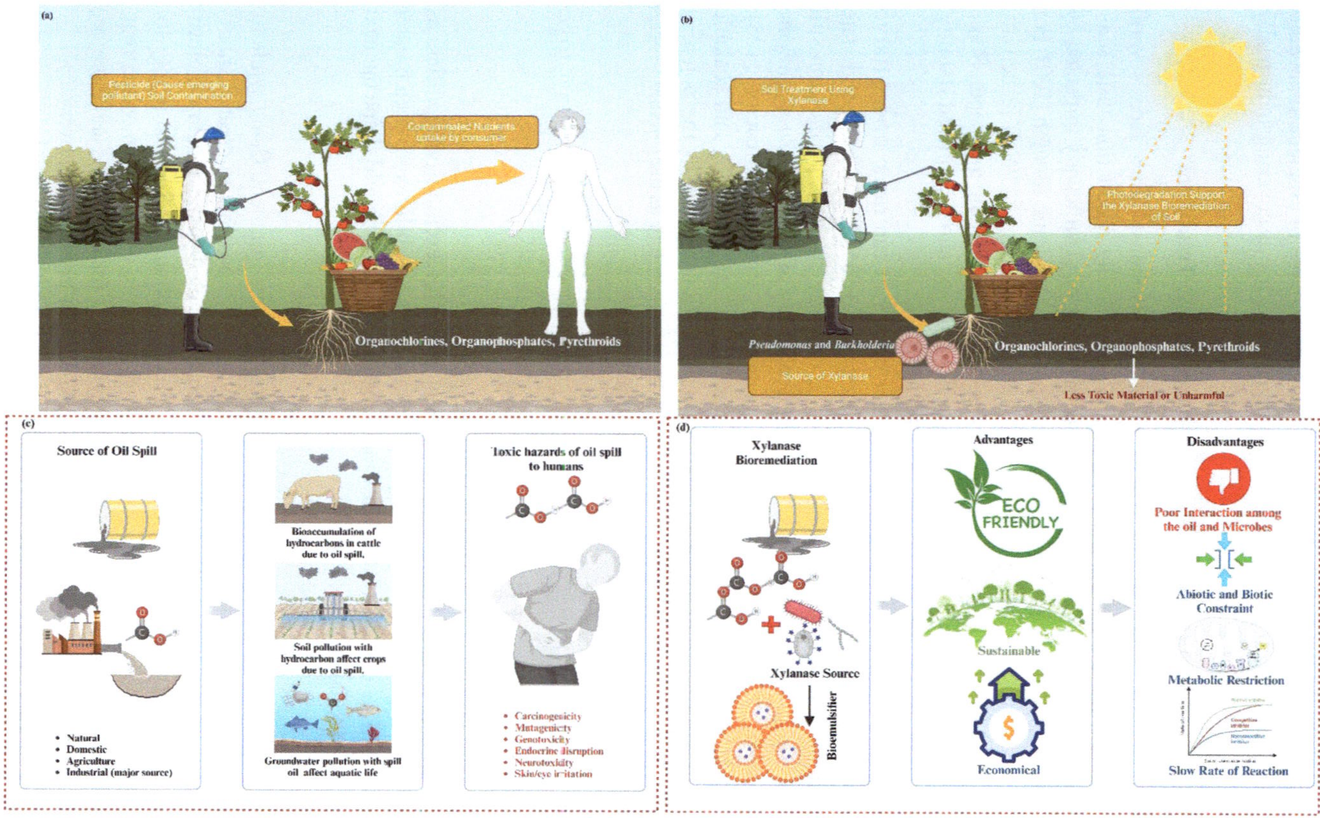

Fig. 5 **a** Impact of pesticides on soil and its exposure to human, **b** Degradation of pesticides via xylanases (fungi, bacteria), **c** Impact of oil spill on ecosystem as well human, **e** Treatment of contaminated environment and its pros and cons

60 °C for PW-xyl9 and 55 °C for PW-xyl37. Their specific activities were 470.1 U/mg and 113.7 U/mg, respectively, with Km values of 8.02 and 18.8 g/L. This research highlights PPWT microbiota as a valuable, untapped source of efficient lignocellulase genes, showcasing the potential of metagenomics for discovering industrially relevant enzymes [104].

Challenges and Limitations in Xylanase-based Waste Treatment

Xylanase has great potential in waste treatment, but its industrial use faces several challenges [110]. These include issues with enzyme stability, production costs, scalability, and microbial resistance. Xylanase activity can be significantly impacted by harsh conditions such as extreme pH, temperature, and high ionic strength [100]. For instance, xylanase from *Thermomyces lanuginosus* retains only 45% of its activity at 55 °C for 2 h [127], limiting its efficiency in industries like pulp and paper. Some xylanases, such as recombinant variants from Bacillus sp., show improved stability, retaining 70% of activity at 60 °C for an hour [88].

The cost of xylanase production is another barrier. It can range from $100 to $500 per kg, depending on the production method [42, 55, 73]. While solid-state fermentation with agricultural waste can lower costs by 20–30%, purification still accounts for 50–60% of the total cost [59]. Scalability also poses issues; enzyme yields drop by 20–30% when scaled up from laboratory to industrial reactors [128]. These challenges highlight the need for more stable, cost-effective, and scalable xylanase solutions for large-scale waste treatment. Figure 6 reveals the potential challenges of xylanase utilization for the waste treatment where as the outside blocks also gives the potential solutions of these challenges.

7 Innovations and Future Directions

The future of xylanase-based waste treatment is full of exciting possibilities, as researchers continue to refine and expand the enzyme's capabilities for even greater efficiency and sustainability. Here are some key directions that could shape the next generation of xylanase applications:

- Smart Enzyme Systems
- Imagine enzymes that can "think" and adapt to their environment. Using synthetic biology and other advances in nanotechnology, it is possible to create "smart" xylanases that respond to environmental conditions on demand, e.g., upon elevation of temperature or upon depressurization by pollutants. Such an ability for precision will ensure that xylanases are involved more efficiently in waste treatments, thereby lowering the amount of enzyme used for treatments and therefore achieving a greater level of sustainability AI-Driven Enzyme Discovery.

  The integration of bioinformatics with artificial intelligence enables new vistas of exploration for whatever liberal xylanases get distributed from truly strange places-such as from the deepest ocean or extreme desert ecosystems. As possible,

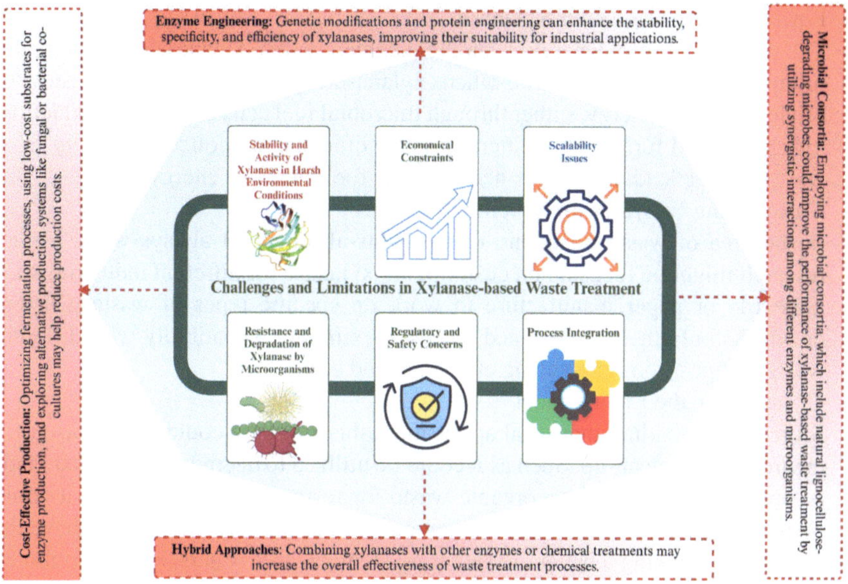

Fig. 6 Challenges associated with the Xylanase based waste treatment approaches and its potential solutions

AI sorts through large databases to even tell us of much more efficient and stable enzymes meant for work on some of the toughest challenges in waste treatment.

- Collaborative Enzyme Teams
  Rather than employing one enzyme alone, it is imagined that a team of enzymes would be employed together to break down wastes more effectively. Keeping in view such microbial consortia producing xylanases along with few others such as cellulases and ligninases will facilitate a significant improvement in the breakdown of complex materials such as plant waste, which can be speedier and more efficient.
- Recycling Enzymes for Long-Term Use
  Enzymes are really expensive; what if they could be reused? This entails immobilizing xylanase on biochar and/or silica beads, which would allow this enzyme to hold up through multiple applications in an industrial process. All the savings, wastes reduced, and much greener life through enzyme application for waste treatment.
- Turning Waste into Valuable Products
  Xylanases might just prove to be instrumental in the actual conversion of agricultural and industrial waste into economically important products such as biodegradable plastics or biofuels. Bringing xylanase into a circular economic model would not mean just treating wastes but also creating new, sustainable resources from them, thus converting something unwanted into something useful.

- Powering Bioenergy with Xylanase
  The production of bioenergy is another segment where xylanase could be really promising. Imagine a scenario where xylanase degrades the organic wastes for producing clean energy. Either through microbial fuel cells or biofuel production, xylanase could further aid in increasing the efficiency of collecting energy from wastes, hence leading towards new avenues for renewable energy sources.
- Customizing Enzymes for Specific Industries
  In the area of waste treatment, one-size-fits-all does not always work. Future research might be focusing on customizing xylanase for different industries such as textile or paper manufacture to work on specific types of waste. Custom enzymes could then be designed to give maximum compatibility with different waste compositions, enhancing efficiency and cost.
- Cleaning Up the Environment
  More than providing industrial application, this xylanase could also be used for environmental clean-up. Such as it could be utilized to degrade plant-based kinds of pollution or decompose organic waste for restoring soil quality. As it could be incorporated into other bioremediation approaches, xylanase should have the potential of making a better cleaner ecosystem, cleaner waters, and less polluted environments.

These simply mark the onset of innovation. With the continued research, xylanase will be an important future tool for sustainable waste management, energy production, and environmental restoration. With each further innovation, xylanase takes us one step closer to a cleaner, more sustainable future.

8 Conclusion

In the era of environmental concerns confronting human society with the degradation of complex organic matter, while reducing chemical use, and converting wastes into useful by-products, xylanase-based waste treatment provides an eco-friendly alternative. This chapter discusses the various applications of xylanase in waste treatment, biofuel production, and waste valorization. Subsequent research and case studies affirm xylanase's environmental and economic gains in industrial waste management systems while outlining obstacles for its wider incorporation.

An important conclusion is its role in facilitating the biodegradation of lignocellulosic materials, such as xylan, hemicellulose, and lignin, commonly found in agricultural and industrial waste streams. This enzymatic action potentially helps in reducing chemical oxygen demand (COD) and biochemical oxygen demand (BOD), thereby increasing the efficiency and cost-effectiveness of wastewater treatment. In the pulp and paper, textile, and agricultural industries, xylanase provides less use of toxic chemicals and better efficiency in waste treatment processes. For example, its application in pulp mills has reduced chemical use by 30% and energy consumption during pulping by 20%.

Yet, there are still challenges. These include the instability of xylanase under rough industrial conditions, high production costs, and the non-scalability of production processes—key setbacks. Efforts are now underway to develop more thermostable enzymes and optimize their production processes. Due to the many potential applications of xylanase, it will readily become a critical tool for achieving responsibilities toward circular economy objectives, especially within zero-waste strategies. Future studies should hence aim at improving thermostability, cost-effectiveness, and enzyme synergism toward industrial-scale implementation.

References

1. E.A. Cazier, T.-N. Pham, L. Cossus, M. Abla, T. Ilc, P. Lawrence, Exploring industrial lignocellulosic waste: sources, types, and potential as high-value molecules. Waste Manage. **188**, 11–38 (2024). https://doi.org/10.1016/j.wasman.2024.07.029
2. A. Blasi, A. Verardi, C.G. Lopresto, S. Siciliano, P. Sangiorgio, Lignocellulosic agricultural waste valorization to obtain valuable products: an overview. Recycling **8**(4), 61 (2023). https://doi.org/10.3390/recycling8040061
3. S. Singh et al., Innovations and advances in enzymatic deconstruction of biomass and their sustainability analysis: a review. Renew. Sustain. Energy Rev. **189**, 113958 (2024). https://doi.org/10.1016/j.rser.2023.113958
4. S. Hassan, *Development of Novel Pectinase and Xylanase Juice Clarification Enzymes via a Combined Biorefinery and Immobilization Approach* (2020). https://doi.org/10.21427/G2EY-HK88
5. R. Chaudhary et al., Current status of xylanase for biofuel production: a review on classification and characterization. Biomass Conv. Bioref. **13**(10), 8773–8791 (2023). https://doi.org/10.1007/s13399-021-01948-2
6. M.A.D. Souza, I.T. Vilas-Boas, J.M. Leite-da-Silva, P.D.N. Abrahão, B.E. Teixeira-Costa, V.F. Veiga-Junior, Polysaccharides in agro-industrial biomass residues. Polysaccharides **3**(1), 95–120 (2022). https://doi.org/10.3390/polysaccharides3010005
7. A. Panja, S. Paul, P. Jha, S. Ghosh, R. Prasad, Waste and their polysaccharides: Are they worth bioprocessing? Bioresour. Technol. Rep. **24**, 101594 (2023). https://doi.org/10.1016/j.biteb.2023.101594
8. V. Kamperidou, P. Terzopoulou, Anaerobic digestion of lignocellulosic waste materials. Sustainability **13**(22), 12810 (2021). https://doi.org/10.3390/su132212810
9. A.N. Arzami, T.M. Ho, K.S. Mikkonen, Valorization of cereal by-product hemicelluloses: fractionation and purity considerations. Food Res. Int. **151**, 110818 (2022). https://doi.org/10.1016/j.foodres.2021.110818
10. A. Kalinin, R. Rudnik, A. Tsvetov, K. Bondarenko, A. Shuranova, Emerging markets decoded 2024. SSRN J. (2024). https://doi.org/10.2139/ssrn.4862785
11. S.A. Aladejare, S. Bobbo, Brown vs green energy sources and resource productivity: the role of human capital and technology transfer in developing economies. Sustain. Econ. **2**(4), 416 (2024). https://doi.org/10.62617/se416
12. S.K. Paul et al., Potentials of enzyme biotechnology in urban solid waste management: an assessment, in *Enzyme Biotechnology for Environmental Sustainability* (Elsevier, 2024), pp. 383–414. https://doi.org/10.1016/B978-0-443-22072-2.00010-3
13. G. Venkatesh, Circular bio-economy—paradigm for the future: systematic review of scientific journal publications from 2015 to 2021. Circ. Econ. Sust. **2**(1), 231–279 (2022). https://doi.org/10.1007/s43615-021-00084-3

14. V. Jegatheesan, L. Shu, L. Jegatheesan, Producing fit-for-purpose water and recovering resources from various sources: an overview. Environ. Qual. Mgmt **31**(2), 9–28 (2021). https://doi.org/10.1002/tqem.21780
15. A. Gupta, R. Gupta, Treatment and recycling of wastewater from pulp and paper mill, in *Advances in Biological Treatment of Industrial Waste Water and their Recycling for a Sustainable Future*, eds. by R.L. Singh, R.P. Singh (Springer, Singapore, 2019), pp. 13–49. https://doi.org/10.1007/978-981-13-1468-1_2
16. M. Mujtaba et al., Lignocellulosic biomass from agricultural waste to the circular economy: a review with focus on biofuels, biocomposites and bioplastics. J. Clean. Prod. **402**, 136815 (2023). https://doi.org/10.1016/j.jclepro.2023.136815
17. K. Robertson, K. Flanagan, *Reducing Food Loss And Waste Setting a Global Action Agenda*, World Resource Institute. [Online]. Available: https://web.archive.org/web/20211117130643id_/. https://wriorg.s3.amazonaws.com/s3fs-public/reducing-food-loss-waste-global-action-agenda_0.pdf
18. A. Pandey, Food wastage: causes, impacts and solutions. Sci. Herit. J. **5**(1), 17–20 (2021). https://doi.org/10.26480/gws.01.2021.17.20
19. B. Koul, M. Yakoob, M.P. Shah, Agricultural waste management strategies for environmental sustainability. Environ. Res. **206**, 112285 (2022). https://doi.org/10.1016/j.envres.2021.112285
20. G. G et al., Valorization of agricultural residues: Different biorefinery routes. J. Environ. Chem. Eng. **9**(4), 105435 (2021). https://doi.org/10.1016/j.jece.2021.105435
21. M.F.S. Khan, M. Akbar, Z. Xu, H. Wang, A review on the role of pretreatment technologies in the hydrolysis of lignocellulosic biomass of corn stover. Biomass Bioenerg. **155**, 106276 (2021). https://doi.org/10.1016/j.biombioe.2021.106276
22. M.M. Urugo et al., A comprehensive review of current approaches on food waste reduction strategies. Comp. Rev. Food Sci. Food Safe **23**(5), e70011 (2024). https://doi.org/10.1111/1541-4337.70011
23. Y. Jaouhari et al., From industrial food waste to bioactive ingredients: a review on the sustainable management and transformation of plant-derived food waste. Foods **12**(11), 2183 (2023). https://doi.org/10.3390/foods12112183
24. R.J. Whitehurst, M. Van Oort (eds.), *Enzymes in Food Technology*, 1st edn. (Wiley, 2009). https://doi.org/10.1002/9781444309935
25. A. Sharma et al., Worldwide pesticide usage and its impacts on ecosystem. SN Appl. Sci. **1**(11), 1446 (2019). https://doi.org/10.1007/s42452-019-1485-1
26. S.H. Bai, S.M. Ogbourne, Glyphosate: environmental contamination, toxicity and potential risks to human health via food contamination. Environ. Sci. Pollut. Res. **23**(19), 18988–19001 (2016). https://doi.org/10.1007/s11356-016-7425-3
27. R. Datta, Enzymatic degradation of cellulose in soil: A review. Heliyon **10**(1), e24022 (2024). https://doi.org/10.1016/j.heliyon.2024.e24022
28. P. Chandra, Enespa, Fungal enzymes for bioremediation of contaminated soil, in *Recent Advancement in White Biotechnology Through Fungi*, eds. by A.N. Yadav, S. Singh, S. Mishra, A. Gupta, in Fungal Biology (Springer International Publishing, Cham, 2019), pp. 189–215. https://doi.org/10.1007/978-3-030-25506-0_7
29. V. Corina, An analysis of maritime accidents involving oil tankers from a structural point of view. SBNA **XIX**(1), 71–76 (2018). https://doi.org/10.21279/1454-864X-18-I1-010
30. S.K. Ahuja, G.M. Ferreira, A.R. Moreira, Utilization of enzymes for environmental applications. Crit. Rev. Biotechnol. **24**(2–3), 125–154 (2004). https://doi.org/10.1080/07388550490493726
31. M. Narayanan, S.S. Ali, M. El-Sheekh, A comprehensive review on the potential of microbial enzymes in multipollutant bioremediation: mechanisms, challenges, and future prospects. J. Environ. Manage. **334**, 117532 (2023). https://doi.org/10.1016/j.jenvman.2023.117532
32. L. Porto De Souza Vandenberghe, K. Kley Valladares-Diestra, G. Amaro Bittencourt, A.F. Murawski De Mello, C.R. Soccol, Xylan, in *Biomass, Biofuels, Biochemicals* (Elsevier, 2021), pp. 129–161. https://doi.org/10.1016/B978-0-12-821888-4.00008-3

33. S. Malgas, M.S. Mafa, L. Mkabayi, B.I. Pletschke, A mini review of xylanolytic enzymes with regards to their synergistic interactions during hetero-xylan degradation. World J. Microbiol. Biotechnol. **35**(12), 187 (2019). https://doi.org/10.1007/s11274-019-2765-z
34. K. Šuchová, C. Fehér, J.L. Ravn, S. Bedő, P. Biely, C. Geijer, Cellulose- and xylan-degrading yeasts: Enzymes, applications and biotechnological potential. Biotechnol. Adv. **59**, 107981 (2022). https://doi.org/10.1016/j.biotechadv.2022.107981
35. G. Malhotra, S.S. Chapadgaonkar, Production and applications of xylanases—an overview. bta **99**(1), 59–72 (2018). https://doi.org/10.5114/bta.2018.73562
36. M. Mendonça, M. Barroca, T. Collins, Endo-1,4-β-xylanase-containing glycoside hydrolase families: characteristics, singularities and similarities. Biotechnol. Adv. **65**, 108148 (2023). https://doi.org/10.1016/j.biotechadv.2023.108148
37. N.P. Marques et al., Cellulases and xylanases production by endophytic fungi by solid state fermentation using lignocellulosic substrates and enzymatic saccharification of pretreated sugarcane bagasse. Ind. Crops Prod. **122**, 66–75 (2018). https://doi.org/10.1016/j.indcrop.2018.05.022
38. V. Juturu, J.C. Wu, Microbial exo-xylanases: a mini review. Appl. Biochem. Biotechnol. **174**(1), 81–92 (2014). https://doi.org/10.1007/s12010-014-1042-8
39. K. Taniguchi, S. Karita, M. Umekawa, Synergistic xylan decomposition by a reducing-end xylose-releasing exo-oligoxylanase with other xylanolytic enzymes derived from *Paenibacillus xylaniclasticus* strain TW1. Biosci. Biotechnol. Biochem. **88**(12), 1479–1486 (2024). https://doi.org/10.1093/bbb/zbae130
40. C. Ramos Santos, C.C. Polo, J.M. Corrêa, R.D.C.G. Simão, F.A.V. Seixas, M.T. Murakami, The accessory domain changes the accessibility and molecular topography of the catalytic interface in monomeric GH39 β-xylosidases, *Acta Crystallogr D Biol Crystallogr*, vol. 68, no. 10 (2012), pp. 1339–1345. https://doi.org/10.1107/S0907444912028491
41. N. Rasmussen, N. Mohieddin Abukhdeir, V.C.A. Ward, Hydrogel-based photobioreactor for solid-state cultivation of *Chlorella vulgaris*. Bioresour. Technol. **408**, 131105 (2024). https://doi.org/10.1016/j.biortech.2024.131105
42. N. Bhardwaj, B. Kumar, P. Verma, A detailed overview of xylanases: an emerging biomolecule for current and future prospective. Bioresour. Bioprocess. **6**(1), 40 (2019). https://doi.org/10.1186/s40643-019-0276-2
43. X. Li, Y. Zheng, Lignin-enzyme interaction: Mechanism, mitigation approach, modeling, and research prospects. Biotechnol. Adv. **35**(4), 466–489 (2017). https://doi.org/10.1016/j.biotechadv.2017.03.010
44. N. Kango, U.K. Jana, R. Choukade, Fungal enzymes: sources and biotechnological applications, in *Advancing Frontiers in Mycology & Mycotechnology*, eds. by T. Satyanarayana, S.K. Deshmukh, M.V. Deshpande (Springer, Singapore, 2019), pp. 515–538. https://doi.org/10.1007/978-981-13-9349-5_21
45. S. Ramasamy, Utilizing aerobic and anaerobic mesophilic fungi for the production of multienzymes, in *Bioprospecting of Multi-tasking Fungi for a Sustainable Environment*, eds. by K.B. Uppuluri, R. Selvasembian (Springer Nature, Singapore, 2024), pp. 49–67. https://doi.org/10.1007/978-981-97-4113-7_3
46. K. Sharma, A. Thakur, A. Goyal, Xylanases for food applications, in *Green Bio-processes*, eds. by B. Parameswaran, S. Varjani, S. Raveendran, in Energy, Environment, and Sustainability (Springer, Singapore, 2019), pp. 99–118. https://doi.org/10.1007/978-981-13-3263-0_7
47. M. Kumar, Y. Sun, R. Rathour, A. Pandey, I.S. Thakur, D.C.W. Tsang, Algae as potential feedstock for the production of biofuels and value-added products: Opportunities and challenges. Sci. Total Environ. **716**, 137116 (2020). https://doi.org/10.1016/j.scitotenv.2020.137116
48. A. Bala, B. Singh, Cellulolytic and xylanolytic enzymes of thermophiles for the production of renewable biofuels. Renew. Energy **136**, 1231–1244 (2019). https://doi.org/10.1016/j.renene.2018.09.100
49. M. Sohail et al., Cellulolytic and xylanolytic enzymes from yeasts: properties and industrial applications. Molecules **27**(12), 3783 (2022). https://doi.org/10.3390/molecules27123783

50. J.-K. Lee, S.-Y. Kim, Y.-W. Ryu, J.-H. Seo, J.-H. Kim, Purification and characterization of a novel erythrose reductase from *Candida magnoliae.* Appl. Environ. Microbiol. **69**(7), 3710–3718 (2003). https://doi.org/10.1128/AEM.69.7.3710-3718.2003
51. S. Zhang, J. Wang, H. Jiang, Microbial production of value-added bioproducts and enzymes from molasses, a by-product of sugar industry. Food Chem. **346**, 128860 (2021). https://doi.org/10.1016/j.foodchem.2020.128860
52. D.K. Patel, K. Patel, D. Patel, G. Dave, Engineering of thermostable phytase–xylanase for hydrolysis of complex biopolymers, 3 Biotech, **11**(8), 390 (2021). https://doi.org/10.1007/s13205-021-02936-z
53. L.P. Fernandes, R.Z. Ventorim, M.G. De Oliveira, L.F. Almeida, V.M. Guimarães, G.P. Maitan-Alfenas, Xylooligosaccharides from pretreated rice bran produced by immobilized xylanase. Bioenerg. Res. **17**(4), 2236–2245 (2024). https://doi.org/10.1007/s12155-024-10770-3
54. D. Kour et al., Extremophiles for hydrolytic enzymes productions: biodiversity and potential biotechnological applications, in *Bioprocessing for Biomolecules Production*, 1st edn., eds. by G. Molina, V. Gupta, B. Singh, N. Gathergood (Wiley, 2019), pp. 321–372. https://doi.org/10.1002/9781119434436.ch16
55. W. Wang et al., Efficacy of combination of endo-xylanase and xylan-debranching enzymes in improving cereal bran utilization in piglet diet. Anim. Biosci. **35**(11), 1733–1743 (2022). https://doi.org/10.5713/ab.21.0534
56. S. Dhiman, G. Mukherjee, Recent advances and industrial applications of microbial xylanases: a review, in *Fungi and their Role in Sustainable Development: Current Perspectives*, eds. by P. Gehlot, J. Singh (Springer, Singapore, 2018), pp. 329–348. https://doi.org/10.1007/978-981-13-0393-7_19
57. A. Singh, S. Bajar, A. Devi, N.R. Bishnoi, Adding value to agro-industrial waste for cellulase and xylanase production via solid-state bioconversion. Biomass Conv. Bioref. **13**(9), 7481–7490 (2023). https://doi.org/10.1007/s13399-021-01503-z
58. M. Du Toit, L. Engelbrecht, E. Lerm, S. Krieger-Weber, Lactobacillus: the next generation of malolactic fermentation starter cultures—an overview. Food Bioprocess Technol. **4**(6), 876–906 (2011). https://doi.org/10.1007/s11947-010-0448-8
59. I.D.S.V. Rodrigues et al., Production of xylanases by *Bacillus* sp. TC-DT13 in solid state fermentation using bran wheat. Prep. Biochem. Biotechnol. **50**(1), 91–97 (2020). https://doi.org/10.1080/10826068.2019.1663536
60. Ritika et al., Concomitant inhibitor-tolerant cellulase and xylanase production towards sustainable bioethanol production by Zasmidiumcellare CBS 146.36. Fuel **375**, 132593 (2024). https://doi.org/10.1016/j.fuel.2024.132593
61. S.N. Patel, S. Sharma, A.K. Singh, S.P. Singh, Potential microbial bioresources for functional sugar molecules, in *Microbial Bioreactors for Industrial Molecules*, 1st edn., eds. by S.P. Singh, S.K. Upadhyay (Wiley, 2023), pp. 211–236. https://doi.org/10.1002/9781119874096.ch10
62. A. Jayakrishnan et al., Evolving paradigms of recombinant protein production in pharmaceutical industry: a rigorous review. Science **6**(1), 9 (2024). https://doi.org/10.3390/sci6010009
63. V. Juturu, J.C. Wu, Microbial cellulases: engineering, production and applications. Renew. Sustain. Energy Rev. **33**, 188–203 (2014). https://doi.org/10.1016/j.rser.2014.01.077
64. T. Miao et al., Improved production of xylanase in *Pichia pastoris* and its application in xylose production from xylan. Front. Bioeng. Biotechnol. **9**, 690702 (2021). https://doi.org/10.3389/fbioe.2021.690702
65. Alokika, B. Singh, Production, characteristics, and biotechnological applications of microbial xylanases. Appl. Microbiol. Biotechnol. **103**(21–22), 8763–8784 (2019). https://doi.org/10.1007/s00253-019-10108-6
66. P. Dhaver, B. Pletschke, B. Sithole, R. Govinden, Optimization of xylooligosaccharides production by native and recombinant xylanase hydrolysis of chicken feed substrates. IJMS **24**(23), 17110 (2023). https://doi.org/10.3390/ijms242317110

67. N. Laohakunakorn, Cell-free systems: a proving ground for rational biodesign. Front. Bioeng. Biotechnol. **8**, 788 (2020). https://doi.org/10.3389/fbioe.2020.00788
68. P. Dhaver, B. Pletschke, B. Sithole, R. Govinden, Optimization, purification, and characterization of xylanase production by a newly isolated *Trichoderma harzianum* strain by a two-step statistical experimental design strategy. Sci. Rep. **12**(1), 17791 (2022). https://doi.org/10.1038/s41598-022-22723-x
69. D. Matulis, Selective precipitation of proteins, *CP Protein Science*, vol. 83, no. 1, Feb. 2016, https://doi.org/10.1002/0471140864.ps0405s83
70. R. Ravindran, A. Jaiswal, Microbial enzyme production using lignocellulosic food industry wastes as feedstock: a review. Bioengineering **3**(4), 30 (2016). https://doi.org/10.3390/bioengineering3040030
71. D. Sutay Kocabaş, S. Güder, N. Özben, Purification strategies and properties of a low-molecular weight xylanase and its application in agricultural waste biomass hydrolysis. J. Mol. Catal. B: Enzymatic. **115**, 66–75 (2015). https://doi.org/10.1016/j.molcatb.2015.01.012
72. B.C. Saha, Production, purification and properties of xylanase from a newly isolated *Fusarium proliferatum*. Process Biochem. **37**(11), 1279–1284 (2002). https://doi.org/10.1016/S0032-9592(02)00012-2
73. J. Chen et al., Characterization of a novel cold-active β-Xylosidase from Parabacteroides distasonis and its synergistic hydrolysis of beechwood xylan. Int. J. Biol. Macromol. **284**, 137895 (2025). https://doi.org/10.1016/j.ijbiomac.2024.137895
74. X. Yu et al., Preparation and characterization of ferulic oligosaccharides from different sources by cell-free GH10 and GH11 xylanases. Int. J. Biol. Macromol. **282**, 137287 (2024). https://doi.org/10.1016/j.ijbiomac.2024.137287
75. H. Tai et al., A thermostable xylanase hydrolyzes several polysaccharides from *Bacillus altitudinis* JYY-02 showing promise for industrial applications. Carbohyd. Res. **538**, 109080 (2024). https://doi.org/10.1016/j.carres.2024.109080
76. L.E. Rasmussen, H.R. Sørensen, J. Vind, A. Viksø-Nielsen, Mode of action and properties of the β-xylosidases from *Talaromyces emersonii* and *Trichoderma reesei*. Biotechnol. Bioeng. **94**(5), 869–876 (2006). https://doi.org/10.1002/bit.20908
77. S.S. Adav, A. Ravindran, L.T. Chao, L. Tan, S. Singh, S.K. Sze, Proteomic analysis of pH and strains dependent protein secretion of *Trichoderma reesei*. J. Proteome Res. **10**(10), 4579–4596 (2011). https://doi.org/10.1021/pr200416t
78. G. Mamo, B. Mattiasson, Alkaliphiles: the versatile tools in biotechnology, in *Alkaliphiles in Biotechnology*, vol. 172, eds. by G. Mamo, B. Mattiasson, in Advances in Biochemical Engineering/Biotechnology (Springer International Publishing, Cham, 2020), pp. 1–51. https://doi.org/10.1007/10_2020_126
79. G. Tripathi et al., Utilizing novel *Aspergillus* species for bio-flocculation: a cost-effective approach to harvest Scenedesmus microalgae for biofuel production. Curr. Res. Microbial. Sci. **7**, 100272 (2024). https://doi.org/10.1016/j.crmicr.2024.100272
80. G.B. Sperandio et al., Exploring the synergistic secretome: insights from co-cultivation of *Aspergillus brasiliensis* and *Trichoderma reesei* RUT-C30. JoF **10**(10), 677 (2024). https://doi.org/10.3390/jof10100677
81. M. Dixit, D. Chhabra, P. Shukla, Optimization of endoglucanase-lipase-amylase enzyme consortium from *Thermomyces lanuginosus* VAPS25 using multi-objective genetic algorithm and their bio-deinking applications. Biores. Technol. **370**, 128467 (2023). https://doi.org/10.1016/j.biortech.2022.128467
82. G. Alamnie, A. Gessesse, B. Andualem, Production of surfactant-stable keratinolytic protease from *B. subtilis* ES5 and its application as a detergent additive. Biocatal. Agric. Biotechnol. **50**, 102750 (2023). https://doi.org/10.1016/j.bcab.2023.102750
83. P. Cheng et al., *Tribonema* sp. and *Chlorella zofingiensis* co-culture to treat swine wastewater diluted with fishery wastewater to facilitate harvest. Biores. Technol. **297**, 122516 (2020). https://doi.org/10.1016/j.biortech.2019.122516
84. M. Sanjivkumar, A. Brindhashini, M. Deivakumari, A. Palavesam, G. Immanuel, Investigation on saccharification and bioethanol production from pretreated agro-residues using a mangrove

associated actinobacterium streptomyces variabilis (MAB3). Waste Biomass Valor **9**(6), 969–984 (2018). https://doi.org/10.1007/s12649-017-9886-0
85. L.-Y. Liang, Y.-G. Zheng, Y.-C. Shen, Optimization of β-alanine production from β-aminopropionitrile by resting cells of *Rhodococcus* sp. G20 in a bubble column reactor using response surface methodology. Process Biochem. **43**(7), 758–764 (2008). https://doi.org/10.1016/j.procbio.2008.03.002
86. A.B. Nongonierma, M. Lalmahomed, S. Paolella, R.J. FitzGerald, Milk protein isolate (MPI) as a source of dipeptidyl peptidase IV (DPP-IV) inhibitory peptides. Food Chem. **231**, 202–211 (2017). https://doi.org/10.1016/j.foodchem.2017.03.123
87. N. Beukes, H. Chan, R.H. Doi, B.I. Pletschke, Synergistic associations between *Clostridium cellulovorans* enzymes XynA, ManA and EngE against sugarcane bagasse. Enzyme Microb. Technol. **42**(6), 492–498 (2008). https://doi.org/10.1016/j.enzmictec.2008.01.010
88. S. Kumar, I. Haq, J. Prakash, S.K. Singh, S. Mishra, A. Raj, Purification, characterization and thermostability improvement of xylanase from Bacillus amyloliquefaciens and its application in pre-bleaching of kraft pulp. *3 Biotech*, vol. 7, no. 1 (2017), p. 20. https://doi.org/10.1007/s13205-017-0615-y
89. S.C. Wu, C.M. Lee, Correlation between fouling propensity of soluble extracellular polymeric substances and sludge metabolic activity altered by different starvation conditions. Biores. Technol. **102**(9), 5375–5380 (2011). https://doi.org/10.1016/j.biortech.2010.11.093
90. J. Xu, M.H. Thomsen, A.B. Thomsen, Ethanol production from hydrothermal pretreated corn stover with a loop reactor. Biomass Bioenerg. **34**(3), 334–339 (2010). https://doi.org/10.1016/j.biombioe.2009.11.004
91. M. Victorin, Å. Davidsson, O. Wallberg, Characterization of mechanically pretreated wheat straw for biogas production. Bioenerg. Res. **13**(3), 833–844 (2020). https://doi.org/10.1007/s12155-020-10126-7
92. J. Li et al., Ethanol production from xylan-removed sugarcane bagasse using low loading of commercial cellulase. Biores. Technol. **163**, 390–394 (2014). https://doi.org/10.1016/j.biortech.2014.04.106
93. X. Chen, J. Shekiro, R. Elander, M. Tucker, Improved xylan hydrolysis of Corn Stover by deacetylation with high solids dilute acid pretreatment. Ind. Eng. Chem. Res. **51**(1), 70–76 (2012). https://doi.org/10.1021/ie201493g
94. N. Dutta, A. Mukhopadhyay, A. Kr. Dasgupta, K. Chakrabarti, Improved production of reducing sugars from rice husk and rice straw using bacterial cellulase and xylanase activated with hydroxyapatite nanoparticles. Bioresour. Technol. **153**, 269–277 (2014). https://doi.org/10.1016/j.biortech.2013.12.016
95. M.G. Tabka, I. Herpoël-Gimbert, F. Monod, M. Asther, J.C. Sigoillot, Enzymatic saccharification of wheat straw for bioethanol production by a combined cellulase xylanase and feruloyl esterase treatment. Enzyme Microb. Technol. **39**(4), 897–902 (2006). https://doi.org/10.1016/j.enzmictec.2006.01.021
96. Mohd Khairul Hakimi Bin Abdul Wahab, *Crosslinking Immobilisation of Surface-Engineered Xylanase and Its Hydrolysis Performance of Oil Palm Empty Fruit Bunches for Xylooligosaccharide Production.* [Online]. Available: https://eprints.utm.my/102125/1/MohdKhairulPSChE2021.pdf
97. A. Hamid et al., Enzymatic hydrolysis of low temperature alkali pretreated wheat straw using immobilized β-xylanase nanoparticles. RSC Adv. **13**(2), 1434–1445 (2023). https://doi.org/10.1039/D2RA07231A
98. S.S.M. Sukri, A.M. Mimi Sakinah, Production of high commercial value xylooligosaccharides from meranti wood sawdust using immobilised xylanase. Appl. Biochem. Biotechnol. **184**(1), 278–290 (2018). https://doi.org/10.1007/s12010-017-2542-0
99. A. Esmaeeli et al., A comprehensive review on pulp and paper industries wastewater treatment advances. Ind. Eng. Chem. Res. **62**(21), 8119–8145 (2023). https://doi.org/10.1021/acs.iecr.2c04393
100. G.K. Gupta, M. Dixit, R.K. Kapoor, P. Shukla, Xylanolytic enzymes in pulp and paper industry: new technologies and perspectives. Mol. Biotechnol. **64**(2), 130–143 (2022). https://doi.org/10.1007/s12033-021-00396-7

101. O.B. Chukwuma, M. Rafatullah, H.A. Tajarudin, N. Ismail, Lignocellulolytic enzymes in biotechnological and industrial processes: a review. Sustainability **12**(18), 7282 (2020). https://doi.org/10.3390/su12187282
102. P. Immerzeel, J. Fiskari, Synergism of enzymes in chemical pulp bleaching from an industrial point of view: a critical review. Can. J. Chem. Eng. **101**(1), 312–321 (2023). https://doi.org/10.1002/cjce.24374
103. J. Wang, S. Liu, J. Ma, X. Piao, Changes in growth performance and ileal microbiota composition by xylanase supplementation in broilers fed wheat-based diets. Front. Microbiol. **12**, 706396 (2021). https://doi.org/10.3389/fmicb.2021.706396
104. J. Wang, J. Liang, Y. Li, L. Tian, Y. Wei, Characterization of efficient xylanases from industrial-scale pulp and paper wastewater treatment microbiota. AMB Expr. **11**(1), 19 (2021). https://doi.org/10.1186/s13568-020-01178-1
105. T. Jana et al., Quality improvement of banana fiber through sequential enzymatic treatment. Syst. Microbiol. Biomanuf. **4**(4), 1284–1297 (2024). https://doi.org/10.1007/s43393-024-00265-9
106. S. Sood, K. Li, C. Sand, L. Pal, M.A. Hubbe, Fruit wastes: a source of value-added products, in *Adding Value to Fruit Wastes* (Elsevier, 2024), pp. 3–48. https://doi.org/10.1016/B978-0-443-13842-3.00001-0
107. S. Chavan, B. Yadav, A. Atmakuri, R.D. Tyagi, J.W.C. Wong, P. Drogui, Bioconversion of organic wastes into value-added products: a review. Biores. Technol. **344**, 126398 (2022). https://doi.org/10.1016/j.biortech.2021.126398
108. S. Chatterjee, S. Venkata Mohan, Fungal biorefinery for sustainable resource recovery from waste, *Bioresource Technology*, vol. 345 (2022), p. 126443. https://doi.org/10.1016/j.biortech.2021.126443
109. M. Tian, A. Wai, T.K. Guha, G. Hausner, Q. Yuan, Production of endoglucanase and xylanase using food waste by solid-state fermentation. Waste Biomass Valor **9**(12), 2391–2398 (2018). https://doi.org/10.1007/s12649-017-0192-7
110. B.S. Pereira, C. De Freitas, J. Contiero, M. Brienzo, Enzymatic production of xylooligosaccharides from xylan solubilized from food and agroindustrial waste. Bioenerg. Res. **15**(2), 1195–1203 (2022). https://doi.org/10.1007/s12155-021-10373-2
111. A. Sachidhanandham, A.P. Periyasamy, Environmentally friendly wastewater treatment methods for the textile industry, in *Handbook of Nanomaterials and Nanocomposites for Energy and Environmental Applications*, eds. by O.V. Kharissova, L.M. Torres-Martínez, B.I. Kharisov (Springer International Publishing, Cham, 2021), pp. 2269–2307. https://doi.org/10.1007/978-3-030-36268-3_54
112. A. Dukare et al., Microbial xylanase aided biobleaching effect on multiple components of lignocelluloses biomass based pulp and paper: a review. Nord. Pulp Pap. Res. J. **38**(3), 459–480 (2023). https://doi.org/10.1515/npprj-2023-0005
113. V. Kumar, P. Verma, A critical review on environmental risk and toxic hazards of refractory pollutants discharged in chlorolignin waste of pulp and paper mills and their remediation approaches for environmental safety. Environ. Res. **236**, 116728 (2023). https://doi.org/10.1016/j.envres.2023.116728
114. W.J. Chung, J. Shim, B. Ravindran, Application of wheat bran based biomaterials and nano-catalyst in textile wastewater. J. King Saud University Sci. **34**(2), 101775 (2022). https://doi.org/10.1016/j.jksus.2021.101775
115. S. Rasool, T. Rasool, K.M. Gani, A review of interactions of pesticides within various interfaces of intrinsic and organic residue amended soil environment. Chem. Eng. J. Adv. **11**, 100301 (2022). https://doi.org/10.1016/j.ceja.2022.100301
116. T.G. Ambaye et al., Remediation of soil polluted with petroleum hydrocarbons and its reuse for agriculture: recent progress, challenges, and perspectives. Chemosphere **293**, 133572 (2022). https://doi.org/10.1016/j.chemosphere.2022.133572
117. S. Covino et al., Assessment of degradation potential of aliphatic hydrocarbons by autochthonous filamentous fungi from a historically polluted clay soil. Sci. Total Environ. **505**, 545–554 (2015). https://doi.org/10.1016/j.scitotenv.2014.10.027

118. M. Dixit, G.K. Gupta, Z. Usmani, M. Sharma, P. Shukla, Enhanced bioremediation of pulp effluents through improved enzymatic treatment strategies: a greener approach. Renew. Sustain. Energy Rev. **152**, 111664 (2021). https://doi.org/10.1016/j.rser.2021.111664
119. D.P. Prendergast, P.M. Gschwend, Assessing the performance and cost of oil spill remediation technologies. J. Clean. Prod. **78**, 233–242 (2014). https://doi.org/10.1016/j.jclepro.2014.04.054
120. T. Kuthiala, K. Thakur, D. Sharma, G. Singh, M. Khatri, S.K. Arya, The eco-friendly approach of cocktail enzyme in agricultural waste treatment: a comprehensive review. Int. J. Biol. Macromol. **209**, 1956–1974 (2022). https://doi.org/10.1016/j.ijbiomac.2022.04.173
121. A. Reyes-César, Á.E. Absalón, F.J. Fernández, J.M. González, D.V. Cortés-Espinosa, Biodegradation of a mixture of PAHs by non-ligninolytic fungal strains isolated from crude oil-contaminated soil. World J. Microbiol. Biotechnol. **30**(3), 999–1009 (2014). https://doi.org/10.1007/s11274-013-1518-7
122. S.A. Aitkeldiyeva, E.R. Faizulina, L.G. Tatarkina, A.V. Alimbetova, O.N. Auezova, G.A. Spankulova, Prospects for the use of thermotolerant microorganisms for remediation of oil-contaminated soils. Eurasian J. Appl. Biotechnol. **2** (2020). https://doi.org/10.11134/btp.2.2020.4
123. O.V. Olagoke, G.G. Babaniyi, F.J. Afolabi, B.R. Babaniyi, Soil microbial enzymes and mitigation of heavy metal uptake by plants, in *Ecological Interplays in Microbial Enzymology*, eds. by N.R. Maddela, A.S. Abiodun, R. Prasad, in Environmental and Microbial Biotechnology (Springer Nature, Singapore, 2022), pp. 215–246. https://doi.org/10.1007/978-981-19-0155-3_11
124. E. Priyadarshini, S.S. Priyadarshini, B.G. Cousins, N. Pradhan, Metal-fungus interaction: review on cellular processes underlying heavy metal detoxification and synthesis of metal nanoparticles. Chemosphere **274**, 129976 (2021). https://doi.org/10.1016/j.chemosphere.2021.129976
125. Z. Qiu, J. Tang, J. Chen, Q. Zhang, Remediation of cadmium-contaminated soil with biochar simultaneously improves biochar's recalcitrance. Environ. Pollut. **256**, 113436 (2020). https://doi.org/10.1016/j.envpol.2019.113436
126. C. Liang et al., Improving β-glucosidase and xylanase production in a combination of waste substrate from domestic wastewater treatment system and agriculture residues. Biores. Technol. **318**, 124019 (2020). https://doi.org/10.1016/j.biortech.2020.124019
127. S. Hati, S. Mandal, J. Prajapati, Novel starters for value added fermented dairy products. Curr Res Nutr Food Sci **1**(1), 83–91 (2013). https://doi.org/10.12944/CRNFSJ.1.1.09
128. B. Aslanbay Güler, İ. Deniz, Z. Demirel, E. İmamoğlu, "Evaluation of different scale-up strategies for *Haematococcus pluvialis* cultivation in airlift photobioreactors. ase **0**(0), 0–0 (2022). https://doi.org/10.26650/ASE202211036078

Chapter 9
Industrial-Scale Production of Xylanase

Kamran Ali, Abdul Sattar jatoi, Saim Yar Khan, Muhammad Asim khan, and Nehal Ahmed

Abstract Because xylanase is one of the few hemicellulolytic enzymes able to hydrolyze xylan, it is now widely used in industry. Due to rising usage in pulp and paper, animal feed, food and beverages, textile and biofuel sectors, xylanase manufacturing has improved greatly. Under conditions of submergency or solid-state cultures, large-scale fermentation depends on filamentous fungi (e.g., *Aspergillus* spp.) and bacteria (e.g., *Bacillus* spp.). Enhancements in the yield and stability of xylanases have resulted from improvements in fermentation conditions, better use of substrates (for example, agro-industrial materials) and changes in the DNA of microbial strains. Processes developed downstream have been improved to allow for economical enzyme recovery and formulation. In addition, new trends in bioprocess monitoring, bioreactor development and scale-up methods make it easier to produce xylanase sustainably. This chapter explains how industrial xylanase production has developed and what challenges remain, also providing directions for better and more extensive use in the future.

Keywords Industrial scale · Xylanase

1 Introduction

Xylanase helps biotechnology worldwide by converting xylan, an important part of hemicellulose, into its sugars such as xylose. Producing xylanase on a large scale has accelerated because it has many uses in bleaching paper and pulp, processing animal feed, making biofuel and working in food and textiles [1]. This chapter discusses the techniques, problems and improvements found in making xylanase on a large scale for industrial reasons.

Xylan is broken down by the enzyme into xylo-oligosaccharides and sugar by the breakdown of xylan in plant cell walls. This action by enzymes is necessary for

K. Ali · A. S. jatoi (✉) · S. Y. Khan · M. A. khan · N. Ahmed
Department of Chemical Engineering, Dawood University of Engineering and Technology Karachi, Karachi, Pakistan
e-mail: abdul.sattar@duet.edu.pk

A. Sattar Jatoi (ed.), *Xylanase from Microorganisms*,
https://doi.org/10.1007/978-981-95-0824-2_9

making the best use of plant biomass in various businesses. Xylanase improves the paper-bleaching process, making it possible to use small amounts of chlorine and therefore helping protect the environment in the paper and pulp industry. By using an enzyme, the industry can turn lignocellulosic biomass into fermentable sugars and make second-generation bioethanol [2].

Because microbial sources are efficient, affordable and can be scaled up, most xylanase is produced through industrial processes involving microbes. Fungi and bacteria are used in the production of xylanase. Enzymes from *Trichoderma reesei* and *Aspergillus niger* are popular in industry since they are easy to produce in large amounts and survive throughout different types of fermentation. Just as fungi have, *Bacillus subtilis* and *Streptomyces* spp., types of bacteria, are gaining interest because of their prompt growth and the convenience of genetically altering them (Fig. 1) [3].

Even with a lot of progress, making xylanase industrially has its own set of problems. Such steps involve improving fermentation, finding low-cost ways to process the products and addressing how enzyme activity changes according to what it is processing. Furthermore, higher production should meet demands for sustainability, lowering harm to nature and preserving profitability.

Xylanase can greatly enhance industrial work by endorsing sustainability and making resources more efficient. Large-scale manufacturing of the enzyme is proof of how well microbiology, biochemical engineering and being environmentally responsible can work together. It introduces and explores the techniques, advances and problems in producing xylanases, supporting the examination of how they are used industrially.

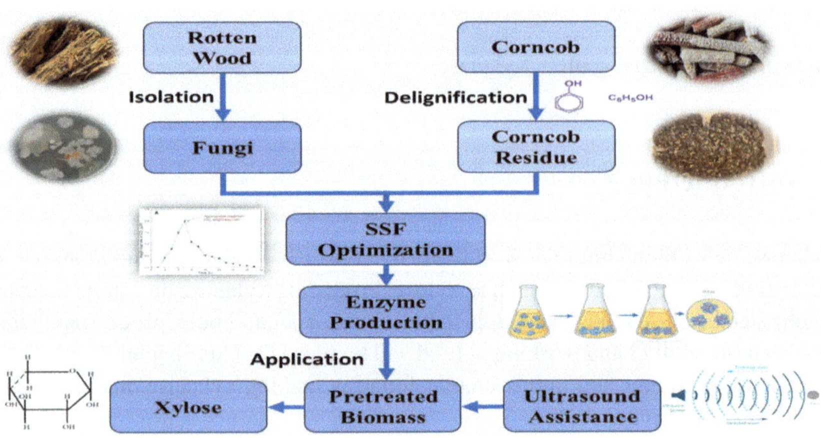

Fig. 1 Schematic representation of fungal enzyme production and application for biomass pretreatment [1]

2 Xylanase Production Process

The production of xylanase on an industrial scale requires a multifaceted approach, integrating microbial selection, substrate utilization, fermentation techniques, and process optimization [4]. Each step plays a critical role in determining the yield, activity, and cost-effectiveness of the final enzyme product. This section explores these components in detail [2].

2.1 Microbial Sources of Xylanase

Microorganisms are important because they naturally produce and secrete xylanases whenever they are grown on xylan-based media. Xylan is the main part of hemicellulose found throughout plant cell walls and xylanases are enzymes that speed up its breakdown. Because xylanase is useful in several sectors, including paper and pulp, food and feed, textile, biofuel and bioremediation, its microbial creation is valued industrially. How well xylanase is produced and how suitable it is for different uses mainly depends on which microorganism is used, since this determines its stability under various temperature and pH ranges and the kind of substrate it breaks down [3]. Among all microorganisms, researchers have preferred fungi because they are used extensively for making commercial xylanase. Both *Aspergillus niger* and *Trichoderma reesei* are distinguished by their ability to make a lot of extracellular xylanase. Xylan is what fungi eat, as it is part of the lignocellulosic biomass they live on, so they naturally create xylan-destroying enzymes. Being effective in acidic environments is a big advantage of fungal xylanases, particularly for the paper industry, where softening wood during pulp bleaching is necessary. The ability of fungal xylanases to withstand acidic conditions lets them be directly used in industrial processes which makes operations simpler and less expensive. Xylanases from fungi often differ in both substrate affinity and molecular weight which allow them to be used in many different roles. *Trichoderma reesei* is attractive for biofuel production as it is able to produce xylanases and cellulases together in high volume.

At the same time, microbe strains have attracted notice since they can produce thermally stable proteins called xylanases. *Bacillus subtilis*, *Bacillus licheniformis* and all members of Streptomyces are appreciated for their high-temperature enzyme stability, a feature important for use in harsh industries. Retaining their catalytic activity at high temperatures enables the enzymes to perform well in processes such as biobleaching, making animal feed and bioethanol making. Since these bacterial xylanases show high activity at higher pH, they add strength to the existing acidic fungal enzymes and allow for applications over a wider range. Because they work well in high temperatures and an alkaline environment, bacterial xylanases can easily be added to industrial processes that pre-treat biomass, making biomass hydrolysis more effective. Fungi take a longer time to multiply than bacteria, so using bacteria can lead to an increase in enzyme production and decrease in how long and how much it costs to ferment. Bacterial xylanases may also have unique traits, for example, being able to withstand protease action and several chemical inhibitors, providing them additional value in mixed industrial conditions. Most of the enzyme xylanase is produced by fungi and bacteria, though scientists are now studying how different yeasts might do this well too. Xylanases produced by species such as Cryptococcus are unique in their chemistry and might be applied in biotransformation and biocatalysis. Yeast-derived xylanases are less thoroughly examined and made available in smaller amounts than fungi and bacterial enzymes, but their unique affinities and processes may be applied to selectively modify materials made from xylan. Because yeasts can be genetically changed and are simple to grow, they are excellent options for future investigation and use in industry. Moreover, for many industrial applications, yeast xylanases might work in ways that require only a few or no changes which could be useful for making unique enzyme cocktails. Many industries depend on the usefulness of microbial xylanases. Using xylanases, the paper and pulp industry performs biobleaching of pulp, allowing for less use of chemicals that harm the environment. For these reasons, fungal xylanases are preferred because they enhance total pulp brightness and strength while controlling chlorine needs. In baking, adding xylanases to wheat flour helps improve how dough is handled and how good the bread turns out by modifying arabinoxylans and making the bread last longer. Which organism is selected matters because the mildly acidic and complex dough affects enzyme activity [5]. Breaking down the plants in livestock feed through thermostable bacterial xylanases increases the ease with which animals can access and use nutrients. To hydrolyze lignocellulosic biomass and liberate hemicellulose sugars, production of biofuel relies on microbial xylanases. Because some microbes are thermotolerant and can work in a harsh environment, selecting such xylanase producers can allow for fewer enzymes and higher yields in bioethanol production. In addition, xylanases from microorganisms help in retting and bleaching of plant fibers in the textile industry, use in waste management to convert agricultural residues and in the production of xylooligosaccharides with prebiotic use. Thanks to recent breakthrough sin molecular biology and biotechnology, xylanase-producing microbes can be alteredto create more enzymes and modify them to satisfy various industrial needs. Using recombinant DNA technology, scientists can make xylanase genes produce more in suitable organisms which leads to higher titters and less costly production. Using metagenomic technology, biologists have revealed new xylanase genes from previously unknown microbes in difficult environments [6]. The natural xylan degradation ability and adaptability to different working conditions make microbial sources important for xylanase production. Because *Aspergillus niger* and *Trichoderma reesei* are contenders for being highly active in extracellular enzymes and being acid-resistant, their xylanase isoforms are useful for various purposes. Bacteria such as *Bacillus subtilis*, *Bacillus licheniformis* and *Streptomyces* spp. provide enzymes that can still function at high temperatures and basic pH, so they are utilized in strong industrial settings. Even though yeasts are less popular, they produce unique xylanases with possible uses for certain applications. Still, engineers continuously search for new microbial xylanase producers which could increase their function and reduce costs for sustainable business processes [7].

2.2 Substrates for Xylanase Production

The substrate chosen is important because it supplies needed nutrients and triggers the making of xylanase. Because wheat bran, corn cob, sugarcane bagasse, rice husk and sawdust are available, cheap and high in xylan, these are commonly used to make xylanase. In contrast, using xylose or commercial xylan, while being more expensive, is practiced in studies and small-scale applications because the activity can be precisely studied. Applying steam explosion, acid hydrolysis or an enzymatic hydrolysis treatment to substrates can enhance microbial access to xylans, resulting in a stronger yield of xylanase.

2.3 Fermentation Techniques

The production of xylanase at large scales relies on fermentation, affecting both the success and practicality of making the product [3]. Depending on a microorganism's species and the needs of the application, both Submerged Fermentation (SmF) and Solid State Fermentation (SSF) have certain benefits and drawbacks. In Submerged Fermentation, microorganisms are grown in liquid medium where all important conditions such as pH, heating, aeration and shaking can be carefully controlled [5]. This degree of control of the process is crucial for improving the living conditions of xylanase-producing microbes, specifically *Bacillus subtilis* and *Aspergillus niger*. The compatibility of SmF with bioreactor systems makes it suitable for big commercial operations that can be closely watched, managed automatically and easily scaled. Moreover, SmF makes it easier to get reliable results and to process them further because the fermentation broth has lower solid content. SmF is popular yet comes with drawbacks, like higher costs for water and energy and an enzyme yield that is much lower per grain of substrate when compared to SSF. On top of that, if agitation is too high, problems such as foaming and shear sensitivity related to microbes can make process optimization more difficult. Unlike Submerged Fermentation, Solid-State Fermentation uses materials such as wheat bran, sugarcane bagasse and similar lignocellulosic residues both as nutrients and as a physical support for microbes. The process better resembles the natural environment of many filamentous fungi that are known to make large amounts of xylanase. The technology is known to decrease operational costs, produce less wastewater and increase the production amount when used. Using this technique with cheap, local industrial waste from farming makes it popular in developing or small-scale agricultural areas. Also, the harsh setting mimicking real substrate colonization within SSF can lead to the generation of extra enzymes. Still, there are certain obstacles that stop SSF from being widely applied in industry. It is challenging to keep both temperature and moisture throughout the matrix uniform and stable when working with large substrates. Environmental instability may cause microbial growth to decrease, enzyme formation to vary and expose the product to contamination. Because these products require strict control of humidity and temperature, these fermenters need to be carefully designed and built for the job. So, in order to improve upon SmF and SSF, people are paying more attention to hybrid types of fermentation. The goal is to gain both high machine control and scalability from SmF as well as improved productivity and enzyme levels from SSF. If pre-fermentation is used in SSF, with enzyme extraction afterward and then microbial growth in SmF, productivity goes up while costs stay low. As a result, using hybrid systems, surveillance tools and automation is helping establish better and more sustainable methods for making industrial xylanase [6].

2.4 *Process Optimization Parameters*

Getting production processes just right maximizes the amount and activity of xylanase in production. Microbes from different sources have different needs, with fungal xylanases needing acidic conditions (pH 4.5–6) and bacterial xylanases being stable in neutral or alkaline environments (pH 7–9). In particular, thermophilic strains can live and grow best at temperatures between 50 and 70 °C. The proper balance between organic materials containing carbon and nitrogen is an important thing to consider. Carbon in microbial cultures often comes from xylan-rich agro-residues, whereas peptone, yeast extract and inorganic sources influence both microbial growth and enzyme output. Aeration and agitation must be balanced, mainly for aerobic species, since direct oxygen transfer is necessary; they should not be conducted so vigorously as to impair the cells. Additionally, inducers such as xylan or xylose or various surfactants, can strongly enhance the release of enzymes. Metal ions such as magnesium or calcium are also sometimes included to help stabilize how the enzyme looks [3]. In order to optimize these complicated factors, researchers now make use of techniques like response surface methodology (RSM) and artificial neural networks (ANN) which help them quickly determine the best settings while decreasing the work involved in doing experiments.

3 Mass Transfer Operation for Production of Xylanase

Maximizing the yield of xylanase requires efficient mass transfer of needed nutrients, gases and substrates by microbes for their growth and activity. Microbial activity in aerobic fermentation systems depends heavily on how efficiently oxygen is transferred which is influenced by the amount of oxygen transfer, the speed of agitation and the oxygen transfer coefficient (kLa). If airflow and agitation are suitable, dissolved oxygen is available and oxygen isn't unevenly distributed; if agitation is stronger than necessary, it can injure bacteria and lessen the digester's productivity [3]. Making sure kLa is high in large fermenters keeps the oxygen balance effective. Microbial growth and the secretion of xylanase depend on internal diffusion and available nutrients. A big difference between the medium and cells tends to increase diffusion, whereas thickness or haze in the medium, caused by plenty of biomass, can interfere with the transfer of matter in the solution. To reduce viscosity, options such as dilution or enzymatic processes are used and in solid-state fermentation (SSF), limiting mass transfer in the substrate can be avoided by breaking the substrate into smaller pieces and mixing it periodically [5]. Microbial metabolism may be interrupted when metabolic by-products, including organic acids, are not promptly eliminated. It can be reached by aerating and agitating properly which allows volatiles to escape and stop the pH from changing repeatedly. Besides, the way substances transfer at the boundaries between gas, liquid and solid has a major impact on process efficiency. Putting in effort to have small bubbles and better contact of solids with the liquid, can attract more microbes to theSSF. In addition, moving mass transfer experiments from small to industrial scale needs more resources and therefore CFD models are employed to support and adjust mass transfer parameters and ensure efficient mass production [6].

4 Industrial Applications of Xylanase

There has been a great deal of interest in xylanase in industry because it is effective at splitting hemicellulose, an important plant fiber. It is useful across a range of industries, allowing for healthier chemical alternatives that boost product quality, control spending and protect the environment [3]. Xylanase is most often used in the pulp and paper industry as a biological bleaching agent. Xylanase partly limits xylan within wood, making it easier to take out lignin during the bleaching process. Thanks to this, the use of tough chemicals such as chlorine and chlorine oxide is less necessary which reduces waste and makes the environment cleaner. In addition, using enzymatic pre-bleaching brightens the paper, improves its strength and makes it of higher quality while also helping to save energy during processing. Xylanase is being used in paper pulp production because it supports both sustainability and low-cost processes [5].

Xylanase is used in fabric processing to make these steps gentler on the environment and more productive. It is mainly necessary in biopolishing and scouring to break down hemicellulosic materials on natural fibers including cotton and flax. Thanks to the enzymes, the clothing softens and is more pleasant for consumers to wear. It is well known that xylanase helps make the surface of fiber more uniform which leads to a better and more even uptake of dye, making the colors on dyed textiles brighter. Such positive results are reached by avoiding rough alkaline processes, helping to protect water, chemicals and nature during textile production [6]. Application of xylanase is highly beneficial to the food and beverage industry. Adding xylanase to wheat flour in baking helps to break down arabinoxylans and increases dough elasticity, traps more air in the dough, raises the loaf volume and gives products a better texture and longer shelf life. With the help of xylanase, it becomes easier to extract juice from plants and clearer juices and alcohol can be made. Because xylanase reduces thickness and keeps processing clean, it improves both product and processing qualities. Enzymes play a key role in food processing, proving their usefulness with both cleaner labels and outstanding functionality [7].

Producing biofuels, particularly converting lignocellulosic biomass to bioethanol, benefits from the use of xylanase. Before any lignocellulosic materials like agricultural residues and woody biomass can be turned into ethanol, their complex carbohydrates, mainly cellulose and hemicellulose, must be changed into fermentable sugars. Thanks to xylanase, the hemicellulose content, determined as xylan, is decomposed into simple sugars xylose and others that can immediately be fermented by microorganisms. Because of this, breaking down biomass into ethanol takes less time for better results. Using xylanase means less demanding preprocessing which reduces both the costs and environmental impact of making biofuels [8].

Xylanase is used in both animal feed and agriculture to boost the nutritional value and how digestible plant-based feeds are which frequently contain a lot of NSPs, mainly xylan as higlighted in Fig. 2. NSPs can raise the stickiness of food inside the intestines and prevent proper absorption of nutrients by poultry and pigs. Dividing xylan into simple sugars with xylanase reduces the thickness of the feed and releases locked nutrients, making all nutrients (proteins, carbohydrates and minerals) more accessible [9]. Animals that receive enzyme supplements tend to grow faster, eat less feed and enjoy better overall health. It is now possible to use xylanase to degrade crop residues faster which speeds up the reuse of organic materials in the soil. It benefits the soil, makes crops more fertile and aids sustainable farming by using fewer chemicals and getting better growth from crops [10].

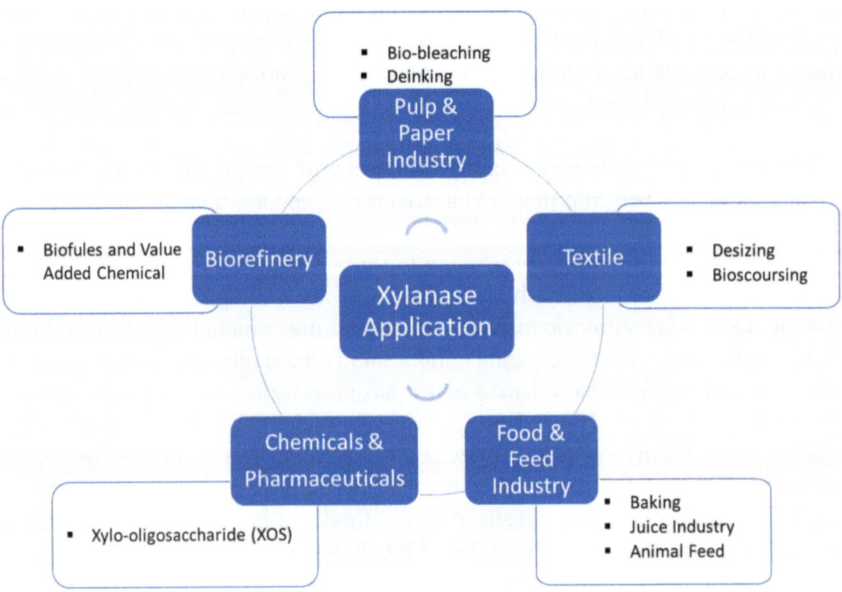

Fig. 2 Application of xylanase for various sector [3]

5 Economic and Sustainability Considerations

For commercial production of xylanase to be appropriate for industry, a detailed examination of its economic and environmental impact is necessary. Such points are key to assessing both how efficiently xylanase can be produced and how the technology affects the environment over a long period. Evaluating such production is mainly based on the production cost, whose main contributors include the types of raw materials, microbial strains, fermentation techniques and handling after fermentation. Agricultural waste such as wheat bran, rice husks and corn stover are broadly available and inexpensive, making them perfect sources of carbon for growing microorganisms. The type of microbe used for production such as Aspergillus or Bacillus, matters a lot since it affects how productive and easy the process is. It is possible to improve the way things are made by optimizing the pH, temperature, nutrients found and the level of air in fermentation. Fed-batch and solid-state fermentation make enzymes more productive and lower the amount of energy needed for their production [5]. These techniques help cut expenses and speed up enzyme manufacturing, avoiding much purification. Enzyme yield and productivity are vital for ensuring that xylanase is appealing for industrial use as well. Thanks to genetic engineering and strain improvement in biotechnology, high-yielding microbial strains are now used in production. Commonly, these contrasting strains show greater fitness and give rise to more enzyme when grown under the best conditions. Making industry more productive lowers production expenses which helps keep costs competitive and encourages companies to purchase more industrial hardware. In addition, xylanase manufacturing should be done sustainably and have as small an impact on the environment as possible. By using agro-industrial waste as a feedstock, both disposal matters and renewable options for fermentation are improved. Additionally, xylanase adds value to green industry practices. In the pulp and paper business, xylanase cuts down chlorine use which means emissions of harmful substances from the process are lower. Using xylanase in biofuel production allows lignocellulosic biomass to be turned into sugar which encourages a circular economy and lowers dependence on fossil fuels. Furthermore, the need for xylanase and competitive costs influence if it is reasonable to manufacture the product. Because sustainable technologies are receiving more attention, xylanase—an enzyme—is in higher demand for food processing, bioenergy and animal nutrition. To stay on track with market expectations, enzyme producers must introduce new innovations, make products suitable for particular uses and operate their systems more efficiently to control costs. Energy usage and efficiency play important roles in making xylanase production financially successful and environmentally friendly [6]. The key steps of fermentation and downstream processing use a lot of energy for agitation, controlling temperature and aeration which affects both the company's expenses and the environment. Some ways to solve this are by adopting bioreactors that conserve energy, improving processing conditions and using solar or wind energy to power production. Furthermore, handling downstream tasks with crude enzyme solutions rather than pure ones can greatly lower both energy usage and the related costs. Combining these approaches reduces emissions during enzyme production and also supports the worldwide move toward environmentally-friendly manufacturing. Ultimately, to make xylase production possible and sustainable, multiple actions are required such as cutting costs, growing productivity, recycling valuable materials, using energy effectively and meeting what the market requires. To show xylanase works well, this approach is crucial for demonstrating its nature as an environmentally friendly solution for industry [7].

6 Challenges and Future Prospects

Xylanase is produced at a large scale today, yet still encounters many technical, economic and environmental difficulties. Even so, the outlook is hopeful as major progress is being made in microbial biotechnology, enzyme engineering and improving processes. Strain development and optimization present one of the biggest problems in this field. Although several strains of fungi and bacteria like Aspergillus and Bacillus have been identified as xylanase producers, obtaining steady and high xylanase production using these microbes is still a challenge industrially. Not only should they produce high yields, but their sturdiness, lasting stable properties and resistance to industrial changes in pH and temperature are also needed. Adapting aeration, nutrient mixture and inoculum numbers increases the difficulties and the cost of the process. Still, the field looks promising with recent advancements in genetic engineering and synthetic biology. The use of CRISPR-Cas9 and modeling metabolic pathways is helping us create strains that produce more while tolerating a wide range of stresses. Additionally, new bioreactor designs and improved process controls should help to finely monitor fermentation and make its dynamics more efficient. It is also important to mention that high production costs are frequently a result of requiring pure sugars and costly nutrient mixtures. Handling aeration, sterilization and temperature control within fermentation also greatly adds to production costs. The new strategies plan to use cheap and renewable feedstocks, like wheat bran, corn stover or leftovers from food processing. Not only do these lignocellulosic materials lower the cost of raw materials, but they also help achieve sustainability by transforming waste material. In addition to this, interest is growing in solid-state fermentation (SSF) which helps save water, energy and uses simpler equipment, while increasing product strength. Still, these strategies must be improved to become more scalable and reproducible. Stability and proper storage are important issues because xylanase has a tendency to lose activity in industrial applications. If enzymes are exposed to temperature changes, impact from grinding or the wrong pH, they may require both cooling and special stabilizers which add to the cost and make things less convenient. Various emerging methods which include encapsulating enzymes, linking proteins together and immobilization, are being used to help increase xylanase stability. Scientists are applying site-directed mutagenesis and directed evolution to create versions of xylanase that are stable at high temperatures, in different pHs and last longer in industry. Commercialization of xylanase is further complicated by steps taken to process and purify the product. On an industrial scale, separating xylanase from the fermenter takes several processes—filtration, centrifugation, precipitation and chromatography—and can be very expensive and result in lost product. Future steps are taken to make purification easier by separating waste with membranes, using special tags for pulling out needed proteins and combining the steps of manufacturing and purification into one setup. The potential improvements could make the process quicker and reduce the expense of making each part. The ecological effect of large-scale enzyme production is an additional important aspect. Intense use of energy, large amounts of water and producing side products can actually damage the environment, undoing any benefits from biocatalyst use. In the future, setting up renewable energy and waste management systems inside the industrial process is vital for sustainable fermentation. Turning all the components of biomass into valuable substances such as enzymes, fuels and chemicals is considered a highly promising way to increase sustainability and use resources more efficiently. Furthermore, the market for enzymes is still constrained by rising early costs, challenges from other chemicals or enzymes and unawareness among possible customers. Although xylanase helps reduce costs and preserve the environment, industries are extremely reluctant to make the change from their current approach. It is believed that this situation will improve as the world's industries become more sustainable and follow new environmental rules. If green technologies are supported by investments and education, plus backed by enzyme producers, adoption is likely to increase. Also, partners between biotechnology and sectors such as pulp and paper, animal feed, biofuel help create xylanase formulas adapted to their needs, opening up the market for biotechnology firms. In short, while xylanase hasn't experienced much industry use, the combination of advances in technology, better planning and concern for the environment can encourage it to succeed.

7 Discussion

A major difficulty in industrial xylanase production involves picking and improving the microbial strains needed. Even though numerous microorganisms such as bacteria and fungi, have been found to make xylanase, very few do so in a reliable and high-yielding way during industrial production. In addition, getting the conditions right for microbe and enzyme action and stability in fermentation is very important.

Furthermore, xylanase is often produced on a large scale using tough media which leads to high costs for the product. Because agricultural residues are cheap but have varied chemistry, it is difficult to create a standard process for production. Such variability may also cause variations in the amount and activity of enzymes, so manufacturers find it hard to raise their yield while maintaining quality.

A big problem, too, is making enzymes stable. Just like other enzymes, xylanase is affected by things like pH level, temperature and ionic strength in its environment. For the enzyme to be effective in industrial use, it must maintain its activity both while producing and while stored. Furthermore, after fermentation, purity and separation steps increase complexity and demand a lot of energy which leads to higher cost of producing cell culture medicine.

Being Highly Mindful of How Our Actions Affect the Environment.

Xylanase production is getting a lot of focus because it may reduce the negative effect on the environment. Processes that use agricultural residues and lignocellulosic waste help to reduce waste and are sustainable. Moreover, the pulp and paper industry uses xylanase which means less need for hazardous chemicals and a better, more environmentally friendly way to produce products.

In addition, breaking down lignocellulosic materials into ethanol with the assistance of enzymes promotes both more effective and environmentally responsible production of biofuels. When xylanase makes it easier for plants to break down organic matter into fermentable sugars, it greatly supports growing the bioenergy sector.

Still, commercial-scale xylanase production brings certain problems related to the environment. Fermentation takes up a lot of energy and leaving the waste behind can damage the environment if not handled right. For this reason, achieving sustainable xylanase production calls for methods that save energy, keep waste down and maximize the use of resources.

The Importance of Budget and Cutting Cost.

Making xylanase in large quantities is expensive mainly because special resources and the complicated nature of large-scale fermentation add to the initial outlay. Recently, efforts in biotechnology have been aimed at lowering production costs by adding agricultural waste as a substrate and boosting how efficiently fermentation takes place.

Modifying the genes of microbes to boost enzyme production and make them more stable has been a leading subject of interest. Combining strain engineering with SSF and fed-batch can greatly bring down the production cost per unit of xylanase. Moreover, if the principles of biorefinery are used—so that multiple items are processed from a single input—it could increase the profitability of xylanase production.

When manufacturers need greener solutions, enzymes like xylanase should see rising competition. As the use of environmentally friendly approaches increases, companies in textiles and pulp and paper are expected to adopt more xylanase technology. Yet, if xylanase is to be used in countries around the world, it must become cheaper to produce and purify.

Future Prospects.

In the future, several new advancements in xylanase production give us the hope of solving present problems. Further advancements in synthetic biology, altering enzymes and enhancing production are expected to guide industrial xylanase production in the future.

Mechanical techniques employed today can strengthen microbial cultures and help make sure the xylanase is produced in higher quantities with reliable stability. Innovations in bioprocessing, including continuous fermentation, could decrease expenses and increase how well enzymes are made. The introduction of sustainable power sources and responsible waste management into production can improve both the energy use and environmental safety of making xylanase.

Creating new uses for xylanase is another important area for further study. As a result, since many prefer eco-friendly cleaners, xylanase can be an alternative to traditional chemical cleaners used at homes and in businesses. Also, as biocatalysis and green chemistry attract additional interest, there is potential to research xylanase for use in many chemical changes, enabling its broader adoption in industry.

8 Conclusion

Xylanase which is now big business due to new synthesis techniques, has the ability to enhance animal feed, food and beverage, textiles, pulp and paper and production of biofuels. The enzyme plays a crucial role in improving industrial sustainability by turning complex plant material into forms that can be used in several applications.

Despite the difficulties of strain development, maintaining the stability of enzymes, the high financial cost of production and the complicated nature of downstream processing, genetic engineering, advancements in fermentation and enzyme stabilization are offering answers. Making xylanase more sustainable and affordable is possible using inexpensive renewable resources and adopting better production techniques. Xylanase is playing an important role in helping industry choose more environmentally friendly solutions. The importance of this enzyme is clear when we use it in waste processing, biofuel generation and safe manufacturing. Apart from its current functions in brewing, baking and textiles, xylanase is now used for biocatalysis and eco-friendly detergent formulas, meaning more people are needing this enzyme.

Despite some difficulties, the manufacture of xylanase looks set for a bright future. There is a growing likelihood that xylanase will be significant for several industrial applications because ongoing studies, technology and dedication to sustainable practices are supporting its use. A range of new options can be developed through better organization of the process, reducing costs and making the enzyme perform better.

References

1. B. Alriksson, S.H. Rose, W.H. van Zyl, A. Sjode, N.-O. Nilvebrant, L.J. Jonsson, Cellulase production from spent lignocellulose hydrolysates by recombinant *Aspergillus niger*. Appl. Environ. Microbiol. **75**, 2366–2374 (2009)
2. S. Sharma, P. Verma, K. Agrawal, Harnessing the potential of fungal xylanases: an insight into its application and technological advancements. *Industrial Crops and Products*, vol. 222, 2024/12/15/ (2024), p. 119967
3. C. Pandey, P. Sharma, N. Gupta, Engineering to enhance thermostability of xylanase: for the new era of biotechnology. J. Appl. Biol. Biotechnol. **11**, 41–54 (2023)

4. H. Yang, D. Sun, Y. Li, W. Zhao, Z. Jiang, Q. Yan, et al., Development of a green, efficient and renewable biorefinery strategy for the production of xylooligosaccharides from corncobs based on a novel specific xylanase. *Carbohydrate Polymer Technologies and Applications*, vol. 10. 2025/06/01/ (2025), p. 100858
5. G. Msigwa, J.O. Ighalo, P.-S. Yap, Considerations on environmental, economic, and energy impacts of wind energy generation: projections towards sustainability initiatives. *Science of the Total Environment*, vol. 849. 2022/11/25/ (2022), p. 157755
6. E. Santoyo-Castelazo, A. Azapagic, Sustainability assessment of energy systems: integrating environmental, economic and social aspects. *Journal of Cleaner Production*, vol. 80, 2014/10/01/ (2014), pp. 119–138
7. A. Castillo-Landero, A.P. Ortiz-Espinoza, A. Jiménez-Gutiérrez, A process intensification methodology including economic, sustainability, and safety considerations. Ind. Eng. Chem. Res. **58**, 6080–6092 (2018)
8. T. Abena, A. Simachew, A review on xylanase sources, classification, mode of action, fermentation processes, and applications as a promising biocatalyst. BioTechnologia **105**(3), 273 (2024)
9. A. Sarangi, H. Thatoi, Xylanase as a promising biocatalyst: a review on its production, purification and biotechnological applications. Proc. Nat. Acad. Sci. India Sect. B: Biol. Sci. **2024**, pp. 1–16
10. C. Qiu, N. Liu, X. Diao, L. He, H. Zhou, W. Zhang, Effects of cellulase and xylanase on fermentation characteristics, chemical composition and bacterial community of the mixed silage of king grass and rice straw. Microorganisms **12**(3), 561 (2024)

Chapter 10
Metagenomics and Discovery of Novel Xylanases

Abdul Sattar Jatoi, Bazil Owais Nasir, Saim Akram, and Owais Baig

Abstract The practical applications of xylanases have made them highly sought after in pulp and paper, biofuel, animal feed and the food and textile industries. Most of the world's environmental microorganisms cannot be cultured under current enzyme discovery methods. Scientists are using metagenomics, a technique that focuses on all the genes in an environment without culturing, to discover novel xylanases with different ways of catalyzing reactions. Metagenomic techniques utilized in finding and characterizing unique xylanases are fully explained in this chapter. The methods cover the creation of metagenomic libraries, screening of xylanase genes through simulations, to cloning, expressing and purifying those genes. A list of important methods is presented for evaluating how stable, active and useful for industry an enzyme is. The chapter also examines the different uses of metagenome-producing xylanases and looks at their effects on making different industries more sustainable. Importance is given to combining bioinformatics, synthetic biology and high-throughput functional screening to address current problems and help find better xylanases. Here, we show that metagenomics is essential for improving the variety of xylanases and inspiring changes in enzyme bioprocessing.

Keywords Xylanase · Metagenomics · Biocatalyst discovery · Functional screening

1 Introduction

Growing interest in more sustainable biotechnologies has made lignocellulose-degrading enzymes, mainly xylanases, important, as they assist in the breakdown of hemicellulose which makes up a major part of plant cell walls [1]. In industries such as biofuel, paper and pulp, animal feed and food and beverage, xylanases (specifically endo-1,4-β-xylanases) help break down xylan into shorter chains and xylose,

A. S. Jatoi (✉) · B. O. Nasir · S. Akram · O. Baig
Department of Chemical Engineering and Technology Karachi, Karachi, Pakistan
e-mail: abdul.sattar@duet.edu.pk

A. Sattar Jatoi (ed.), *Xylanase from Microorganisms*,
https://doi.org/10.1007/978-981-95-0824-2_10

making them necessary [2]. Traditional methods of enzyme discovery, which rely on cultivating microorganisms from natural environments, are severely limited by the fact that over 99% of microbial species are uncultivable under standard laboratory conditions [3]. Consequently, a vast majority of potentially useful enzymatic diversity remains inaccessible through conventional microbiological techniques. The advent of metagenomics, defined as the direct extraction and analysis of DNA from environmental samples, revolutionized this field by allowing access to the collective genetic material of entire microbial communities, including those comprised of uncultured organisms [4]. This culture-independent approach enables researchers to bypass the need for microbial isolation, instead permitting the cloning, sequencing, and functional analysis of genes directly from environmental DNA (eDNA) [5]. In the context of xylanase discovery, metagenomics has proven especially powerful, facilitating the identification of novel glycoside hydrolase (GH) families and unique catalytic mechanisms that would otherwise remain hidden [6]. There are two primary approaches to metagenomics: sequence-based and function-based. Sequence-based metagenomics involves high-throughput DNA sequencing followed by bioinformatics analysis to identify putative xylanase genes based on homology to known GH families, such as GH10, GH11, GH30, and GH43, often using databases like CAZy, Pfam, and NCBI BLAST [7]. With this system, large numbers of xylanase genes can be swiftly found, but their function must be proven by testing. Unlike DNA-based metagenomics, function-based metagenomics makes use of suitable expression vectors to insert DNA from the environment and grow them in E. coli host cells [8]. After screening with a xylan-Congo red plate, functional activity is confirmed by clear areas formed around colonies containing xylanases. Because of this method, we can discover enzymes with surprising functions, even when their sequences differ from those of known enzymes and it resulted in identifying xylanases with features like favorable temperature and pH, strong heat resistance and tolerance to harmful materials—all desirable for use in industry [9]. Many genes for xylanase have been found in soil, in rotting plant matter, hot springs, sea beds, termite stomachs and the stomach of cows or related ruminants, thanks to metagenome studies [10]. In pertinent cases, xylanases from thermophilic sites are functional at around 80 °C, allowing their use in industrial processes executed in high temperatures. In pulp bleaching areas, where the pH is very high, soda lake xylanases have proved useful [11]. A range of enzyme characteristics and uses proves how well microbial communities have adapted to different ecosystems. Developments in NGS, single-cell genomics and synthetic biology have boosted both the power and relevance of metagenomics by offering better information and advanced tools for mining genes [12]. In addition, applying machine learning to metagenomic analyses is starting to make the identification of industrial candidates easier. Nevertheless, metagenomic genes still face several obstacles, with codon bias and problems with protein folding or no post-translational modifications in other hosts, so researchers are now studying Bacillus subtilis, Pichia pastoris and cell-free expression systems [13]. Identified xylanases are tested thoroughly to learn about their kinetic parameters (Km and Vmax), action across different temperatures and pH values, suitable substrates and condition-dependent stability. Often, novel xylanases found by metagenomics include extra activities, working with arabinofuranosidases, acetyl xylanse esterases or ferulic acid esterases which help them decompose hemicellulose, making them suited for use in integrated biorefineries [14]. Besides, discoveries through metagenomics have revealed different folding patterns and active sites in enzymes, enhancing our knowledge of enzyme growth and mechanisms [15].Thanks to our approach to a bio-based economic system, more robust and varied enzyme toolkits will be required [16]. Metagenomics is examined in this chapter, along with its significance for discovering new xylanases and biocatalysts. Cases are discussed where metagenomics has been useful and the current challenges and future opportunities in this area are also highlighted.

2 Metagenomics

A key aspect of using metagenomics is that scientists can examine all types of microbes in an entire community without having to culture and study them one at a time which would be very difficult if not impossible [17]. This culture-independent strategy was developed to address the fundamental limitation of traditional microbiology, which can access only a small fraction—estimated at less than 1%—of the microbial diversity present in nature. Metagenomics overcomes this by extracting total DNA, also referred to as environmental DNA (eDNA), from diverse sources such as soil, marine sediments, compost, hot springs, animal guts, and wastewater systems [18]. Metagenomics mainly uses two types of methods: those based on sequences and those based on function. Environmental DNA is sequenced rapidly and in bulk using new methods and the resulting data is analyzed by computers to see what types of genes are present, what metabolic pathways exist and what species are present [19]. It depends largely on Illumina, Oxford Nanopore and PacBio technologies, as their data allows scientists to dive into the details of microbial genomes. Once the sequencing is done, experts use computer software to bring together the small reads, identify genes and investigate their resemblance to NCBI, KEGG and CAZy databases [20]. Tools like MetaGeneMark, Prodigal, and MEGAHIT are commonly used for gene prediction and metagenome assembly, while taxonomic classifiers like Kraken2 or Kaiju assign reads to phylogenetic lineages [21]. Through sequence-based metagenomics, researchers can identify homologs of known enzymes—including xylanases—by looking for conserved domains such as glycoside hydrolase families (e.g., GH10, GH11, GH30, GH43), often using hidden Markov models (HMMs) for sensitive detection. This has led to the discovery of putative novel xylanases with unusual or enhanced catalytic properties, even when they bear little sequence similarity to previously known proteins [22]. However, one limitation of this method is that gene prediction does not confirm functional activity; hence, experimental validation is essential to confirm the enzymatic potential of predicted sequences. In contrast, function-based metagenomics involves cloning environmental DNA fragments into suitable expression vectors, transforming them into a host organism—commonly *Escherichia coli*—and screening the resulting clones for enzymatic activities of interest. This approach does not require any prior sequence knowledge and can identify completely novel enzymes, including those with no homology to known genes. Functional screening is typically carried out using activity-based assays, such as colorimetric or fluorometric substrate cleavage tests, plate-based halo assays (e.g., Congo red-xylan agar), or chromogenic indicators. Libraries for function-based metagenomics can be constructed in plasmids for small inserts or in fosmids, cosmids, and bacterial artificial chromosomes (BACs) for capturing larger genomic fragments, which may include regulatory elements that enhance expression [23]. Function-based metagenomics has successfully uncovered novel xylanases with desirable industrial traits such as thermostability, alkaliphilicity, and tolerance to solvents, inhibitors, or extreme pH—properties often absent in laboratory strains. However, this approach is not without challenges [24]. *E. coli* may not be able to perform heterologous expression for reasons such as wrong codon use, misfolded proteins or the missing of essential post-translational procsesses or cofactors. Moreover, metagenomics researchers often have to go through significant amounts of data because rare and specific target activities often need complex substrates.

Heterologous expression in *E. coli* can fail due to codon usage bias, misfolding, or the absence of required post-translational modifications or cofactors. Additionally, screening large libraries can be labor-intensive and time-consuming, particularly when target activities are rare or require complex substrates [25]. Despite these challenges, combining both sequence-based and function-based strategies offers a powerful, complementary approach to enzyme discovery [26]. Now, machine learning algorithms are being taught on big datasets to guess enzyme tasks from sequence clues, enabling researchers to focus on the genes that should be studied next. Using synthetic biology tools, scientists are able to make expression systems that help boost how much metagenomic proteins work [27]. Additionally, shotgun metagenomics can be paired with transcriptomics (metatranscriptomics) and proteomics (metaproteomics) to provide a more dynamic view of microbial community function, revealing not only which genes are present but which are actively transcribed and translated under specific conditions [28]. These omics layers deepen our understanding of microbial interactions and the ecological context of enzyme function. Importantly, metagenomics has also advanced the concept of the "microbial dark matter"—the vast majority of microbial life that remains uncultured and poorly understood. By shedding light on this hidden diversity, metagenomics enables the exploration of novel catalytic functions and enzyme mechanisms that expand our biochemical knowledge and open doors to innovative biotechnological applications [29]. In the field of biomass conversion and bioenergy, for example, metagenomically derived xylanases contribute to the enzymatic toolbox needed for the efficient breakdown of plant hemicellulose into fermentable sugars, a key step in the production of second-generation biofuels [30]. These enzymes are also being tailored for use in biorefineries, animal nutrition, food processing, textile manufacturing, and paper bleaching. In summary, metagenomics represents a transformative paradigm in enzyme discovery, particularly for the identification of novel xylanases from diverse and extreme environments [31]. By enabling researchers to explore the full genetic and functional potential of microbial communities, this approach has not

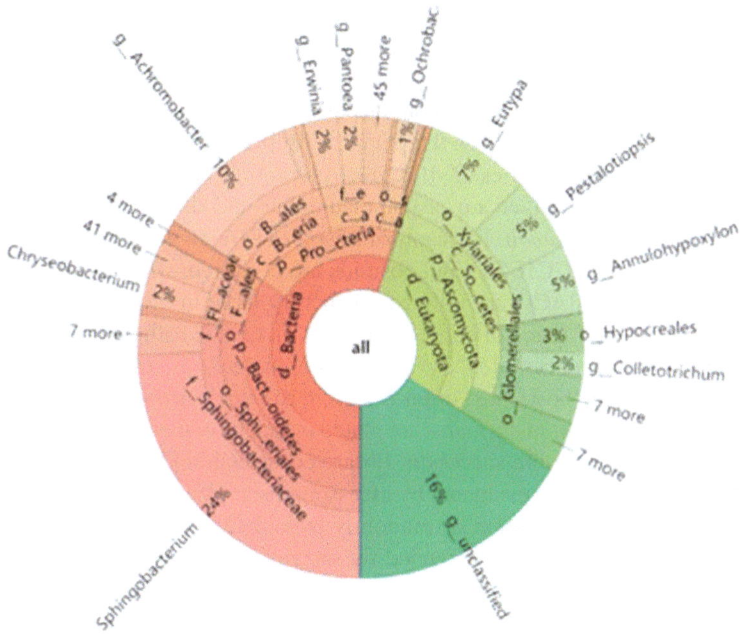

Fig. 1 Metagenomics of the xylanase enzyme [33]

only expanded the known diversity of xylanases but also provided critical insights into microbial ecology, enzyme evolution, and industrial biotechnology [32]. As sequencing technologies continue to evolve and computational tools become more powerful and accessible, the potential of metagenomics to uncover novel biocatalysts with tailored properties is virtually limitless. This section has provided a foundational understanding of the principles, methodologies, strengths, and challenges of metagenomics, setting the stage for deeper exploration into specific applications in the discovery of xylanases and related enzymes (Fig. 1).

3 Construction of Metagenomic Libraries

Creating metagenomic libraries is essential in function-based metagenomics that aim to discover new xylanases from hard-to-culture microorganisms found in many environments such as soil, compost material, marine sediments, hot springs and the guts of animals [34]. Initially, scientists need to separate high-molecular-weight environmental DNA (eDNA) from the environment, as its quality must ensure both cloning and long-term stability in the organism hosting it. Typically, extraction methods destroy the microbial cells in the sample with mechanical and chemical agents and enzymes, then remove contaminating substances like humic substances, polysaccharides and phenolics to ensure these do not affect the later process [35]. Once apart, the eDNA is measured by its size, usually cutting out fragments anywhere from 2 kb to more than 40 kb and repaired to make sure it can pair with the chosen vector ends. The vector system selected is based on the planned insert size and the screening method used [36]. Small-insert libraries (2–10 kb) are commonly constructed using plasmid vectors such as pUC19, which are easy to manipulate and allow high transformation efficiency in *Escherichia coli*. For capturing larger DNA fragments (up to ~ 40 kb), fosmid vectors (e.g., pCC1FOS) and cosmids are preferred, whereas bacterial artificial chromosomes (BACs) are utilized when very large inserts (100–300 kb) are desired, allowing the capture of entire operons or metabolic pathways [37]. Ligation of the prepared environmental DNA into the chosen vector is followed by transformation into a suitable host, most often *E. coli*, through electroporation or chemical methods. The resulting transformants are plated on selective media containing antibiotics and screened to ensure successful library construction [38]. The diversity and coverage of the library are evaluated by assessing insert sizes and redundancy, often by PCR or restriction digestion of randomly picked clones. Once a metagenomic library is established, it can be screened for enzymatic activities using a variety of high-throughput assays [39]. In the case of xylanases, functional screening typically involves plating transformants on agar containing xylan (e.g., birchwood or oat spelt xylan) and a chromogenic dye like Congo red. Active clones secrete xylanase enzymes that degrade the xylan substrate, leading to clear halos around colonies after dye staining and destaining—an easy and visual indicator of enzymatic activity [40]. Other screening methods may use fluorogenic substrates, colorimetric reactions, or plate-based zymograms for detecting activity under specific conditions (e.g., temperature, pH, or salt concentration) [41]. To improve expression of environmental genes, metagenomic libraries may be constructed using expression vectors that [42] binding sites tailored for efficient translation in the host. However, not all environmental genes are readily expressed in *E. coli* due to issues like codon bias, absence of post-translational machinery, or improper protein folding. As such, alternative hosts like *Streptomyces*, *Bacillus subtilis*, *Pseudomonas putida*, or *Pichia pastoris* are increasingly employed to broaden the host range and improve heterologous expression of metagenomic genes [43]. Dual-host shuttle vectors compatible with both *E. coli* and Gram-positive bacteria or yeast are also being developed to facilitate screening in multiple systems [44]. For screening rare activities or low-expression enzymes, enrichment strategies can be employed, such as culturing library clones in xylan-rich media to select for functionally active candidates, or using selective pressures that enrich for clones with particular traits (e.g., thermotolerance, halotolerance, solvent resistance) [45]. Additionally, robotics and microfluidic technologies are being integrated into high-throughput screening platforms to increase the efficiency and scale of functional screening [46]. In parallel, function-based metagenomic screening is increasingly complemented by sequence-based approaches wherein library clones are sequenced and analyzed for the presence of conserved motifs or catalytic domains characteristic of xylanases. This hybrid approach enables the prioritization of clones for expression based on bioinformatic predictions, reducing the screening burden while still capturing novel functional diversity [47]. Importantly, metagenomic libraries can be preserved and archived for future screening under different conditions or against alternative substrates, providing a valuable and renewable resource for enzyme mining [48]. Researchers have found that using functional metagenomic libraries for xylanase discovery has introduced new GH enzymes, for example GH10 and GH11 and, more rarely, GH30, GH5 and GH43, each featuring different functions and activities [49]. Unlike other enzymes, xylanases have extraordinary thermostability, functioning above 90 °C and some can survive either in acidic or alkaline conditions, allowing them to be involved in biofuels, paper bleaching, food processing and improving animal feed, among other industries [50]. Additional carbohydrate-binding modules (CBMs) or additional types of domains have been identified in some of these enzymes, making them attract complex hemicellulosic substrates better [51]. Modern techniques in synthetic biology and protein engineering allow experts to optimize xylanases found using metagenomics, making it easy to design chimeric xylanases that work better, use more substrates and remain stable [52]. Furthermore, the combination of metagenomic library construction with directed evolution techniques—whereby random mutations are introduced into xylanase genes followed by functional selection—offers a powerful strategy for tailoring enzymes to meet specific industrial requirements [53]. Besides, constructing metagenomic libraries and applying directed evolution where mutations are added randomly to xylanase genes give us an effective approach for adjusting the enzyme to meet various industrial demands. Therefore, constructing metagenomic libraries is a key aspect of functional metagenomics, connecting gene potential with useful enzyme activity and supports the search for xylanases with important and unique traits [54]. As our research goes further and molecular tools advance, metagenomic libraries will always have a crucial role in any effort to learn from nature's enzymes for creating sustainable technology [34–36, 49–52, 54] (Fig. 2).

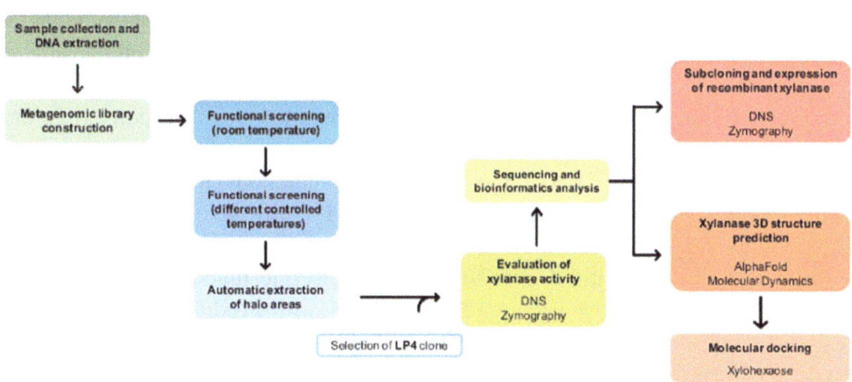

Fig. 2 Construction of metagenomic libraries of xylanase enzyme [55]

4 In Silico Identification of Xylanase Gene

The evolution of NGS and high-throughput bioinformatics has improved xylanase discovery, mainly by enabling computers to find xylanase genes in complex natural samples [56]. Within silico techniques, raw DNA data from metagenomic samples can be used to predict and mark up genes, specifically those for xylanases, without the use of prior experiments [57]. You begin by getting good metagenomic data using Illumina (illumina), Oxford Nanopore or PacBio (long reads), then process the raw data using Trimmomatic, FASTQC and BBTools to eliminate low-quality reads and sequences that have adapters [58]. Contigs are then made by joining the cleaned reads using MEGAHIT, SPAdes or MetaSPAdes which are designed to work best on big and varied metagenomic datasets [59]. The resulting contigs undergo gene prediction using algorithms such as Prodigal, MetaGeneMark, or Glimmer-MG, which identify open reading frames (ORFs) likely to encode proteins. Once predicted, the ORFs are annotated using functional databases and domain-search tools like BLASTp (against NCBI NR), HMMER (against PFAM or TIGRFAMs), InterProScan, and eggNOG-mapper, allowing researchers to infer the function of genes based on homology to known proteins [60]. Specifically for xylanases, in silico identification relies heavily on the detection of conserved catalytic domains and motifs associated with glycoside hydrolase (GH) families, particularly GH10, GH11, GH30, GH43, and GH5, which are known to include xylan-degrading enzymes [17]. The Carbohydrate-Active enZYmes (CAZy) database serves as a cornerstone for this type of functional classification, cataloging enzymes involved in the synthesis, modification, and degradation of carbohydrates, and providing GH family annotations based on both sequence similarity and catalytic mechanism [29]. HMM profiles from CAZy or dbCAN (a tool built upon CAZy data) are frequently employed to scan metagenomic ORFs for glycoside hydrolase domains, identifying candidate xylanase genes even when they share low sequence identity with characterized enzymes [61]. Moreover, signal peptide prediction tools like SignalP or Phobius are used to determine if the predicted proteins are secreted, a key trait for industrial applications where extracellular activity is preferred [62]. Additional bioinformatic analysis often includes the prediction of physicochemical properties such as molecular weight, isoelectric point (pI), and the presence of disulfide bonds or glycosylation sites using tools like ProtParam, NetNGlyc, and NetPhos. Structural and functional predictions can be enhanced using protein modeling platforms such as AlphaFold2, I-TASSER, or SWISS-MODEL, which allow for the visualization of active sites, catalytic residues, and substrate-binding pockets, providing deeper insights into enzyme functionality [63]. After splitting, contigs enter the gene-prediction phase, with Prodigal, MetaGeneMark or Glimmer-MG marking possible open reading frames in the helping sequence. Predicted ORFs, once found, are examined using functional databases and tools and eggNOG-mapper to see if they are similar to known proteins. Today, several DNA-sequencing techniques and phylogenetic tools such as MEGA and PhyML, allow researchers to find out the positions of xylanase genes in evolutionary lineages [64]. This comparative analysis is particularly valuable for identifying enzymes with potentially unique catalytic activities or environmental adaptations [65]. Additionally, people are more often using computational docking tools including AutoDock, SwissDock and PyMOL-based plugins to examine interactions between xylanases and xylan, estimate binding strength and see how well these xylanases fit to xylan or tolerate various industrial settings [66]. A major benefit of working with in silico methods is that screening thousands of genes can be finished within a few days, costing much less and requiring less effort than traditional screening would [67]. These approaches are especially useful for analyzing metagenomic data from places like hot springs, deep-sea vents and alkaline lakes since the organisms there often make unique enzymes with special properties. Using machine learning (ML) and deep learning techniques, scientists can now improve the accuracy of prediction and find hidden aspects of large metagenomic data in in silico experiments. Tools such as DeepEC, EnzymeMiner, and EFICAz use trained algorithms to classify proteins and predict enzymatic function based on complex sequence features, outperforming traditional homology-based methods in many cases [68]. These approaches can identify distant homologs or detect entirely new enzyme families with no close match in current databases. Once high-confidence xylanase candidates are identified, codon optimization tools like JCat or OPTIMIZER are often used to adapt the gene sequences for expression in heterologous hosts like *E. coli*, *Bacillus subtilis*, or *Pichia pastoris*, thereby streamlining the downstream processes of gene synthesis, cloning, expression, and functional characterization [69]. Moreover, the integration of in silico predictions with metatranscriptomic and metaproteomic data can provide functional validation, confirming that the identified xylanase genes are actively transcribed and translated in the native microbial community under specific environmental conditions [70]. This layer of confirmation adds robustness to the candidate selection and helps avoid the pursuit of genes that are merely present but not expressed. Ultimately, the in silico identification of xylanase genes offers a powerful, cost-effective, and comprehensive strategy for mining metagenomic data, dramatically expanding our understanding of microbial enzymatic capabilities and enabling the discovery of novel enzymes with tailored industrial applications. As computational resources continue to improve and bioinformatics tools become more sophisticated, the precision and efficiency of in silico enzyme discovery will only increase, playing an indispensable role in the future of industrial biotechnology, biorefining, and sustainable bioenergy production (Fig. 3).

5 Cloning, Expression, and Purification of Xylanases

To check if the xylanases discovered by metagenomic methods are functional and useful, scientists must clone, express and purify them [71]. Whenever novel xylanase genes are found using computer or metagenomic methods, the first task is to use PCR with primers that contain special sequences which makes them suitable for cloning into another organism [72] for downstream cloning. High-fidelity DNA polymerases like Phusion or Q5 are typically used to ensure error-free amplification.

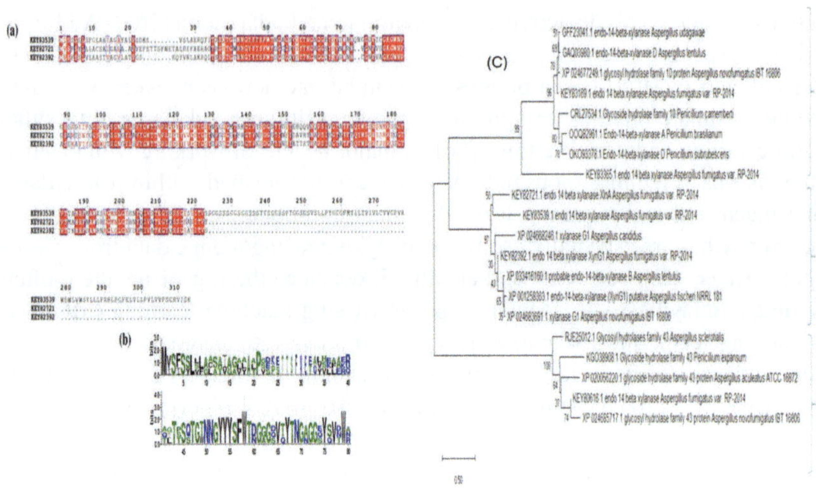

Fig. 3 Identification of xylanase enzyme [59]

The amplified product is then cloned into an appropriate expression vector, which often contains regulatory elements such as a strong promoter (e.g., T7, lac, tac, or arabinose-inducible araBAD), ribosome binding site, selectable marker (e.g., ampicillin or kanamycin resistance), and affinity tags such as His-tag, GST-tag, or MBP-tag for facilitating protein purification [73]. For prokaryotic expression, *Escherichia coli* remains the most widely used host due to its rapid growth, well-understood genetics, and availability of engineered strains like BL21(DE3), Rosetta, or SHuffle, which are designed to enhance the expression of heterologous proteins, improve codon compatibility, or promote disulfide bond formation. Codon optimization of the xylanase gene for *E. coli* is often carried out using tools such as JCat, GeneArt, or online codon usage databases to increase translation efficiency. The recombinant plasmid is then introduced into competent *E. coli* cells via transformation (heat shock or electroporation), and successful transformants are selected on antibiotic-containing media [74]. Induction of protein expression is typically achieved using isopropyl β-D-1-thiogalactopyranoside (IPTG) in T7-based systems or arabinose in pBAD systems, with variables such as inducer concentration, temperature, and duration optimized to maximize yield while minimizing inclusion body formation [75]. In some cases, autoinduction media are used for high-density cultures with gradual and self-regulated induction [76]. After expression, cells are harvested by centrifugation and lysed using sonication, enzymatic lysis (e.g., lysozyme), or high-pressure homogenization in lysis buffers that maintain protein solubility and activity. Soluble and insoluble protein fractions are analyzed by SDS-PAGE to confirm expression and determine solubility [77]. If the xylanase is insoluble, refolding from inclusion bodies or co-expression with molecular chaperones may be necessary. Soluble recombinant xylanase is then purified, typically using affinity chromatography. His-tagged xylanases are purified using nickel-nitrilotriacetic acid (Ni–NTA) or cobalt-based

(Talon) resins, where the target protein binds to the resin and is subsequently eluted with imidazole gradients [11]. Ion-exchange chromatography is sometimes used with DEAE or Q-Sepharose to tidy the protein and size-exclusion chromatography removes any unwanted aggregates for more uniform purification [78]. SDS-PAGE and Western blotting are performed to confirm the purity and molecular weight of the eluted xylanase. Tests for enzymatic activity in a protein include quantifying the liberation of reducing sugars from xylan materials (birchwood or oat spelt) using the DNS method or other similar tests [26]. Other ways to characterize the enzyme include finding the ideal pH and temperature, measuring how stable it stays at different temperatures and identifying which kinds of substrates it acts on. When the chosen xylanaseis not produced well in *E. coli* or is glycosylated and requires further modifications, alternative organisms are chosen, including Pichia pastoris, Bacillus subtilis or Streptomyces species. Extracellular expression in these systems is often made easier by using signaling proteins like the α-factor prepro-leader in *P. pastoris* and PelB in *E. coli* [79]. Moreover, thanks to progress in synthetic biology, Golden Gate and Gibson Assembly have become popular methods to quickly and efficiently clone xylanase genes into several types of expression vectors [80]. Because of their high efficiency and use of microwell formats, robotic systems and microtiter-based methods are becoming preferred for screening different xylanase variants. Moreover, modifications can be done afterward, allowing for enhanced thermal stability, tolerance to acid or alkaline environments or better substrate binding. When a recombinant xylanase is stable and remains active, it can either be immobilized for use or lyophilized for storing and shipping [81]. Ultimately, cloning, expressing and purifying xylanases allow us to use genetic data in practical ways in biofuels, paper and pulp industries, food processing and cleaning up the environment. Great strides in the development of these systems and technologies help researchers make enzymes from difficult organisms useful which broadens what is possible in enzyme biotechnology [82].

6 Biochemical and Biophysical Characterization

Proper characterization of xylanases is required to learn about their enzyme actions, molecular traits and possible role in many industries, particularly if they come from metagenomic sources. Biological attributes of purified recombinant xylanases are measured to assess how well they work, what types of substrates they use, their stability and their overall form under many different environmental conditions. Optimal and thermostability temperature ranges of the enzyme are important measures, so testing for these usually involves incubating them with birchwood, beechwood or oat spelt xylan at various temperatures between 20 and 100 °C and checking for the reducing sugars released with the DNS assay. Thermostability is tested by comparing how much the activity of an enzyme decreases when it is heated for various durations and canting the thermal half-life (T_{50} or $t_{1/2}$) uncovers how well the enzyme can handle heat [83]. In the same manner, using buffer systems with pH values between 3.0 and 11.0 (for example, citrate, phosphate, Tris–HCl or glycine–NaOH), the best pH for activity of the enzyme is determined [84]. For predicting how well an enzyme will function in paper bleaching or being added to animal feed, its activity after incubation in several pH buffers is tested [85]. To determine substrate specificity, researchers commonly compare an enzyme's action on natural materials such as arabinoxylan, glucuronoxylan and carboxymethyl cellulose, as well as a synthetic substrate called p-nitrophenyl-xylopyranoside (pNPX), as these are common xylans found in nature [86]. Kinetic parameters are calculated after conducting enzyme assays at several substrate concentrations and plotting the results using Lineweaver–Burk, Michaelis–Menten or Eadie-Hofstee plots. They tell us about both how effectively enzymes bind to their substrates and how fast they react. Together, biochemistry and biophysics can analyze structural stability, the way proteins fold and how many units linked together they are. Quantitative assessment of the secondary structure and melting temperatures of proteins is done by using CD spectroscopy. Differential scanning calorimetry (DSC) measures changes in how a protein stabilizes between different temperatures, showing when and how energy is released or absorbed during unfolding [29]. Understandably, FTIR and Raman spectroscopy are also applied to investigate the shapes of proteins and their hydrogen bonds at various environmental settings [87]. DLS and SEC provide data on the hydrodynamic radius and size of the enzyme which may influence both its catalytic performance and the stability of the production process. Resolving the atomic frames of potential xylanases may be done through either structural modelling or experimental crystallography. Reliable models can be predicted by AlphaFold, I-TASSER or MODELLER which are often validated by checking the Ramachandran plots or RMSD between the structures calculated and those determined by experiments. Molecular docking simulations using AutoDock, SwissDock or Glide aid in understanding how the substrate binds to the active site, how catalytic residues function and possible inhibition ways. To underline the role of certain amino acids in catalysis and stability, scientists often use site-directed mutagenesis at conserved active site sites like those with glutamate or aspartate in GH10/GH11 xylanases [88]. In addition, enzymes are tested in assays that include metal ions, organic solvents (e.g., ethanol, DMSO, acetone) or are diluted using high ionic strength, useful for industrial purposes under harsh conditions. Metal ion involvement or blocking can be detected by diluting reaction solutions with Ca^{2+}, Mg^{2+}, Mn^{2+}, Zn^{2+}, Cu^{2+}, Na^{+}, K^{+} or EDTA. Some cases involve using polyols, surfactants or stabilizers to determine their effect on enzyme function and structure [89]. Glycosylation patterns for eukaryotic systems like Pichia pastoris are evaluated using SDS-PAGE with glycoprotein staining, mass spectrometry or degrading the modifications with enzymes to check the change in activity and stability. Surface hydrophobicity measurements using ANS-binding assays and isoelectric focusing (IEF) also contribute to understanding protein folding and aggregation behavior. To determine the suitability of the enzyme for biotechnological deployment, activity is tested under simulated industrial process conditions—e.g., in the presence of lignocellulosic biomass hydrolysates, under high temperatures, or extreme pH values, or during co-incubation with cellulases or ligninases to assess synergistic action in biomass degradation. Enzyme immobilization

Table 1 Overview of biochemical characteristics of xyalanse

Parameter	Typical range/value	Source organism(s)	Reference(s)
Optimum pH	5.0–7.0	*Aspergillus niger*	[83]
Optimum temperature (°C)	50–60	*Trichoderma reesei*	[84]
pH stability range	4.0–9.0	*Bacillus subtilis*	[85]
Thermal stability (°C)	Up to 70 °C; moderate stability	*Thermomyces lanuginosus*	[86]
Molecular weight (kDa)	20–45	*Streptomyces sp., Bacillus sp.*	[92]
Michaelis constant (Km, mg/mL)	1.8–5.2	*Aspergillus awamori, Bacillus pumilus*	[87]
Maximum velocity (Vmax, μmol/min/mg)	100–850	*Bacillus circulans, Thermotoga maritima*	[88]
Isoelectric point (pI)	4.2–6.8	*Aspergillus spp., Bacillus spp.*	[89]
Substrate specificity	Birchwood xylan > oat spelt xylan	*Multiple fungal and bacterial strains*	[93]
Metal Ion influence	Mn^{2+}, Ca^{2+}—activators; Hg^{2+}—strong inhibitor	*Bacillus licheniformis*	[94]

studies may also be conducted, using supports such as alginate beads, silica, chitosan, or magnetic nanoparticles to improve reusability and stability, and the activity and recovery are measured over multiple cycles [90]. Collectively, these biochemical and biophysical characterizations provide a comprehensive profile of novel xylanases, linking sequence-based discovery with functional and structural validation. Such exhaustive profiling ensures that only the most promising enzymes are selected for further scale-up, engineering, or commercialization. Moreover, this rigorous approach facilitates the development of tailor-made biocatalysts suited to specific industrial niches such as bioethanol production, pulp biobleaching, food and beverage processing, textile treatment, and animal feed formulation, further underscoring the value of metagenomics in biotechnological innovation [91] (Table 1).

7 Applications of Novel Xylanases

Finding novel xylanases, especially those from metagenomic libraries, has introduced exciting chances for their use in various industries because of their unique chemical properties such as tolerating heat, alkaline conditions and different kinds of substrates [95]. Xylan-hydrolyzing enzymes are key in converting lignocellulosic biomass and are fundamental to different green biotechnological procedures. One of the most impactful applications lies in the **pulp and paper industry**, where xylanases are used in environmentally friendly bioleaching processes. Here, xylanases act by removing hemicellulose-bound lignin residues from wood pulp, reducing the need for harsh chlorine-based chemicals and significantly lowering the generation of toxic chlorinated effluents [96]. Thermostable and alkaline-active xylanases, often discovered through metagenomic approaches, are especially favored for integration into kraft pulping and chlorine dioxide bleaching sequences. In the **biofuel sector**, xylanases are vital components of enzymatic cocktails used for the saccharification of agricultural residues like wheat straw, corn stover, sugarcane bagasse, and hardwoods, which contain a high proportion of xylan [97]. These enzymes increase the yield of fermentable xylose and other sugars from hemicellulose, thereby improving the efficiency of second-generation bioethanol and biobutanol production. Novel xylanases exhibiting resistance to inhibitors like furfural or high concentrations of salts and solvents—conditions typical of pretreated biomass—have shown promise in enhancing overall biomass conversion [80]. In the **animal feed industry**, xylanases improve the nutritional value and digestibility of cereals and plant-based feed ingredients (e.g., wheat, barley, rye), especially in monogastric animals such as poultry and swine [98]. By breaking down arabinoxylans and other non-starch polysaccharides that are otherwise indigestible, xylanases reduce intestinal viscosity, enhance nutrient absorption, and promote gut health. The use of thermostable and acid-stable xylanases—derived from extremophilic microbes via metagenomics—ensures enzyme functionality during feed pelleting and under gastrointestinal conditions. Another important domain is the **food and beverage industry**, where xylanases are employed in baking to improve dough handling and bread quality [99]. They modify the water-binding capacity and molecular structure of arabinoxylans, resulting in softer crumbs, higher loaf volume, and extended shelf life. In brewing, xylanases facilitate the breakdown of non-starch polysaccharides during mashing, enhancing wort clarity and filtration rates. In the processing of coffee, plant juices, and wines, xylanases assist in maceration, clarification, and flavor enhancement by acting synergistically with pectinases and cellulases. In **textile and biofinishing processes**, xylanases contribute to the eco-friendly processing of plant-based fabrics by removing hemicellulosic residues, improving dye uptake, softness, and brightness while minimizing water and energy consumption compared to conventional chemical treatments. In **wastewater treatment and bioremediation**, xylanases aid in the degradation of xylan-rich organic matter in industrial effluents from agro-processing, paper mills, and textile factories. Engineered microbial consortia expressing novel xylanases have been developed to enhance the decomposition of recalcitrant organics and reduce chemical oxygen demand (COD) in polluted waters. In the **pharmaceutical and nutraceutical fields**, xylanase-generated xylo-oligosaccharides (XOS) are emerging as prebiotic ingredients due to their selective fermentation by beneficial gut microbes like *Bifidobacteria* and *Lactobacilli.* Novel xylanases capable of producing short-chain XOS with specific degrees of polymerization are being exploited for the large-scale, enzymatic synthesis of functional food ingredients. In **bioplastics and biopolymers**, xylanase treatment of hemicellulose-rich biomass facilitates the production of xylose, which can be chemically or biologically converted into xylitol, lactic acid, or furfural—key precursors for biobased materials. The adaptability of novel xylanases to extreme environments, as observed in enzymes sourced from hot springs, deep-sea vents, compost piles, and saline soils, makes them highly attractive for harsh processing conditions. Their inclusion in **integrated biorefineries** not only improves sugar recovery but also contributes to the circular bioeconomy by enabling the valorization of agro-waste into high-value chemicals and fuels. Recent developments in synthetic biology and enzyme engineering have further broadened application potential by creating designer xylanases with enhanced catalytic properties, reduced feedback inhibition, and multifunctionality, such as bifunctional xylanase-arabinofuranosidases or xylanase-acetyl xylan esterases. Additionally, immobilized xylanases—using matrices like alginate beads, mesoporous silica, or magnetic nanoparticles—have demonstrated reusability and stability, making them suitable for continuous industrial operations and reducing process costs. Metagenomic exploration has also led to the discovery of **cold-active xylanases**, suitable for low-temperature processes in food preservation or detergent formulations where energy efficiency and mild conditions are crucial. In **diagnostic and analytical applications**, xylanases are used in carbohydrate profiling and structural studies of plant biomass, assisting researchers in understanding plant cell wall architecture and improving biomass pretreatment strategies. Looking ahead, the integration of novel xylanases into **smart bioprocessing systems**—combining enzyme cocktails, engineered microbes, and AI-guided optimization—promises to further revolutionize industries reliant on hemicellulose degradation. Using the metagenomic approach increases the variety of recognized xylanases and makes available biocatalysts that are hard to grow in a laboratory, improving the resources for different types of biotechnology. Novel xylanases play an increasingly important role in supporting sustainability, using resources sensibly and protecting the environment in various industrial areas (Fig. 4).

8 Conclusion

All told, using metagenomics has allowed researchers to find novel xylanases in the genes of microorganisms living in all kinds of environments. Unlike standard ways of culturing, using metagenomics lets researchers take and use DNA from the environment, giving access to a variety of heat-resistant, salt tolerant, acid or alkali tolerant and highly selective xylanolytic enzymes. Changing the focus in enzyme bioprospecting has increased our knowledge of microorganisms and enzymes, as well as led to the discovery of xylanases with greater usefulness in industry. Metagenomic libraries, functional screening and in silico gene mining help quickly find and characterize a wide range of xylanase candidates. The use of modern molecular cloning, heterologous systems and protein purifiers has made the change from finding sequences to studying proteins much faster. With computational tools for modeling, docking and evolution, we have discovered more details about how these enzymes function and what they do. Work on understanding thermal and acid–base profiles, how the enzymes function which substrates they can act on and their robustness in processes has helped make xylanases suitable for certain uses. They have a wide

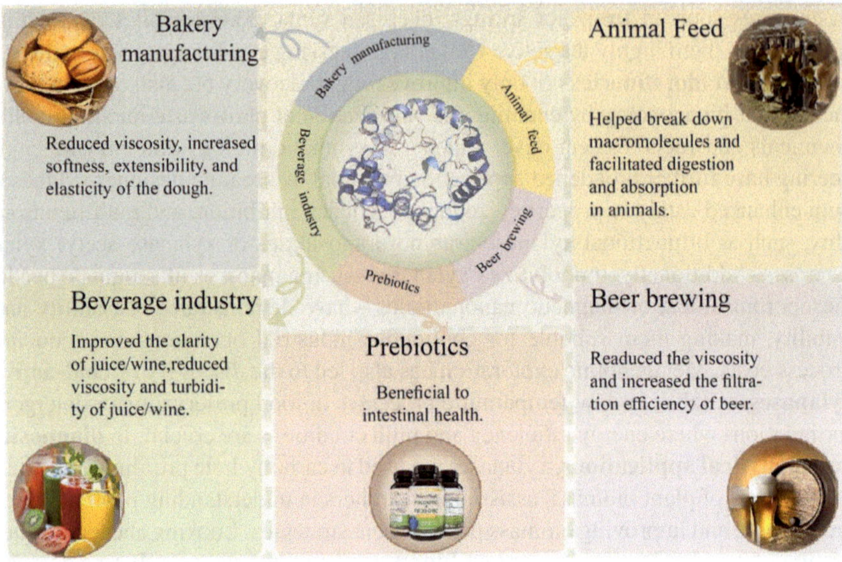

Fig. 4 Application of novel xylanase enzyme [100]

range of important uses, found in pulp and paper, biofuel production, giving extra nutrition to animals, food and beverage industries, textile finishing and pharmaceutical companies. Furthermore, the ability to generate value-added products like xylo-oligosaccharides, xylitol, and biofuels from lignocellulosic waste using metagenome-derived xylanases underscores their importance in promoting sustainable and circular bioeconomies. Despite the remarkable progress, challenges remain, particularly in the expression of full-length functional enzymes from complex metagenomic data, optimization of enzyme stability in non-native hosts, and scalability of enzyme production. However, with the ongoing integration of synthetic biology, machine learning, and high-throughput screening platforms, the efficiency and precision of enzyme discovery and engineering are expected to improve significantly. Looking ahead, the future of metagenomic xylanase research lies in the creation of customized enzyme toolkits, adaptive bioprocesses, and smart biocatalysts that can operate under dynamic industrial conditions with minimal environmental impact. Overall, metagenomics continues to be a powerful and transformative approach in the quest for novel xylanases, fostering innovation in biotechnology and enabling the transition toward greener, more sustainable industries.

References

1. L. Fülöp, The inhibition of xylanase enzymes by oligosaccharides produced during the degradation of biopolymers in biomass. *Biomass and Bioenergy,* vol. 195, 2025/04/01/ (2025), p. 107693
2. X. yan Wang, Z. li Ding, Z. sheng Zhang, C. bin Zhu, S. Yang, H. Fei, Application of two crucial hemicellulolytic enzymes in fish diet: xylanase and mannanase. *Animal Feed Science and Technology,* vol. 319, 2025/01/01/ (2025), p. 116178
3. Y. Xie, X. Ji, Z. Tian, Y. Wang, X. Mo, F. Zhang, et al., Extraction of high-purity lignin from the kraft pulping black liquor by enzyme purification process with alkaline-resistant xylanase and cellulose. Int. J. Biol. Macromol. **295**, p. 139574, 2025/03/01/ (2025)
4. N. Kanwal, N. Rashid, M. Irfan Majeed, H. Nawaz, A. Amber, M. Zohaib, et al., Surface-enhanced Raman spectroscopy for the characterization of xylanases enzyme. *Spectrochimica Acta Part A: Molecular and Biomolecular Spectroscopy,* vol. 325, 2025/01/15/ (2025), p. 125065
5. R.M. Maas, F. Kokou, M.C.J. Verdegem, J.W. Schrama, Enzyme supplementation (phytase and xylanase) improves low quality diets in Nile tilapia. *Aquaculture Reports,* vol. 40, 2025/03/15/ (2025), p. 102650
6. J.I. Vargas, J.P. Gulizia, J.R. Hernandez, C.T. Simões, E.G. Guzman, J. Lee, et al., Dose response of xylanase enzyme on ileal viscosity, digestibility of nutrients, and performance of broiler chickens fed wheat-based diets. J. Appl. Poultry Res. **33**, 100479, 2024/12/01/ (2024)
7. N.T. Tinh, L.M. Bao, D.D. Nhan, P.P. Bao, T.M. Hoang, N.T.H. Nam, et al., Green process prorogue by xylanase enzyme and cellulose-derived sulfonated graphene oxide-like reinforcement in xylose and furfural production from corncob biomass. Renew. Energy **232**, 121039, 2024/10/01/ (2024)
8. I. Tyagi, P. Yadav, P.R. Gogate, Intensification of activity of xylanase enzyme using ultrasound and its application for brown rice polishing. Chem. Eng. Process. Process Intensification **196**, 109683, 2024/02/01/ (2024)
9. K. Zhang, X. Qi, N. Feng, Y. Wang, H. Wei, M. Liu, Antioxidant capacity of xylooligosaccharides generated from beechwood xylan by recombinant family GH10 *Aspergillus niger* xylanase A and insights into the enzyme's competitive inhibition by riceXIP. Enzyme Microb. Technol. **179**, 110456, 2024/09/01/ (2024)
10. A.A. Nascimento, É.S. de Macêdo, G.S. Gonçalves, T.P. da Cruz, B. Wernick, V.R.B. Furuya, et al., Xylanase and β-glucanase in tandem improve performance, digestive enzymes activity and digestibility in juvenile Nile tilapia fed corn distillers' dried grains with solubles-added diet. Animal Feed Sci. Technol. **306**, 115816, 2023/12/01/ (2023)
11. O. Ulucay, A. Gormez, C. Ozic, For biotechnological applications: purification and characterization of recombinant and nanoconjugated xylanase enzyme from thermophilic *Bacillus subtilis*. Biocatal. Agric. Biotechnol. **44**, 102478, 2022/09/01/ (2022)
12. H. Nadeem, S.Z.H. Shah, M. Fatima, Maryam, M.M. Shahzad, Effects of exogenous xylanase supplementation in plant-based diets on growth performance, nutrient digestibility and digestive enzyme activities of juvenile mori (*Cirrhinus mrigala*). Ani. Feed Sci. Technol. **291**, 115391, 2022/09/01/ (2022)
13. N. Morgan, M.M. Bhuiyan, R. Hopcroft, Non-starch polysaccharide degradation in the gastrointestinal tract of broiler chickens fed commercial-type diets supplemented with either a single dose of xylanase, a double dose of xylanase, or a cocktail of non-starch polysaccharide-degrading enzymes. Poult. Sci. **101**, 101846, 2022/06/01/ (2022)
14. J. Kaushal, M. Khatri, G. Singh, S.K. Arya, A multifaceted enzyme conspicuous in fruit juice clarification: an elaborate review on xylanase. Int. J. Biol. Macromol. **193**, 1350–1361, 2021/12/15/ (2021)
15. R.L. Miorin, L.H.C. Batista, F.A. Nascimento, L.F. Costa e Silva, A. Koontz, J.E. Pettigrew, et al., Effect of supplementation strategies and the use of exogenous xylanase enzyme on ruminal fermentation, digestibility, animal performance, and carcass characteristics of Nellore bulls grazing during dry season. Anim. Feed Sci. Technol. **290**, 115373, 2022/08/01/ (2022)

16. L. Du, X. Cui, H. Li, Y. Wang, L. Fan, R. He, et al., Enhancing the enzymatic hydrolysis efficiency of lignocellulose assisted by artificial fusion enzyme of swollenin-xylanase. Ind. Crops Prod. **173**, 114106, 2021/12/01/ (2021)
17. H.-B. Kim, K.-T. Lee, M.-J. Kim, J.-S. Lee, K.-S. Kim, Identification and characterization of a novel KG42 xylanase (GH10 family) isolated from the black goat rumen-derived metagenomic library. Carbohydr. Res. **469**, 1–9, 2018/11/01/ (2018)
18. Z. Li, X. Li, T. Liu, S. Chen, H. Liu, H. Wang, et al., The critical roles of exposed surface residues for the thermostability and halotolerance of a novel GH11 xylanase from the metagenomic library of a saline-alkaline soil. Int. J. Biol. Macromol. **133**, 316–323, 2019/07/15/ (2019)
19. C. Qian, N. Liu, X. Yan, Q. Wang, Z. Zhou, Q. Wang, Engineering a high-performance, metagenomic-derived novel xylanase with improved soluble protein yield and thermostability. Enzyme Microb. Technol. **70**, 35–41, 2015/03/01/ (2015)
20. S. Ariaeenejad, H. Lanjanian, E. Motamedi, K. Kavousi, A.A. Moosavi-Movahedi, G. Hosseini Salekdeh, The stabilizing mechanism of immobilized metagenomic xylanases on bio-based hydrogels to improve utilization performance: computational and functional perspectives. Bioconjugate Chem. **31**, 2158–2171, 2020/01/01/ (2020)
21. C.-C. Yang, Y.-J. Lin, F.-S. Wen, Novel xylanase activity and biofuel production in Lacrimispora xylanolytica isolated from termite guts symbionts. Bioresour. Technol. **429**, 132489, 2025/08/01/ (2025)
22. P. Gao, J. Ma, R.A. Basit, X. Lou, W. Zhang, H. Song, et al., Investigation of the enzymatic characteristics of thermotolerant xylanase McXyn0243 derived from *Malbranchea cinnamomea*, and its application in the degradation of three agricultural residues for the production of xylooligosaccharides. Int. J. Biol. Macromol. 143973, 2025/05/05/ (2025)
23. M. OuYang, C. Liu, P. Sheng, Q. Qiu, K. OuYang, Y. Li, et al., Characterization of acidophilic cellulase with β-glucanase and xylanase activity from the rumen metagenome and in vitro evaluation of barley digestibility. Int. J. Biol. Macromol. **306**, 141141, 2025/05/01/ (2025)
24. V.P. Patel, S. Rothe, Chapter 14—Metagenomics: a mining enzymes from microbial cells for biofuel production, in *Genetic and Genome-Wide Microbial Insights: Bioenergy*, eds. by J.A. Parray, N. Singh, D. Egamberdieva (Academic Press, 2025), pp. 233–247
25. Y. Ma, H. Pan, Z. Feng, P. Li, M. Fu, X. Yi, et al., Metagenomic analysis of a top-down enriched straw biodegradation synthetic bacterial consortium (StrBsyn) facilitates anaerobic rice straw degradation. Process Saf. Environ. Prot. **192**, 1260–1270, 2024/12/01/ (2024)
26. X. Qi, X. Liu, Y. Zhang, J. Hao, H. Rao, D. Zhao, et al., A novel wheat arabino-xylanase from *Trichoderma gamsii*: study on biochemical characterization, substrate recognition mechanism and its application in wheat flour dough and Chinese bread. LWT **219**, 117550, 2025/03/01/ (2025)
27. W.-W. Zhang, Y.-X. Guo, Q.-J. Chen, Y.-Y. Wang, Q.-Y. Wang, Y.-R. Yang, et al., Metagenomic insights into the lignocellulose degradation mechanism during short-term composting of peach sawdust: core microbial community and carbohydrate-active enzyme profile analysis. Environ. Technol. Innov. **37**, 103959, 2025/02/01/ (2025)
28. J. D'aes, M.-A. Fraiture, B. Bogaerts, Y. Van Laere, S.C.J. De Keersmaecker, N.H.C. Roosens, et al., Metagenomics-based tracing of genetically modified microorganism contaminations in commercial fermentation products. Food Chem. Mol. Sci. **10**, 100236, 2025/06/01/ (2025)
29. S. Ariaeenejad, E. Hosseini, M. Maleki, K. Kavousi, A.A. Moosavi-Movahedi, G.H. Salekdeh, Identification and characterization of a novel thermostable xylanase from camel rumen metagenome. Int. J. Biol. Macromol. **126**, 1295–1302, 2019/04/01/ (2019)
30. Y. Yang, M.-S. Niu, H. Yu, W. Shi, L.-J. Chai, Z.-M. Lu, et al., Exploring the contribution of temperature-adapted microbiota to enzyme profile of saccharification in Daqu using metagenomics and metaproteomics. LWT **197**, 115916, 2024/04/01/ (2024)
31. J.-J. Escuder-Rodríguez, M.-E. DeCastro, M. Becerra, E. Rodríguez-Belmonte, M.-I. González-Siso, 21—Advances of functional metagenomics in harnessing thermozymes, in *Metagenomics (Second Edition)*, M. Nagarajan, ed. by (Academic Press, 2025), pp. 493–516

32. Y. Huang, D. Li, Y. Mu, Z. Zhu, Y. Wu, Q. Qi, et al., Exploring the heterogeneity of community and function and correspondence of "species-enzymes" among three types of Daqu with different fermentation peak-temperature via high-throughput sequencing and metagenomics. Food Res. Int. **176**, 113805, 2024/01/01/ (2024)
33. Y. Lin, C. Li, C. Wei, H. Lin, L. Zhang, Mining, identification, and characterization of three xylanases from the microbiota of *T. fuciformis* with its companion strains. Catalysts **14**, 15 (2024)
34. T.S. Crofts, A.G. McFarland, E.M. Hartmann, Mosaic ends tagmentation (METa) assembly for highly efficient construction of functional metagenomic libraries. mSystems **6**, 2021/05/11/ (2021)
35. J. Cheng, L. Pinnell, K. Engel, J.D. Neufeld, T.C. Charles, Versatile broad-host-range cosmids for construction of high quality metagenomic libraries. J. Microbiol. Methods. **99**, 27–34, 2014/04/01/ (2014)
36. S. Pandey, S. Gulati, E. Goyal, S. Singh, K. Kumar, L. Nain, et al. Construction and screening of metagenomic library derived from soil for β-1, 4-endoglucanase gene. Biocatal. Agric. Biotechnol. **5**, 186–192, 2016/01/01/ (2016)
37. H.E. Biyik, M. Menekse, N.C. Balci, M. Tuter, M. Wehrl, N.G. Karaguler, Construction of pre-enriched metagenomic library for isolating novel hydrolase enzymes for LCO2/SCCO2. Curr. Opinion Biotechnol. **22**, S86, 2011/09/01/ (2011)
38. H. Bertrand, F. Poly, V.T. Van, N. Lombard, R. Nalin, T.M. Vogel, et al., High molecular weight DNA recovery from soils prerequisite for biotechnological metagenomic library construction. J. Microbiol. Methods. **62**, 1–11, 2005/07/01/ (2005)
39. V. Tsopanakis, E. Anastasiadou, M.D. Mikkelsen, A.S. Meyer, I.V. Pavlidis, Identification and characterization of a novel thermostable PL7 alginate lyase from a submarine volcanic metagenomic library. Enzyme Microb. Technol. **180**, 110486, 2024/10/01/ (2024)
40. F.-M. Chang, Y.-H. Chen, P.-S. Hsu, T.-H. Wu, I. H. Sung, M.-C. Wu, et al., RNA metagenomics revealed insights into the viromes of honey bees (*Apis mellifera*) and *Varroa mites* (*Varroa destructor*) in Taiwan. J. Invertebr. Pathol. **211**, 108341, 2025/07/01/ (2025)
41. R. Kogoj, M. Bosilj, A. Celar Šturm, M. Korva, K. Strašek Smrdel, E. Kvas, et al., Bioinformatic challenges in metagenomic next generation sequencing data analysis while unravelling a case of uncommon campylobacteriosis. J. Biomed. Inf. **166**, 104841, 2025/06/01/ (2025)
42. A. Alma'abadi, H. Behzad, M. Alarawi, D. Conchouso, Y. Saito, M. Hosokawa, et al., Identification of lipolytic enzymes using high-throughput single-cell screening and sorting of a metagenomic library. New Biotechnol. **70**, 102–108, 2022/09/25/ (2022)
43. Y. Yang, Y. Deng, L. Liu, X. Yin, X. Xu, D. Wang, et al., Establishing reference material for the quest towards standardization in environmental microbial metagenomic studies. Water Res. **245**, 120641, 2023/10/15/ (2023)
44. A.K. Murray, L. Zhang, J. Snape, W.H. Gaze, Functional metagenomic libraries generated from anthropogenically impacted environments reveal importance of metabolic genes in biocide and antibiotic resistance. Curr. Res. Microb. Sci. **4**, 100184, 2023/01/01/ (2023)
45. O. Coskuner-Weber, S. Alpsoy, O. Yolcu, E. Teber, A. de Marco, S. Shumka, Metagenomics studies in aquaculture systems: Big data analysis, bioinformatics, machine learning and quantum computing. Computat. Biol. Chem. **118**, 108444, 2025/10/01/ (2025)
46. S. Chai, X. Zhang, Y. Gao, L. Nie, C. Xiang, M. Cao, et al., Identification and characterization of a novel endo-β-1,4-glucanase from a soil metagenomic library. Carbohydr. Res. **510**, 108460, 2021/12/01/ (2021)
47. C.A. Gaulke, E.R. Schmeltzer, M. Dasenko, B.M. Tyler, R. Vega Thurber, T.J. Sharpton, Evaluation of the effects of library preparation procedure and sample characteristics on the accuracy of metagenomic profiles. mSystems **6**, 2021/09/14/ (2021)
48. Q. Peng, L. Quan, H. Zheng, J. Li, G. Xie, Analyzing the contribution of top-down and bottom-up methods to the construction of synthetic microbial communities in Jiuyao. Food Microbiol. **129**, 104759, 2025/08/01/ (2025)
49. J. Liu, Z. Jia, S. Li, Y. Li, Q. You, C. Zhang, et al., Identification and characterization of a chitin deacetylase from a metagenomic library of deep-sea sediments of the Arctic Ocean. Gene **590**, 79–84, 2016/09/15/ (2016)

50. Z. Mai, L. Wang, Q. Zeng, Characterization of a novel isoflavone glycoside-hydrolyzing β-glucosidase from mangrove soil metagenomic library. Biochem. Biophys. Res. Commun. **569**, 61–65, 2021/09/10/ (2021)
51. A.K. Wani, F. Qadir, N. Elboughdiri, F. Rahayu, Saefudin, D. Pranowo, et al., Metagenomics and plant-microbe symbioses: microbial community dynamics, functional roles in carbon sequestration, nitrogen transformation, sulfur and phosphorus mobilization for sustainable soil health. Biotechnol. Adv. **82**, 108580, 2025/09/01/ (2025)
52. Y. Xu, L. Yao, S. Song, M. Sun, H. Wang, C. Yu, et al., Revealing the microbial diversity and volatile flavor formation in finger citron kombucha by metagenomic and GC-MS analysis. Food Biosci. **59**, 104087, 2024/06/01/ (2024)
53. Y.M. Kang, M.K. Kim, J.M. An, M.A. Haque, K.M. Cho, Metagenomics of un-culturable bacteria in cow rumen: construction of cel9E–xyn10A fusion gene by site-directed mutagenesis. J. Mol. Catal. B Enzym. **113**, 29–38, 2015/03/01/ (2015)
54. Y. Shen, N. Liu, Z. Wang, Recent advances in the culture-independent discovery of natural products using metagenomic approaches. Chin. J. Nat. Med. **22**, 100–111, 2024/02/01/ (2024)
55. J. Sousa, C. Santos-Pereira, J.S. Gomes, Â.M.A. Costa, A.O. Santos, R. Franco-Duarte, et al., Heterologous expression and structure prediction of a xylanase identified from a compost metagenomic library. Appl. Microbiol. Biotechnol. **108**, 329, 2024/05/10 (2024)
56. X. Qi, Y. Wang, N. Feng, Y. Kong, K. Zhang, X. Ouyang, et al., Competitive inhibition of family GH5 *Aspergillus niger* endo-glucanase A (AnglA5) by rice xylanase inhibitor (RIXI): experimental and in silico insights. Int. J. Biol. Macromol. **307**, 142011, 2025/05/01/ (2025)
57. S.A. Nour, G.M. El-Sayed, H.A.A. Taie, M.T.H. Emam, A.F. El-Sayed, R.G. Salim, Safe production of *Aspergillus terreus* xylanase from *Ricinus communis*: gene identification, molecular docking, characterization, production of xylooligosaccharides, and its biological activities. J. Genet. Eng. Biotechnol. **20**, 121, 2022/12/01/ (2022)
58. B. Dutta, A. Banerjee, P. Chakraborty, R. Bandopadhyay, In silico studies on bacterial xylanase enzyme: structural and functional insight. J. Genetic Eng. Biotechnol. **16**, 749–756, 2018/12/01/ (2018)
59. S.R. Dodda, M. Hossain, B.S. Kapoor, S. Dasgupta, V.P.R.B, K. Aikat, et al., Computational approach for identification, characterization, three-dimensional structure modelling and machine learning-based thermostability prediction of xylanases from the genome of *Aspergillus fumigatus*. Computat. Biol. Chem. **91**, 107451, 2021/04/01/ (2021)
60. G. Elliott, A. Durand, R.K. Hughes, P.A. Kroon, R. D'Ovidio, N. Juge, Isolation and characterisation of a xylanase inhibitor Xip-II gene from durum wheat. J. Cereal Sci. **50**, 324–331, 2009/11/01/ (2009)
61. S. Ariaeenejad, M. Maleki, E. Hosseini, K. Kavousi, A.A. Moosavi-Movahedi, G.H. Salekdeh, Mining of camel rumen metagenome to identify novel alkali-thermostable xylanase capable of enhancing the recalcitrant lignocellulosic biomass conversion. Bioresour. Technol. **281**, 343–350, 2019/06/01/ (2019)
62. S. Ariaeenejad, K. Kavousi, A.S.A. Mamaghani, S.F.S. Motahar, H. Nedaei, G.H. Salekdeh, In-silico discovery of bifunctional enzymes with enhanced lignocellulose hydrolysis from microbiota big data. Int. J. Biol. Macromol. **177**, 211–220, 2021/04/30/ (2021)
63. S. Miyauchi, V.S. Te'o, P.L. Bergquist, K.M.H. Nevalainen, Expression of a bacterial xylanase in *Trichoderma reesei* under the egl2 and cbh2 glycosyl hydrolase gene promoters. New Biotechnol. **30**, 523–530, 2013/06/25/ (2013)
64. M. Paul, H. Thatoi, Chapter 14—microbial xylanases, their structural characteristics, and industrial applications: a biotechnological advancement, in *Innovations in Fermentation and Phytopharmaceutical Technologies*, H. Thatoi, S. Mohapatra, S.K. Das, eds. by (Elsevier, 2022), pp. 315–339
65. S. Kishishita, M. Yoshimi, T. Fujii, L.E. Taylor, S.R. Decker, K. Ishikawa, et al., Cellulose-inducible xylanase Xyl10A from Acremonium cellulolyticus: Purification, cloning and homologous expression. Protein Express. Purif. **94**, 40–45, 2014/02/01/ (2014)

66. C. Bouiche, N. Boucherba, S. Benallaoua, J. Martinez, P. Diaz, F.I.J. Pastor, et al., Differential antioxidant activity of glucuronoxylooligosaccharides (UXOS) and arabinoxylooligosaccharides (AXOS) produced by two novel xylanases. Int. J. Biol. Macromol. **155**, 1075–1083, 2020/07/15/ (2020)
67. N.S. Betts, H.M. Collins, N.J. Shirley, J.A. Cuesta-Seijo, J.G. Schwerdt, R.J. Phillips, et al., Identification and spatio-temporal expression analysis of barley genes that encode putative modular xylanolytic enzymes. Plant Sci. **308**, 110792, 2021/07/01/ (2021)
68. C.d.A. Guerreiro, L.A.D. Andrade, C. Fernández-Lainez, L.N. Fraga, G. López-Velázquez, T.M. Marques, et al., Bioactive arabinoxylan oligomers via colonic fermentation and enzymatic catalysis: evidence of interaction with toll-like receptors from in vitro, in silico and functional analysis. Carbohydr. Polym. **352**, 123175, 2025/03/15/ (2025)
69. M.F.S. Sam-on, S. Mustafa, A.M. Hashim, M.T. Yusof, S. Zulkifly, M.A.H. Roslan, Determination of prebiotic utilisation capability of potential probiotic *Bacillus velezensis* FS26 through in silico and in vitro approaches. Food Biosci. **53**, 102566, 2023/06/01/ (2023)
70. T. Furukawa, Y. Shida, N. Kitagami, K. Mori, M. Kato, T. Kobayashi, et al., Identification of specific binding sites for XYR1, a transcriptional activator of cellulolytic and xylanolytic genes in *Trichoderma reesei*. Fungal Genet. Biol. **46**, 564–574, 2009/08/01/ (2009)
71. D. Driss, F. Bhiri, R. ghorbel, S.E. Chaabouni, Cloning and constitutive expression of His-tagged xylanase GH 11 from *Penicillium occitanis* Pol6 in *Pichia pastoris* X33: purification and characterization. Protein Express. Purif. **83**, 8–14, 2012/05/01/ (2012)
72. D.M. Tran, T.U. Huynh, T.H. Nguyen, T.O. Do, A.D. Nguyen, Data on gene cloning, expression, purification, and characterization of the glycoside hydrolase family 11 from *Bacillus velezensis*. Data Brief. **52**, 109834, 2024/02/01/ (2024)
73. H. Karaoglu, K.M.A. Ramadan, S.A. Al Hashedi, A. AlShoaibi, Z. Iqbal, R. Aydın, et al., Selection, heterologous production, and functional characterization of a thermostable xylanase from anoxybacillus for dough and bread quality enhancement. Int. J. Biol. Macromol. **312**, 144000, 2025/06/01/ (2025)
74. A. Hamid, A. Zafar, I. Liaqat, M.S. Afzal, L. Peng, M.K. Rauf, et al. Effective utilization of magnetic nano-coupled cloned β-xylanase in saccharification process. RSC Adv. **12**, 6463–6475, 2022/02/22/ (2022)
75. N. Bhardwaj, V.K. Verma, V. Chaturvedi, P. Verma, Cloning, expression and characterization of a thermo-alkali-stable xylanase from *Aspergillus oryzae* LC1 in *Escherichia coli* BL21(DE3). Protein Express. Purif. **168**, 105551, 2020/04/01/ (2020)
76. P. Bajpai, Chapter 7—Molecular cloning and expression of xylanase gene, in *Microbial Xylanolytic Enzymes*, P. Bajpai, ed. by (Academic Press, 2022), pp. 109–116
77. U. Cakmak, N. Saglam Ertunga, Gene cloning, expression, immobilization and characterization of endo-xylanase from *Geobacillus* sp. TF16 and investigation of its industrial applications. J. Mol. Catal. B Enzymatic. **133**, S288–S298, 2016/11/01/ (2016)
78. A. Saleem, S. Waris, T. Ahmed, R. Tabassum, Biochemical characterization and molecular docking of cloned xylanase gene from Bacillus subtilis RTS expressed in *E. coli*, Int. J. Biol. Macromol. **168**, 310–321, 2021/01/31/ (2021)
79. P. Boonchuay, S. Takenaka, A. Kuntiya, C. Techapun, N. Leksawasdi, P. Seesuriyachan, et al., Purification, characterization, and molecular cloning of the xylanase from *Streptomyces thermovulgaris* TISTR1948 and its application to xylooligosaccharide production. J. Mol. Catal. B Enzym. **129**, 61–68, 2016/07/01/ (2016)
80. Y. Lai, J. Wang, N. Xie, G. Liu, D.C. Lacap-Bugler, Identification of a novel forkhead transcription factor MtFKH1 for cellulase and xylanase gene expression in *Myceliophthora thermophila* (ATCC 42464). Microbiol. Res. **294**, 128097, 2025/05/01/ (2025)
81. M. Shrivastava, A. Aishwarya, C.M.G.A. Fontes, A. Goyal, A novel bifunctional type I α-l-arabinofuranosidase of family 43 glycoside hydrolase (BoGH43_35) from Bacteroides ovatus with endo-β-1,4-xylanase activity. Carbohydr. Res. **552**, 109432, 2025/06/01/ (2025)
82. D. Verma, T. Satyanarayana, Cloning, expression and applicability of thermo-alkali-stable xylanase of *Geobacillus thermoleovorans* in generating xylooligosaccharides from agro-residues. Bioresour. Technol. **107**, 333–338, 2012/03/01/ (2012)

83. M.L.T.M. Polizeli, A.C.S. Rizzatti, R. Monti, H.F. Terenzi, J.A. Jorge, D.S. Amorim, Xylanases from fungi: properties and industrial applications. Appl. Microbiol. Biotechnol. **67**, 577–591, 2005/06/01 (2005)
84. T. Matsuzawa, S. Kaneko, K. Yaoi, Improvement of thermostability and activity of Trichoderma reesei endo-xylanase Xyn III on insoluble substrates. Appl. Microbiol. Biotechnol. **100**, 8043–8051, 2016/09/01 (2016)
85. B. Battan, J. Sharma, S.S. Dhiman, R.C. Kuhad, Enhanced production of cellulase-free thermostable xylanase by *Bacillus pumilus* ASH and its potential application in paper industry. Enzyme Microb. Technol. **41**, 733–739, 2007/11/01/ (2007)
86. S. Singh, A.M. Madlala, B.A. Prior, *Thermomyces lanuginosus*: properties of strains and their hemicellulases. FEMS Microbiol. Rev. **27**, 3–16 (2003)
87. C. Techapun, N. Poosaran, M. Watanabe, K. Sasaki, Thermostable and alkaline-tolerant microbial cellulase-free xylanases produced from agricultural wastes and the properties required for use in pulp bleaching bioprocesses: a review. Process Biochem. **38**, 1327–1340, 2003/04/30/ (2003)
88. T. Yu, S. Anbarasan, Y. Wang, K. Telli, A.S. Aslan, Z. Su, et al., Hyperthermostable *Thermotoga maritima* xylanase XYN10B shows high activity at high temperatures in the presence of biomass-dissolving hydrophilic ionic liquids. Extremophiles. **20**, 515–524, 2016/07/01 (2016)
89. K. Choengpanya, S. Arthornthurasuk, P. Wattana-amorn, W.-T. Huang, W. Plengmuankhae, Y.-K. Li, et al., Cloning, expression and characterization of β-xylosidase from *Aspergillus niger* ASKU28. Protein Express. Purif. **115**, 132–140, 2015/11/01/ (2015)
90. R. Fonseca-Maldonado, L.F. Ribeiro, G.P. Furtado, L.M. Arruda, L.P. Meleiro, J.S. Alponti, et al., Synergistic action of co-expressed xylanase/laccase mixtures against milled sugar cane bagasse. Process Biochem. **49**, 1152–1161, 2014/07/01/ (2014)
91. P. Vishwakarma, E. Sachdeva, A. Thakur, A.S. Ethayathulla, A. Goyal, P. Kaur, Deciphering the structural and biochemical aspects of xylosidase from *Pseudopedobacter saltans*. Int. J. Biol. Macromol. **291**, 139042, 2025/02/01/ (2025)
92. S. Subramaniyan, P. Prema, Biotechnology of microbial xylanases: enzymology, molecular biology, and application. Crit. Rev. Biotechnol. **22**, 33–64, 2002/01/01 (2002)
93. T. Collins, C. Gerday, G. Feller, Xylanases, xylanase families and extremophilic xylanases. FEMS Microbiol. Rev. **29**, 3–23 (2005)
94. M. Kapoor, L. M. Nair, and R. C. Kuhad, Cost-effective xylanase production from free and immobilized *Bacillus pumilus* strain MK001 and its application in saccharification of *Prosopis juliflora*. Biochem. Eng. J. **38**, 88–97, 2008/01/15/ (2008)
95. X. Li, L. Zhang, Z. Jiang, L. Liu, J. Wang, L. Zhong, et al., A novel cold-active GH8 xylanase from cellulolytic myxobacterium and its application in food industry. Food Chem. **393**, 133463, 2022/11/01/ (2022)
96. X. Liu, Y. Liu, Z. Jiang, H. Liu, S. Yang, Q. Yan, Biochemical characterization of a novel xylanase from *Paenibacillus barengoltzii* and its application in xylooligosaccharides production from corncobs. Food Chem. **264**, 310–318, 2018/10/30/ (2018)
97. Y. Zhang, X. Liu, M. Liu, L. Han, D. Zhao, H. Rao, et al., Enzymatic modification of whole wheat dough gluten matrix development and bread quality by a novel wheat arabino-xylanase from *Podospora comata* with its properties and substrate specificity mechanism. Int. J. Biol. Macromol. **310**, 142860, 2025/05/01/ (2025)
98. V. Kumar, T. Satyanarayana, Thermo-alkali-stable xylanase of a novel polyextremophilic *Bacillus halodurans* TSEV1 and its application in biobleaching. Int. Biodeterior. Biodegradation. **75**, 138–145, 2012/11/01/ (2012)
99. C.K. Kim, H.S. Choi, S.J. Lee, J.H. Lee, J.H. Lee, H.Y. Yoo, et al., Production of xylanase from a novel engineered Pichia pastoris and application to enzymatic hydrolysis process for biorefinery. Process Biochem. **65**, 130–135, 2018/02/01/ (2018)
100. D. Mu, P. Li, T. Ma, D. Wei, M. Montalbán-López, Y. Ai, et al., Advances in the understanding of the production, modification and applications of xylanases in the food industry. Enzym. Microb. Technol. **179**, 110473, 2024/09/01/ (2024)

Chapter 11
Applications of Xylanase in Food Processing

Pei-Gee Yap and Chee-Yuen Gan

Abstract Xylanase belongs to the glycosidase (GH) family which cleaves β-1,4-xylosidic linkages in xylan, a hemicellulose that forms the major component in plant cell walls into xylose and xylan oligosaccharides. This enzyme has broad applications in the food industry and the enzyme market shows robust potential due to increasing consumer demand for more natural food products. Case studies on the role of xylanase in various food processing sectors, such as baking, beverage production, prebiotics and animal feed are discussed. Various factors affecting xylanase activity, such as pH, temperature and activating/inhibiting agents are discussed and analyzed. Additionally, this chapter covers the microbial sources, production and purification methods of xylanase, with special highlights on the pros and cons of each method. This chapter also emphasizes the commercialization of xylanase, addressing safety considerations and regulatory aspects. Despite its broad utility, challenges related to optimizing enzyme stability and activity in harsh food processing conditions remain. Future research should focus on enhancing xylanase's functional properties to meet industrial requirements. This chapter concludes by emphasizing trends and opportunities for further research, focusing on developing more cost-effective production techniques and expanding xylanase's application for more sustainable and efficient food production processes.

Keywords Industrial applications · Xylanase activity · Enzyme stability · Enzyme production · Enzyme safety

P.-G. Yap (✉) · C.-Y. Gan (✉)
Analytical Biochemistry Research Centre (ABrC), University Innovation Incubator Building, SAINS@USM Campus, Universiti Sains Malaysia, Bayan Lepas, Penang, Malaysia
e-mail: pgyap@usm.my

C.-Y. Gan
e-mail: cygan@usm.my

A. Sattar Jatoi (ed.), *Xylanase from Microorganisms*,
https://doi.org/10.1007/978-981-95-0824-2_11

1 Introduction

Enzymes are biological catalysts that have been used in food production for thousands of years, since our ancestors accidentally discovered cheese when they stored milk in containers made from ruminant stomachs. Enzymes are systematically grouped into seven different classes based on the type of reaction they catalyze, namely hydrolases, oxidoreductases, lyases, transferases, ligases, isomerases and translocases according to the International Union of Biochemists and Molecular Biology (IUBMB). Hydrolases appear as the largest group of commercial enzymes in the world. Among them, endo-1,4-β-xylanase or sometimes referred as xylanase (EC 3.2.1.8), belongs to the glycoside hydrolase (GH) family which hydrolyze glycosidic bonds between two or more carbohydrates or between a carbohydrate and a non-carbohydrate moiety [1]. Specifically, xylanases cleave β-1,4-xylosidic linkages in xylan, a hemicellulose that forms the major component in plant cell walls into xylose and xylan oligosaccharides. According to the Carbohydrate-Active enZYmes (CAZy) database (http://www.cazy.org/), various GH families are associated with xylanase activity including 5, 7, 8, 9, 10, 11, 16, 26, 30, 43, 44, 52, and 62 [2]. Each family features different physicochemical properties, three-dimensional folds, substrate specificities and catalytic mechanisms, giving rise to two different catalytic mechanisms, i.e., retention and inversion of the anomeric configuration at the glycosidic bond [1, 2].

Initially discovered in 1955, xylanases are recognized for their role in degradation of lignocellulose biomass for diverse applications including biofuel production, pulp bleaching for paper production and desizing and scouring of textiles [1, 3, 4]. In the food industry, xylanases are widely used in baking, beverage processing, prebiotics production and animal feed processing. Growing consumer preference for more natural food products and acceptance to enzyme-based applications have nevertheless driven the market growth of food grade xylanase, which is forecast to reach USD 33 billion by 2033, growing at a compound annual growth rate (CAGR) of 5.4% from 2023. It is therefore crucial to have a closer look into the current status of research and application of xylanase in the food processing industry. This chapter explores the industrial applications of xylanase in food processing, including case studies focusing on the enzyme's role in enhancing bakery products, clarifying beverages, producing prebiotics and improving animal feed digestibility. Additionally, the chapter delves into factors influencing xylanase activity, such as pH, temperature and activators/inhibitors which are crucial during industrial process design to support optimum enzyme performance. An overview of xylanase production techniques is discussed, followed by extraction and purification techniques for enhanced enzyme production. Finally, real-world commercial xylanase products for food processing are elaborated, with special highlights on safety considerations and future research directions.

2 Industrial Applications and Case Studies

Due to the inherent ability to breakdown hemicellulose (specifically xylan) into simpler sugars, xylanases have become increasingly more essential in the food processing industry. Application of xylanases could effectively enhance product quality, reduce processing costs, increase productivity and more importantly, minimize the use of chemical additives that raise safety concerns in consumers. This versatile biological agent offers wide functionalities, i.e., improving dough quality in baking, facilitating juice clarification and beer processing, producing prebiotics for gut health and boosting nutrient digestibility in animal feed. In this section, key examples of food products and processes utilizing xylanase are discussed. Case studies of food processing leveraging this enzyme from different sources are summarized in Table 1.

2.1 Bakery

In the baking industry, xylanase is added to enhance dough properties by modifying the arabinoxylan present in wheat flour. Arabinoxylans are non-starch polysaccharides although present in minute amount in wheat flour (generally 2–3%), unfavorable for bread making. It is classified as soluble water-extractable arabinoxylan (WE-AX) and insoluble water-unextractable xylan (WU-AX). The former is crucial to stiffen the dough and stabilize gas cells to improve loaf volume to produce airy and soft bread. Conversely, the latter has a strong water holding capacity and competes with gluten for water, which is detrimental to proper gluten matrix formation. High content of WU-AX hence gives rise to sticky, inconsistent and less elastic dough that is hard to knead and shape. In addition, the baking process involves moisture loss and starch retrogradation which affects the final texture of bread. During the baking process, the temperature of the dough rises. Starch granules gelatinize to set the structure of bread crumb. Moisture loss from the surface forms the crust while water retained by the gelatinized starch network provides a soft and spongy texture. As the bread cools after baking, the starch molecules begin to retrograde (i.e., realign and recrystallize), contributing to the firmness of bread. This is where xylanase comes in place to hydrolyze the linear arabinoxylan backbones to reduce the amount of WU-AX to improve bread crumb structure, loaf volume, elasticity, extensibility and water absorption, making it a key component in bread-making formulations.

Yegin et al. [12] reported that a xylanase from *Aureobasidium pullulans* when supplemented at 125 U/100 g wheat flour, improved the bread volume by 30% and maintained the moisture content within the ideal range of 35–40%. The xylanase also significantly reduced the firmness and gumminess by ~75% and increased the springiness by 111%. These effects are more pronounced than commercial xylanases. In another study, 5 ppm of *Chaetomium* sp. xylanase improved the rheological properties of bread which is indicated by reduction of torque values of water absorption

Table 1 Applications summary in the food processing industry of xylanases from different sources (2013–2024)

Application	Microbial origin	Molecular weight (kDa)	Expression host	Production method	Enzyme activity (U/mL)	Optimal pH	Optimal temperature	Activating agent	Inhibiting agent	Reference
Bakery	*Thermotoga maritima*#	ND	*P. pastoris*	SmF	1.52×10^6 U/mL	5.0	100	ND	ND	[5]
	Sorangium cellulosum#	50.8	*E. coli*	SmF	ND	6.0	50	Ba^{2+}, Ca^{2+}, Mg^{2+}, DTT	Zn^{2+}, Cu^{2+}, Co^{2+}, Mn^{2+}, Cr^{3+}, SDS, MeOH, EtOH	[6]
	Chaetomium sp.#	20.2	*A. niger*	SmF	3824 U/mL	7.0	70	–	Ag^{+}, Ni^{2+}, Cu^{2+}	[7]
	Halolactibacillus miurensis#	45	*P. pastoris*	SmF	550.68 U/mL	6.5	45	ND	ND	[8]
	Anaeromyces robustus#	44	*P. pastoris*	SmF	134.49 U/mL	5.5	40	Ba^{2+}	Fe^{2+}, Mn^{2+}, Co^{2+}, Zn^{2+}, Cu^{2+}, SDS, EDTA	[9]
	Orpinomyces sp. PC-2#	25.0	*E. coli*	SmF	7.73–12.8 U/g	3, 5	60	ND	ND	[10]
	Streptomyces sp. S27#	20	*P. pastoris*	SmF	82 U/mL	6.0	75	Al^{3+}, NH^{4+}, Ba^{2+}, Mg^{2+}, β-ME	Cu^{2+}, Hg^{+}, Ag^{+}, Mn^{2+} Zn^{2+}, Fe^{2+}, SDS, Xylose	[11]

(continued)

Table 1 (continued)

Application	Microbial origin	Molecular weight (kDa)	Expression host	Production method	Enzyme activity (U/mL)	Optimal pH	Optimal temperature	Activating agent	Inhibiting agent	Reference
	Aureobasidium pullulans NRRL Y-2311–1	21.6	–	SmF	ND	4.0	30–50	ND	ND	[12]
	Trichoderma stromaticum	22–34	–	SSF	9233.2–35,937.5 U/mg	5.0	50–60	Ag^{2+}, Zn^{2+}, Mn^{2+}, Co^{2+}, β-ME	Fe^{2+}, Hg^{2+}, Mg^{2+}, SDS, EDTA	[13]
	Bacillus halodurans TSEV1	40	–	SmF	69 U/mL	9.0	80	ND	ND	[14]
Brewery	*Pseudothermotoga thermarum*#	130	*E. coli*	SmF	151.9 U/mg	5.5	95	K^+. Ni^{2+}, Ca^{2+}, Co^{2+}, Mg^{2+}	Cu^{2+}, SDS	[15]
	Thermotoga maritima#	40	*E. coli*	SmF	18,000 U/mL	5.5	90	ND	ND	[16]
	Chaetomium sp. CQ31#	44.9	*P. pastoris*	SmF	102 U/mL	6.5	85	ND	ND	[17]
	Neocallimastix sp. GMLF1#	27	*E. coli*	SmF	ND	7.5	55	ND	ND	[18]
	Gloeophyllum trabeum CBS 900.73#	48	*E. coli*	SmF	ND	4.5	75	ND	ND	[19]
	Humicola insolens Y1#	41–44	*E. coli*	SmF	16.8 U/mL	6.0–7.0	70–80	ND	ND	[20]
Juice clarification	*Bacillus altitudinis* JYY-02#	27.98	*E. coli*	SmF	1105 U/mg	6.5	50	Cu^{2+}, Fe^{3+}	K^+, Zn^{2+}, Mg^{2+}	[21]

(continued)

Table 1 (continued)

Application	Microbial origin	Molecular weight (kDa)	Expression host	Production method	Enzyme activity (U/mL)	Optimal pH	Optimal temperature	Activating agent	Inhibiting agent	Reference
	Aspergillus niger GIO	ND	–	SSF	1.35–3.18 U/ml	5.0	40	Ca^{2+}, Fe^{2+}, Na^{+}, K^{+}, SDS, Urea	EDTA	[22]
	Bacillus megaterium	ND	–		1.93–3.13 U/ml	6.2	40	Ca^{2+}, Fe^{2+}, Na^{+}, K^{+}, SDS, Urea	EDTA	
	Bacillus altitudinis	43	–	SmF	82.664 U/mL	5.0	40	Zn^{2+}, Mg^{2+}, Mn^{2+}	Fe^{2+}	[23]
	Bacillus licheniformis	40	–		153.59 U/mL	4.0	40	Zn^{2+}, Mn^{2+}	Fe^{2+}, Fe^{3+}	
	Geobacillus vulcani GS90	40.1	–	SmF	1671.4 U/mg	8.0	55	Mn^{2+}, Co^{2+}, Ca^{2+}, Acetone, Chloroform	Fe^{3+}, Cd^{2+}, SDS, DTT, β-ME, CHAPS	[24]
	Fusarium sp. 21[#]	23–31	*E. coli*	SmF	ND	5.0–6.0	45	Na^{+}, Zn^{2+}, Mn^{2+}, Mg^{2+}, Ni^{+}, Fe^{2+}, Cu^{2+}, Mn^{2+}, Mg^{2+}	Ca^{2+}, Mn^{2+}, Ba^{2+}, β-ME, SDS	[25]
	Pediococcus acidilactici GC25	48.15	–	SmF	318.45 U/mg	7.0	40	–	–	[26]

(continued)

Table 1 (continued)

Application	Microbial origin	Molecular weight (kDa)	Expression host	Production method	Enzyme activity (U/mL)	Optimal pH	Optimal temperature	Activating agent	Inhibiting agent	Reference
	Aspergillus japonicus (UFMS 48.136)	32	–	SmF	448.3 U/mg	5.0–6.0	50–60	Mn^{2+}, β-ME, Tween-20, Tween-80, Triton X-100	SDS	[27]
	Aspergillus niger	36	–	SSF	572 U/mg	5.0	35	ND	ND	[28]
Prebiotics production	*Clostridium boliviensis* E-1[#]	52.9	*E. coli*	SmF	ND	7.0–9.0	65	Mg^{2+}, Fe^{2+}, Ca^{2+}	Cu^{2+}, EDTA	[29]
	Trichoderma asperellum ND-1[#]	24–36	*E. coli*	SmF	78.8–446.2	6.0	50–65	Mg^{2+}, NH^{+}, Ni^{2+}, Li^{+}, K^{+}, Ca^{2+}, Co^{2+}, Ba^{2+}	Cd^{2+}, Al^{3+}, Fe^{3+}, Mn^{2+}, Cu^{2+}, SDS, EDTA	[30]
	Bacillus sp.	30	–	SmF	31.3 U/mg	4.6	40	K^{+}, Na^{+}	Cu^{2+}, Zn^{2+}, Fe^{3+}	[31]
	Bacillus paralicheniformis[#]	–	*E. coli*	SmF	504.12 U/mL	8.0	50	ND	ND	[32]
	Bacillus subtilis KCX006[#]	–	*E. coli*	SmF	ND	6.6	50	ND	ND	[33]
	Aspergillus oryzae LC1	35	–	SmF	432.9 U/mL	5.0	30	Fe^{2+}, Ag^{2+}, Mg^{2+}, Mn^{+}, Co^{+}	Cu^{2+}, Hg^{2+}, Zn^{2+}, SDS, EDTA, DMSO	[34]

(continued)

Table 1 (continued)

Application	Microbial origin	Molecular weight (kDa)	Expression host	Production method	Enzyme activity (U/mL)	Optimal pH	Optimal temperature	Activating agent	Inhibiting agent	Reference
	Pichia stipites	31.6	–	SSF	62.5 U/mg	6.0	50	ND	Hg^{2+}, Co^{2+}, Pb^{2+}, Fe^{3+}, Mn^{2+}, Ag^{+}, Fe^{2+}, SDS, Na^{+}, Mg^{2+}, Zn^{2+}, Ca^{2+}, EDTA	[35]
	Paenibacillus barengoltzii#	114–115	*E. coli*	SmF	292.5 U/mg	6.5	60	Mg^{2+}, Ca^{2+}, Ba^{2+}, Co^{2+}	Hg^{2+}, Fe^{3+}, Fe^{2+}, Cr^{3+}	[36]
	Mycothermus thermophilus#	23.3–29.2	*P. pastoris*	SmF	136–377 U/mL	6.0–6.5	65	Ca^{2+}, Ni^{2+}, Ag^{+}, β-ME	SDS	[37]
	Talaromyces amestolkiae	19.86	–	SmF	118.3 U/mg	3.0–4.0	50	ND	ND	[38]
	Bacillus mojavensis UEB-FK	14	–	SmF	250.02 U/mg	4.0	50	Fe^{2+}, Mg^{2+}, Ca^{2+}	Zn^{2+}, Co^{2+}, Mn^{2+}, Cu^{2+}, Hg^{2+}, SDS, EDTA	[39]
Animal feed	*Trichoderma koningii*	ND	–	SSF	31.17 U/mL	4.5	50	ND	ND	[40]
	Trichoderma pseudokoningii	ND	–		57.98 U/mL	4.5	55	ND	ND	
	Trichoderma piluliferum	ND	–	SSF	30.18 U/mL	4.5	50	ND	ND	[41]
	Trichoderma viride	ND	–		67.18 U/mL	6.0	45	ND	ND	

Note #, Recombinant expression; –, Not relevant; β-mercaptoethanol, β-ME; EDTA, Ethylenediaminetetraacetic acid; ND, Not determined; SDS, Sodium dodecyl sulfate; SmF, Submerged fermentation; SSF, Solid-state fermentation capacity, protein weakening, starch gelatinization, stability of starch granules and starch retrogradation. These results subsequently minimized the duration required for gluten network establishment and settlement, as well as contributed to anti-staling property of the dough [7]. In terms of texture, the final volume of the xylanase-treated bread increased by 25.6% while the hardness was almost halved compared to the control. In short, xylanases function to hydrolyze WU-AX to induce redistribution of water within the gluten matrix, allowing more extensive water-gluten interactions to form more extensible dough that ultimately contribute to better texture and higher quality breads [10].

2.2 Beverage Processing

Raw fruit juices contain cellulose, hemicellulose, starch, pectin, xylan and other undissolved solids. These complex constituents contribute to the viscosity and turbidity of juices. While the suspended materials are typically removed via centrifugation or filtration, xylanase is utilized to break down hemicellulose and xylan in the fruit pulp to facilitate better filtration and extraction, leading to clearer and more appealing juices. Clarity is usually evaluated by monitoring the decrease in turbidity of juices. The extent of cell wall degradation is more specifically assessed by determining the reducing sugar content in juices after xylanase treatment. Usually, a decrease in juice turbidity is accompanied by an increase in reducing sugar content, as more hemicellulose are degraded into sugar units. This is observed during the clarification of orange, pomegranate, apricot, peach, apple, grape and kiwi juices by *Pediococcus acidilactici* GC25 xylanase. At 0.15 U/mL, the xylanase reduced ~25% turbidity and increased ~22% reducing sugar content of peach juice [26]. In a recent study, *Bacillus altitudinis* JYY-02 xylanase having an optimal pH of 6.5 was employed to clarify fresh apple, orange, grape and nectarine juices. Maximum clarification (87.1%) accompanied by highest increment of reducing sugar content (11.0 μg/mL) was observed in apple juice after 12 h treatment. In contrast, minimum clarification (<60%) with lowest increment of reducing sugar content (<2 μg/mL) was observed in nectarine and orange juices [21]. The difference in results was ascribed to the pH and type of polysaccharides presence in the juices. For instance, apples are rich in hemicellulose which is favorable for xylanase action whereas the relatively higher acidity of nectarines and oranges may fall out of the pH tolerance range of xylanase (pH 5–8), leading to poorer enzyme performance.

Apart from juice clarification, xylanase plays a crucial role in the brewing industry. During the mashing process, xylanase hydrolyses complex arabinoxylans present in barley, wheat, rye or sorghum malt into fermentable sugars in the liquid wort. This step is crucial as a large amount of arabinoxylan negatively impacts the viscosity and turbidity, hinder filtration process and reduce efficiency of mash separation. For instance, 200 U/g malt of *Chaetomium* sp. CQ31 xylanase improved beer production and quality by shortening the filtration time by 42.3% and reducing the wort viscosity by 8.6% [17]. Notably, xylanase inhibiting proteins (XIPs) are found in plants that particularly inhibit fungal xylanases. XIPs are plant-derived proteins that specifically target fungal xylanases and play defensive roles against phytopathogen invasion [42]. Their presence in cereal grains like wheat, rye and barley poses a concern in beer processing. For instance, *Trichoderma harzianum* and *Neocallimastix* sp. GMLF1 xylanases were applied during the malt saccharification process in the presence and absence of 0.4 mg XIP/g malt. The turbidity of mash treated by XIP-resistant *Neocallimastix* sp. GMLF1 xylanase was significantly reduced whereas negligible changes were observed for XIP-susceptible *Trichoderma harzianum* xylanase compared to the control [18]. It is therefore crucial to screen and select XIP-resistant xylanase to ensure consistent yield and quality of beer.

2.3 Prebiotics Production

Xylanase hydrolyses β-1,4-glycosidic bonds in xylan to produce non-digestible xylooligosaccharides (XOS), which has gained attention as prebiotics to promote gut health. XOS was reported to selectively promote proliferation of intestinal microorganisms. For example, *Bacillus paralicheniformis* xylanase was capable of hydrolyzing xylan into xylobiose, xylotriose and xylotetraose. These XOS supported the growth of four probiotic strains including *Levilactobacillus brevis*, *Enterococcus faecium*, *Pediococcus acidilactici* and *Weissella confusa* among the twenty bacteria tested. Probiotic growth was reflected by the changes in culture medium turbidity and pH [32]. While higher microbial growth increased the medium turbidity, a drop in medium pH was attributed to production of organic acids as the final product of bacterial fermentation. To be more specific, short-chain fatty acids (SCFA) such as acetic acid, lactic acid, succinic acid and propionic acid are typical biomarkers for probiotics species. Nieto-Domínguez et al. [38] ascribed the presence of these SCFAs with increased populations and metabolism of *Bifidobacterium* and *Staphylococcus hominis* in the fecal samples of breast-fed babies supplemented with XOS. The former bacteria are well-known probiotics to improve digestive health whereas the latter bacteria produce bacteriocin hominicin that is effective against pathogenic methicillin-resistant and vancomysin-intermediate *Staphylococcus aureus* (MRSA and VISA) strains [43]. XOS administration also reduced the abundance of disease-causing pathogens *Clostridium perfringens*, *Citrobacter freundii* and *Actinomyces radingae*. The XOS with prebiotic effects is therefore of future interests to be incorporated into infant formula to stimulate probiotic growth and concomitantly suppressing the growth of harmful species.

Notably, enzymatic hydrolysis is a popular method for XOS production. Xylanase is more specific in cleaving xylan to reduce undesired by-products compared to autohydrolysis [44]. Enzymatic approach is also versatile in the sense that it may be coupled with other pretreatment methods for more effective XOS production. For instance, corncob xylan that was first subjected to steam explosion using acidic electrolyzed water followed by *Paenibacillus barengoltzii* xylanase hydrolysis, has yielded almost 75% mg XOS/g xylan. Pretreatment with steam explosion has significantly improved XOS production as the high temperature and pressure aided decomposition of lignocellulose structure in corncob xylan, exposing more β-1,4-glycosidic bonds for higher xylanase accessibility [36]. On the other hand, recycling of xylanases has been reported to improve process economy during XOS production. This is done by immobilization of the enzyme onto various solid matrices. For instance, *Bacillus subtilis* KCX006 xylanase was immobilized on activated carbon, silica and zirconia nanoparticles then compared with free xylanase for XOS production. Immobilization was found to improve thermostability of xylanase over 50–65 °C compared to the free enzyme. Significantly higher yield of short chain XOS was achieved by the immobilization xylanase than the free enzyme using sorghum biomass (13–19%) as substrate. Moreover, the immobilized enzymes could be reused up to eight times at 50 °C, 12 h for each batch of process without significant activity loss for XOS production [33]. In short, emerging technologies are expected to support greener and more efficient production of prebiotics by xylanase as more research reveals the health benefits of XOS.

2.4 Animal Feed Processing

Sorghum, wheat, barley, corn and other agro-industrial wastes are common feedstocks for animal husbandry. However, animals are unable to fully digest due to the lack of specific enzymes in their digestive tracts, often leading to intestinal distress and slow growth. Feed supplement with xylanase, has demonstrated enhanced degradation of complex arabinoxylan to reduce digesta viscosity which subsequently boosts feedstock digestibility and nutrient absorption in animals [45]. This is essential because the degradation of WU-AX into WE-AX contributes to digesta bulk that triggers secretion of cholecystokinin and various enzymes to modulate satiety as well as hydrolysis of fats and proteins. For example, *Trichoderma piluliferum* and *T. viride* xylanases improved the in vitro digestibility of hemicellulose-dense forages, i.e., Tifton185 Hay, corn silage and sugarcane silage in bovine feeding by 5–12%. The enzyme breaks down hemicellulose chains to smaller counterparts that are more available for fermentation by ruminal microbiota. XOS is concomitantly generated, and this is beneficial for probiotics colonization and proliferation to strengthen gastrointestinal health [41]. Similarly, *T. koningii* and *T. pseudokoningii* xylanases increased the in vitro digestibility of cellulose, hemicellulose and lignin complex in Mombaça grass by 12.4 and 10.6%, respectively [40].

Besides, xylanase was used in combination with other enzymes or biomolecules to promote livestock health. Duarte et al. [46] studied the effect of xylanase, protease and xylanase-protease combination treatments on newly weaned pig models. After 24 days, results showed that the combined supplementation significantly improved average daily gain and gut morphology compared to using either enzyme alone. Xylanase reduced 13% digesta viscosity and oxidative stress indicated by lower malondialdehyde concentration in jejunum mucosa while protease improved feed efficiency, increased villus height in jejunum and reduced crypt depth. The combination of xylanase with protease enhanced growth performance and gut health in newly weaned pigs. In another study, combined treatment of xylanase and fermented polysaccharides of *Hericium caputmedusae* demonstrated significant improvement in broiler chicken health by enhancing antioxidants (total-antioxidant capacity, superoxide dismutase and glutathione peroxidase) and anti-inflammatory (increased IL-1ra and IL-10; reduced IL-1β, TNF-α) activities. The treatment also boosted beneficial gut microbiota including *Bacillus* sp. and *Lactobacillus plantarum* and reduced prevalence to pathogens such as *Staphylococcus* strains [47].

3 Factors Affecting Xylanase Activity in Food Processing

Enzymatic activity of xylanase determines its efficiency and suitability for different applications in food processing. The activity of xylanase is influenced by several factors, such as molecular weight, pH, temperature, presence of metal ions and substrate specificity, as summarized in Table 1. Understanding these factors is essential to allow optimal enzyme performance in various food matrices.

3.1 pH

Most microbial xylanases are stable in slightly acidic to neutral environments, i.e., pH 4.5–6.5, although some demonstrated maximum activity in more acidic or alkaline conditions. This property is suitable for food processing applications that require enzyme activity in extreme settings. Taking the beverage industry for example, xylanase is deployed to break down xylan-containing hemicellulose in fruit pulp to reduce the viscosity and turbidity of natural fruit juices. Since fruit juices are acidic at pH 3–4, the working enzyme must maintain stability and catalytically active under this acidic condition. Acidophilic xylanases thus have an edge on juice clarification. For instance, the xylanases derived from *B. altitudinis* and *B. lichenformis* have an optimal pH of 4.0 and are stable over pH 3–8. The xylanases reduced the turbidity of apple, orange, lime, pineapple and kiwi juices by 20–80% and concomitantly increased the total reducing sugar and soluble solid contents in the juices [23]. On the other hand, alkalophilic xylanases are stable and active under alkaline conditions. They are useful in processes where a higher pH is required, for example, in baking. Alkalophilic xylanase may assist in improving dough handling and bread quality by modifying the arabinoxylans in the flour. This can lead to enhanced water absorption and better dough stability. In an earlier study, cane molasses was used by *B. halodurans* TSEV1 to produce a thermo-alkaliphilic xylanase which exerted peak activity at pH 9. Supplementation of the enzyme resulted in softer bread with longer shelf life compared to the control. The xylanase-treated bread gave higher reducing sugar and protein contents. Xylooligosaccharides which are beneficial to intestinal health were also detected in the bread as the result of xylanase action on non-starch polysaccharides [14]. These results indicated that xylanase could aid in the enhancement of bread texture, shelf life and nutritional profile.

Notably, alkalophilic and acidophilic xylanases may still be considered for juice clarification and dough improvement, respectively. The xylanase produced by *Geobacillus vulcani* GS90 has an optimum pH of 8.0 and it showed superior performance to a commercially available β-1,4-xylanase in the enrichment and clarification of apple and orange juices. This could be due to the wide pH tolerance of the enzyme as it could maintain >60% activity at pH 3–9 even when challenged at 90 °C and 120 min [24]. Similarly, the fungus *Orpinomyces* sp. PC-2 secreted an acidophilic xylanase with an optimum pH of 3.0 yet it was effective improving the dough quality. The sponge volume almost doubled whereas the hardness level halved in the presence of xylanase compared to the control. The gumminess and chewiness of the whole-wheat sponge were also reduced by 40% [10]. The wide pH stability of the xylanase (>60% activity at pH 3–8) may have contributed to its positive activities in the modification of bread structural and rheological properties.

3.2 *Temperature*

Temperature is another critical factor affecting xylanase performance. Enzymatic activity gradually increases with temperature up to an optimal point, beyond which the activity drops and halts as the enzyme denatures at higher temperatures. Referring to Table 1, most xylanases show optimal activity at mesophilic temperatures between 40 and 60 °C. This is especially advantageous in the feed processing industry since mesophilic xylanases could function well in the gastrointestinal tract of animals. For instance, the xylanase containing extract from *T. koningii* and *T. pseudokoningii* improved the digestibility of cellulose, hemicellulose and lignin complex in an artificial rumen where temperature is maintained at 39 °C by 12.4 and 10.6% compared to the control, respectively [40]. In another similar study, the enzyme extract from *T. piluliferum* and *T. viride* enhanced the digestibility of various forages by 5–12% [41]. These xylanases retained >50% catalytic activity when challenged at 40 °C for 1 h. Moreover, the half-life of *T. koningii* and *T. pseudokoningii* were 27 and 56 min at 55 °C whereas that of *T. piluliferum* and *T. viride* were 25 and 60 min at 45 °C, respectively. According to the authors, supplementation of thermostable xylanase in bovine diets has aided the breakdown of complex xylan into simple sugars, which can be then easily utilized by the ruminal microbiota for synergistic digestion of the plant cell wall.

Apart from mesophilic xylanases, psychrophilic and thermophilic variants were also found. Psychrophilic enzymes maintain high catalytic performance at cold temperatures (0–20 °C) whereas thermophilic enzymes thrive at elevated temperatures (50–80 °C). The temperature preference of the enzymes lies in their structural differences, whereby the former is more flexible characterized by having more exposure of hydrophobic surfaces and less salt bridge interactions. Flexibility of the substrate-binding hotspots are also among the adapted features that allows optimal functioning of psychrophilic xylanase at low temperatures [48]. The application of psychrophilic and thermophilic xylanases has shown positive results in food processing [49, 50]. In the bakery industry, the cold-adapted xylanase from *Sorangium cellulosum*, for example, retained >20% of its maximum catalytic activity at 20 °C. It has significantly boosted the bread volume by 15% at 0.2 mg/kg, which is superior to that of commercial xylanase (9% increment at 30 mg/kg) at lower dosage volume. Enhancement of other textural properties of bread such as hardness, chewiness, springiness, resilience and cohesiveness are also achieved by this enzyme [6]. In contrast, a thermophilic xylanase isolated from the thermophilic bacteria *Thermotoga maritima* exhibited maximum activity 1.52×10^6 U/mL at 100 °C. It has improved bread volume, reduced the hardness, chewiness and gumminess by approximately 55%, and concomitantly maintained the texture and quality of bread during prolonged storage [5]. Supplementation of another thermophilic xylanase from *Streptomyces* sp. S27 at 75–300 U/kg also resulted in softer bread (64–65%) with higher reducing sugar contents (24–58%) compared to the control [11]. In addition, thermophilic xylanases are favored in beer production, as they exhibited optimal temperature of 70–95 °C (Table 1). Since the mashing process typically occurs at elevated temperatures for prolonged duration, i.e., 45 °C for 15 min, 50 °C for 30 min, 60 °C for 30 min and 70 °C for 60 min, thermostable and thermophilic xylanases are more suited to the harsh condition. For instance, *Gloeophyllum trabeum* CBS 900.73 xylanase (80 U) improved the filtration rate and concomitantly reduced the mash viscosity by 31.3 and 12.8%, respectively [19]. At the same dosage, the *Humicola insolens Y1* xylanase improved the filtration efficiency by 25–35% and reduced mash viscosity by 8–12%. These results outperformed that of the commercial enzyme Ultraflo® (28.2 and 9.0%, respectively) that is used to reduce viscosity and turbidity of beers [20].

3.3 Activating and Inhibiting Agents

The presence of metal ions, surfactants and other chemicals during industrial processing may significantly influence xylanase activity. Understanding these effects is essential for optimizing xylanase applications in food processing. The following trends were observed based on Table 1. For metal ions, Ca^{2+} was often found to stimulate xylanase activity. Ca^{2+} plays a role in maintaining the three-dimensional conformation of enzyme active site to ensure proper substrate recognition and catalysis [51]. For instance, the catalytic activity of *Sorangium cellulosum* xylanase supplemented with 5 mM Ca^{2+} has improved by 26% compared to the control [6]. The presence of Ca^{2+} may also have positive effect on thermal stability of the enzyme. This is exemplified by *Clostridium cellulovorans* xylanase whereby administration of 5 mM Ca^{2+} boosted enzyme activity by 240% and the enzyme retained 70% activity when challenged at 70 °C for 30 min [52]. On the other hand, Hg^{2+} and Cu^{2+} tend to strongly inhibit xylanase activity. Both ions completely inactivated *Bacillus mojavensis* UEB-FK xylanase at 5 mM while the *Bacillus amyloliquefaciens* xylanase treated with 10 mM of the ions remained only 12–22% activity [39, 53]. According to the authors, these heavy metal ions may interact with the functional groups of reactive amino acids in the active site (e.g. thiol, indole and carboxyl groups), hence suppressing catalytic performance. It is also worth noting that metal ions such as Mn^{2+}, Co^{2+}, Fe^{2+} and Ag^{+} exhibited both activating and inhibiting effects to xylanases. Limited information is available on the mechanism that gives rise to variation of results, yet it is postulated that the presence of these metal ions induces conformational alterations in the active site that either favor or disfavor xylanase activity depending on the sources of enzyme [54, 55].

Surfactants like β-mercaptoethanol (β-ME) and detergents like sodium dodecyl sulfate (SDS) are used to denature proteins during pretreatment. β-ME was found to be an activating agent of various xylanases, for instance, introduction of 5 and 10 mM β-ME stimulated the activity of *Mycothermus thermophilus* and *Aspergillus japonicus* xylanases by 145.7 and 40.52%, respectively [27, 37]. Since β-ME works by reducing cysteine-cysteine disulfide bridges, it may prevent oxidation of thiol functional groups to maintain or restore the native structure of enzyme active site, leading to enhanced enzyme performance [53, 56]. In contrast, SDS strongly inhibits xylanase. The presence of 5 mM SDS reduced approximately half the activity of *Trichoderma asperellum* ND-1 and *Fusarium* sp. 21 xylanases, as reported by Li et al. [25] and Zheng et al. [30]. The anionic detergent binds and coats proteins with negative charges. As a result, ionic charges and hydrophobic interactions essential to maintain enzyme structure integrity were distorted, leading to the unfold and denaturation of xylanase. Non-ionic surfactants such as Triton and Tween, on the other hand, conferred positive effects to xylanase activity. For example, *Aspergillus japonicus* xylanase activity was improved 1.5–2.0-fold by 10% Triton X-100, Tween-20 and Tween-80 [27]. Next, ethylenediaminetetraacetic acid (EDTA) is a strong metal chelating agent shown to inhibit xylanase activity by binding or removing essential metal ions that is required for catalysis. Addition of 5 mM EDTA reduced the activity of *Aspergillus oryzae* LC1 xylanase by 36.7% and in a more extreme case, no residual activities of *Aspergillus niger* GIO and *Bacillus megaterium* xylanases were detected [22, 34]. EDTA-induced enzyme inhibition is an indirect way indicating xylanases function as metalloenzymes.

4 Sources of Xylanase and Production Methods

Over the past decade, significant research has focused on the exploration and optimization of the xylanase production from different biological origins. This section aims to provide an overview of the sources of enzymes, emphasizing microbial xylanases due to the high efficiency and ease of production over their animal or plant counterparts. Additional discussions on xylanase extraction and purification techniques will also be included in this section. Understanding the origin of xylanase, along with its production methods, could nevertheless help to appreciate its broad applications and inspire the development of commercial formulation that enhance the efficiency of food processing.

4.1 Sources of Xylanase

According to earlier literature, xylanase could be found in prokaryotes and eukaryotes [57–59]. Based on Table 1, fungi represent the most common source of xylanase, with *Aspergillus, Trichoderma* and *Chaetomium* being the most popular genera. The filamentous fungi *A. niger* has been utilized as a bioconversion agent to produce organic acids, alcohol and fermented products in the food industry. With a positive track record in terms of its safety and stability, *A. niger* and *A. niger* products are granted Generally Regarded As Safe (GRAS) status by the Food and Drug Administration (FDA) [60]. On the other hand, *Trichoderma* sp. is known for its ability to produce various lignocellulolytic enzymes in addition to xylanase, such as β-xylosidase, cellulase, β-mannanase and α-L-arabinofuranosidase [61]. It is more popular in the agriculture industry as a biocontrol agent for plant pathogens, plant growth promoter, decomposition agent and bioremediation agent [62]. In the food industry, its secondary metabolite, i.e., 6-pentyl-α-pyrone, gives off a coconut-like aroma which has potential as a flavoring agent or adjuvant in food [63]. *Chaetomium* sp. is a thermophilic fungus that produces thermostable xylanase, whereby it exhibits optimum catalytic activity at 60–80 °C. This is favorable for industrial applications not only because the enzyme can withstand harsh processes but also accelerate catalytic reactions at elevated temperatures [64].

Apart from fungi, bacteria are also common producers of xylanase, such as those from the genera *Bacillus, Halobacillus, Sorangium, Thermotoga, Pseudothermotoga, Geobacillus, Pediococcus, Paenibacillus* (Table 1). *Bacillus* sp. dominated other groups of bacteria as the host for xylanase production, which could be due to several advantageous traits [65]. Firstly, *Bacillus* sp. is a cosmopolitan bacterium that is easy to isolate and identify. It exhibits more rapid growth and shorter fermentation periods in comparison to yeasts and fungi, allowing for more efficient enzyme production in a shorter time frame. Moreover, *Bacillus* sp. produces extracellular enzymes which streamline the extraction and purification processes, thereby reducing the overall production costs. In fact, numerous microbial enzymes including protease, esterase, lyticase, penicillinase, uricase and α-amylase are commercially produced by *Bacillus* sp. for research purposes. Additionally, *Bacillus* enzymes are known for their robustness and adaptability against broad pH and temperature ranges, making them highly suitable for industrial applications [66]. The natural advantages of *Bacillus* sp. could be further enhanced by genetic modification, i.e., introducing novel traits such as resistance to proteolytic degradation, or optimizing the overexpression of target enzyme to ultimately boost overall productivity [67–69].

Compared to fungi and bacteria, the xylanase produced by yeasts are less characterized and utilized for food applications. There is only one report on the use of *Pichia stiptis* xylanase for probiotics production (Table 1). Other xylan-degrading yeasts include *Blastobotrys mokoenaii, Sugiyamaella xylanicola, Apiotrichum sporotrichoides*, *Saitozyma podzolica* and *Candida tropicalis* [70–72] could be potential candidates for food processing. Notably, *P. stiptis* is a native xylose fermenting yeast which appears to be a common host for heterologous expression of xylanase. This is because yeasts are capable of post-translational modification for proper folding of proteins while offering high level of protein secretion [73]. Researchers may introduce multiple target genes, manipulate metabolic pathways and optimize expression levels more easily in yeasts compared to genetic engineering of fungi systems [74, 75]. Moreover, yeasts could thrive under aerobic and anaerobic conditions to achieve high biomass concentration, hence offering flexibility and scalability in production methods. These factors make yeast a more versatile and often more efficient choice for recombinant protein production [76, 77].

4.2 Fermentation Techniques for Xylanase Production

Over the past decade, fungi, bacteria and yeast were fed on lignocellulosic waste such as beechwood, birchwood, oat spelt, barley, wheat bran, corncob, rice straw, maize straw, sugarcane bagasse, cassava pulp, oil palm leaf, peach palm waste and spent coffee waste for sustainable xylanase production. This process is mainly carried out through submerged fermentation (SmF) or solid-state fermentation (SSF). Each method has its unique advantages and disadvantages, depending on the natural growth preference of the microorganism and substrate selected for xylanase production [78, 79].

Generally, SmF is the most common method for industrial-scale enzyme production in which approximately 90% commercial xylanase is produced through this method [58]. During SmF, microorganisms are grown in a liquid medium containing soluble substrate to facilitate nutrient diffusion. This is preferred by bacterial growth since it requires high water content to thrive. The most prominent advantage of SmF is that the inoculum concentration, pH, temperature, aeration, and nutrient concentration of the liquid medium can be precisely monitored, allowing consistent and scalable enzyme production. Response surface methodology (RSM) was employed to optimize the fermentation condition for optimum xylanase production. Using the Box-Behnken design (BBD) design of experiment, Güler & Özçelik [80] achieved 4.21-fold enhanced xylanase production by *Bacillus* sp. under the optimized condition at 30 °C, pH 8 and 127.27 rpm agitation. In another example, Shakir et al. [81] further optimized the medium additives through Plackett–Burman design followed by central composite design (CCD) whereby 1.5% banana peels, 0.3875% sodium nitrate, 0.8% peptone and 0.775% $MgSO_4$ gave the highest xylanase activity of 4.59 U/mL. However, SmF has several disadvantages such as moderate yield, high capital cost, requirement of skilled personnel, extra effort in maintaining sterile condition and generation of large liquid effluent [82].

On the other hand, SSF involves growing microorganisms on solid substrates without free-flowing water. Although less commercially utilized, SSF has several advantages over SmF for enzyme production. Fungi are the most suitable microbe for this method because of low moisture content in the fermentation vessel. This favors mycelial growth and at the same time minimizes the chances of contamination caused by miscellaneous bacteria. Notably, SSF tends to yield higher concentration of xylanase than SmF (Table 2). This is probably because direct fermentation of solid substrates in SSF resembles the natural growth condition of filamentous fungi, leading to enhanced xylanase production. Certain bacteria may have also adapted to grow well on solid substrates, particularly those isolated from soil where they naturally encounter lignocellulosic material [83–85]. Moreover, xylanase secreted by microbe is less diluted during SSF, since the nutrients are concentrated in the solid substrate instead of diluted in a suspension as in SmF. This reduces the overall cultivation cost, purification processes and energy consumption as SSF involves relatively simple instrumentation control and less generation of effluent [28, 35]. The disadvantages of SSF lies in poor reproducibility of the fermentation process and inconsistency of product quality. Scaling up of SSF is also facing the problem of difficulty in maintaining uniform temperature, moisture, and aeration throughout the solid medium.

It is also worth noting that recombinant technology is popular for xylanase production. The recombinant xylanase genes are typically identified and isolated from native xylan-degrading organisms and then inserted into a host microorganism for expression. The genes can be mined in public genetic databases such as GenBank (https://www.ncbi.nlm.nih.gov/genbank/) and NCBI (https://blast.ncbi.nlm.nih.gov/Blast) where the detailed sequence information is available. This allows researchers to explore the gene's structure, study potential modifications and even improve its characteristics, for example, enhanced thermal stability or activity under different

Table 2 Comparison studies on xylanase activity between submerged fermentation (SmF) and solid-state fermentation (SSF)

Microorganism	Substrate	Enzyme activity		Reference
		SmF	SSF	
Bacillus subtilis	Birchwood Wheat bran	2.62 U/mL	15.55 U/mL	[83]
Trichoderma reesei SAF3	Birchwood	4.75 U/mL	219 U/g	[84]
Bacillus pumilus AB-1	Oat spelt Wheat bran	177 U/mL	489.4 U/g	[85]
Bacillus subtilis ATCC 6633	Barley husk Wheat bran	11.646 U/mL	22.071 U/mL	[86]
Aspergillus niger	Wheat straw Rice husk	6658 U/g 6504 U/g	4549 U/g 5127 U/g	[87]
Bacillus arseniciselenatis DSM 15,340	Birchwood Wheat bran	245.78 U/mL	970.49 U/g	[88]

pH conditions to optimize its efficiency for specific industrial processes. After that, the gene fragment is engineered into the plasmid of the host such as *Escherichia coli*, *Pichia pastoris* or *Aspergillus niger* for mass production of xylanase. This method allows more precise control over enzyme production with higher production yield and consistency compared to conventional SmF and SSF. However, recombinant technology requires higher costs and expertise for specialized equipment, genetic engineering and maintaining optimal growth conditions for the host microorganism.

To summarize, SmF allows better control over the fermentation process to ensure product consistency. This aspect is especially crucial to scale up continuous production, whereby approximately 80% of commercial enzymes were produced using this method. For SSF, it is more cost-effective and gives a lower carbon footprint since it consumes less energy and water. SmF and SSF come with their own set of advantages and disadvantages, making them suitable for different production requirements. Ultimately, the choice between SmF and SSF depends on the type of microorganism, the desired enzyme yield and the available substrates and facilities that support xylanase production.

4.3 Extraction and Purification Techniques of Xylanase

Following the completion of fermentation process, xylanase is separated from the microbial mass, residual substrates and other fermentation by-products. Efficient extraction and purification are essential to obtain high yield of enzyme with desired purity. The enzyme may be secreted by the microorganism intracellularly or extracellularly, this gives rise to slightly different extraction processes. A comparison of these methods is summarized in Table 3.

In the case where xylanase is produced intracellularly, the first step involves cell disruption to release the enzyme. This is typically done using mechanical methods

Table 3 Comparison between precipitation, ATPS and chromatographic purification methods [89, 90]

Method	Purity	Yield	Scalability	Cost	Selectivity
$(NH_4)_2SO_4$ precipitation	Low	Moderate	High	Low	Low
Acetone precipitation	Moderate	High	Moderate	Moderate	Moderate
Aqueous two-phase system (ATPS)	Moderate	Very high	Very high	Low to moderate	Moderate
Ion-exchange chromatography (IEC)	High	Moderate	High	Moderate to high	High
Size-exclusion chromatography (SEC)	Very high	Low	Moderate	High	Low
Affinity chromatography (AC)	Extremely high	Moderate to low	Low to moderate	Very high	Extremely high

such as homogenization or sonication. If SSF is employed, water or suitable buffer will be first added while the solid substrate removed from the fermentation vessel. For instance, *A. niger* was resuspended in 50 mM citrate buffer after SSF and then homogenized at 4 °C to physically break open the cell wall. The lysate was then centrifuged and filtered to recover the crude xylanase [28]. If SmF is employed, the fermentation broth will be separated from cell biomass before cell disruption. For instance, Li et al. [15] centrifuged to collect the expression host cell, *Escherichia coli*, from the fermentation vessel. The cell pellet was then sonicated at 4 °C to release the intracellular xylanase. Temperature control during homogenization or sonication is crucial to preserve the enzyme structure and activity, which is typically done by intermittent operation at low temperatures. When dealing with larger fermentation volume, chemical lysis involving zwitterionic detergent and Triton-X to solubilize cell membrane may be considered. However, the process is slow and the effectiveness of these lysis agents is not guaranteed. It is influenced by various conditions including pH, ionic strength, and the ratio between cell biomass and lysis agents. In a recent report, exposure to Triton-X and lysozyme did not improve cell lysis and production of xylanase by *Geobacillus stearothermophilus* [91]. Moreover, introduction of chemicals may jeopardize the subsequent purification steps or negatively impact the enzyme stability. These factors should be carefully considered when designing the purification protocol [92]. On the contrary, extracellular xylanase is secreted by the microorganism into the fermentation broth or vessel. The extraction procedure starts with separating the liquid medium from the biomass for SmF. For example, the fermentation medium may be centrifuged or simply filtered in which the supernatant or filtrate containing crude xylanase is collected for further purification [9, 38]. For SSF, the enzyme produced by *A. niger* GIO using alkaline-pretreated maize straw was dissolved with water followed by filtration to exclude solids [22].

After extraction, the crude xylanase may still contain impurities such as other proteins, small molecules or salt residuals from the fermentation medium. Initial purification step is conducted to concentrate the enzyme and remove contaminants before more refined purification. Two common methods were reported to selectively precipitate proteins, i.e., ammonium sulfate [$(NH_4)_2SO_4$] precipitation and acetone precipitation. The principle of $(NH_4)_2SO_4$ precipitation is based on the salting-out of proteins. Increasing concentration of the salt increases the ionic strength between protein molecules, leading to protein aggregation and precipitation. Using 80% $(NH_4)_2SO_4$, Carvalho et al. [13] and Chauhan et al. [23] obtained 1.3- and 2.3-fold purified xylanase corresponding to 42 and 10% enzyme yield, respectively. Ding et al. [35] saturated the crude xylanase with 20–60% $(NH_4)_2SO_4$ and reported a cumulative purification fold of 11.7 and yield of 88%. It is noted that $(NH_4)_2SO_4$ precipitation is not highly selective and only moderately improves the enzyme purity. Incomplete precipitation or enzyme denaturation is possible during the process, thereby reducing the overall enzyme yield. The salt concentration required for precipitation also varies depending on the target proteins. Therefore, careful optimization is essential to mitigate enzyme yield loss and acceptable purity. Besides, acetone works in a similar manner by reducing the hydration of proteins. It seems to be a milder method compared to $(NH_4)_2SO_4$ as Algan et al. [24] obtained 1.5-fold and 90% yield whereas

Adiguzel et al. [26] obtained 1.6-fold and 96% yield of partially purified xylanase after precipitation by cold acetone. Another advantage of this method is the ease of solvent removal after protein precipitation. Acetone evaporates easily hence no additional steps are required whereas dialysis or ultrafiltration is usually conducted to remove $(NH_4)_2SO_4$ from the enzyme. Overall, $(NH_4)_2SO_4$ precipitation is useful for larger-scale operations and situations where enzyme selectivity is important, while acetone precipitation offers a faster, low-temperature alternative that is particularly beneficial for sensitive enzymes like xylanase.

Following protein precipitation, chromatographic methods or aqueous two-phase system (ATPS) are used for more refined purification of xylanase. Different types of chromatographic separation are available, i.e., ion exchange chromatography (IEC), size-exclusion chromatography (SEC) and affinity chromatography [AC]. These are deployed either individually or in conjunction with each other in multi-step purification, based on the protein characteristics such as charge, molecular size and specific ligand binding. For instance, the thermostable recombinant xylanase from *B. altitudinis* JYY-02 was 11.7-fold purified by Ni–NTA AC with 43.6% yield and gave a specific activity of 1105 U/mg [21]. In another study, the thermo-alkalophilic xylanase from *B. paramycoides* T4 was first precipitated with 80% $(NH_4)_2SO_4$, followed by IEC using Q-Sepharose column and lastly SEC using Sephadex G-75 column. The final enzyme recorded a specific activity of 3321 U/mg, purity factor of 13.5, and 38.5% recovery [93]. As the number of purification steps increases, the purity of enzymes significantly increased but the yield decreased. More purification steps also imply higher cost and longer processing time, which may not be feasible for large scale enzyme production.

To counter the drawbacks of conventional chromatographic purification, ATPS liquid–liquid fractionation method is proposed. This technique relies on the partitioning of biomolecules between two immiscible aqueous phases, which are formed by a polymer such as polyethylene glycol (PEG) or dextran and salt such as citrates, sulfates or phosphates. The target molecule may partition preferentially into one phase based on its size, charge or hydrophobicity, while other undesired contaminants would partition into the opposite phase. This technique showed remarkable efficiency in extraction, purification and enrichment of various biomolecules including xylanase. The performance of ATPS is affected by the choice of polymer and salt for system formation, molecular weight of polymer, as well as concentrations of polymer and salt. Optimization of these factors often results in successful purification [94]. In fact, 11.3% (w/w) PEG 8000/22.5% (w/w) $MgSO_4$ ATPS outperformed $(NH_4)_2SO_4$ precipitation, IEC and SEC in the purification of fungal xylanase from *A. oryae* LC1. The system recovered the highest enzyme yield (86.8%) with the highest purification factor (13-fold). The xylanase enriched in the PEG-rich top phase also exhibited the highest specific activity (2000 U/mg) among other purification methods [34]. Compared to the chromatographic methods, ATPS is generally more cost effective as it does not require organic solvents or specialized equipment. ATPS could also achieve comparable enzyme yield and purity in a single step. What is more, the ATPS phase components could be recycled, which makes the purification more economic, scalable and preferred for industrial use [95, 96].

Overall, the extraction and purification of xylanase involves several techniques. It is crucial to balance between the desired yield, purity, and scalability offered by each method. Initial extraction methods like $(NH_4)_2SO_4$ and acetone precipitation offer moderate to high enzyme recovery and scalability. If a higher purity is desired, chromatographic methods such as IEC, SEC and AF may be employed, although they may lower overall yield and increase costs. ATPS, on the other hand, is a low-cost method that is highly scalable and gives desirable purity and yield in a single-step process. A careful selection of techniques is critical to achieve optimal balance between enzyme purity, cost-efficiency, and industrial scalability.

5 Commercial Xylanases and Safety Considerations

Comprehensive research on xylanases over the decades has brought about its incorporation into various food processing industries. A list of xylanases with real-world commercial applications is summarized in Table 4.

In breadmaking, bacterial xylanase like Panzea® enhances dough stability and gas retention, leading to improved bread volume and crumb structure. VERON® offers multiple products containing either bacterial or fungal xylanases to produce soft and fluffy breads with elastic crumb from all types of flours. Intens, on the other hand, is a product line comprising various solutions to bakers, i.e., Intens Strength features psychrophilic xylanase which is adapted for frozen bakery uses; Intens Short Bite features thermophilic xylanase for short bite effects in medium to rich breads; Intens Soft & Fine features mesophilic xylanase for bread softness with finer and whiter crumbs. BakeZyme®, which is sold under DSM-Firmenich AG, is derived from *Bacillus subtilis*. The enzyme is optimized for modulating viscosity to enhance dough consistency for improved bread texture, volume and shelf life. For beer processing, Novozymes offer multiple products such as Ultraflo®, Ondea® and Ceremix® containing multi-enzyme blend of xylanase, cellulase, α-amylase, β-glucanase, pullulanase, protease and lipase to hydrolyze cell wall components to minimize turbidity and viscosity to maximize wort separation and beer filtration efficiencies. For animal feed supplementation, Econase XT features a thermostable xylanase that can withstand extreme temperatures during feed processing up to 95 °C. The enzyme also comes with additional prebiotic effects by breaking down arabinoxylans into arabinoxylan XOS. Another thermostable xylanase, Xygest™ HT, when administered at 30, 45 and 90 kU xylanase activity per kg of diet has improved growth performance, feed conversion ratio (FCR) and apparent metabolizable energy (AME) of broilers chicken. Xylanase reduced digesta viscosity, improved nutrient digestibility and equaled the performance to that of positive control group [108]. Enzyme cocktails are also common in feed processing. For example, Axtra® XAP contains a mixture of xylanase, amylase and protease to release trapped nutrients in poultry feeds. Endofeed® and Natugrain® TS contain xylanase and β-glucanase to reduce viscosity and improve digestibility of common poultry and swine feeds such as wheat, corn, rye, sorghum and barley to boost nutrient utilization.

Table 4 List of commercial xylanases incorporated in (a) breadmaking and brewery and (b) animal processing industries along with their safety evaluation by European Food Safety Authority (EFSA)

(a)						
Application	Bread making	Bread making	Bread making	Bread making	Beer processing	
Manufacturing company	Puratos N. V., Belgium	DSM-Firmenich AG, Netherlands	AB Enzymes GmbH, Germany	Novozymes, Denmark	Novozymes, Denmark	
Brand name	Intens	BakeZyme®	VERON®	Panzea®	Ultraflo®; Ondea® Pro; Ceremix®	
Source	*Bacillus subtilis**	*Bacillus subtilis* XAN*	*Bacillus subtilis* AR-153*	*Bacillus licheniformis* NZYM-CE*	*Trichoderma reesei* NZYM-ER*	*Aspergillus oryzae* NZYM-FB*
Evaluation by EFSA (i) Safety	No safety concerns; free from viable cells and rDNA	No safety concerns; free from viable cells and rDNA; no risks for antimicrobial resistance genes	No safety concerns; meets QPS requirements; free from viable cells and DNA	No safety concerns; free from viable cells and rDNA	No safety concerns; free from viable cells and rDNA	No safety concerns
(ii) Toxicology	No genotoxicity concerns; NOAEL with a MoE ≥ 85,000	Toxicological tests not required	Toxicity tests considered unnecessary due to QPS status	No genotoxic concern; NOAEL with a MoE ≥ 85,000	No genotoxicity concern; NOAEL of 1051 mg TOS/kg bw/day	NOAEL of 106 mg TOS/kg bw/day
(iii) Dietary exposure	Estimated up to 0.008 mg TOS/kg bw/day	Estimated up to 0.02 mg TOS/kg bw/day	Estimated up to 0.463 mg TOS/kg bw/day	Estimated up to 0.012 mg TOS/kg bw/day	Estimated up to 0.09 mg TOS/kg bw/day	Estimated up to 0.383 mg TOS/kg bw/day
(iv) Allergenicity	No match to known allergens; no expected food allergic reactions	No match to known allergens; low risk of allergic reactions	No match to known allergens; low risk of allergic reactions	No match to known allergens; low risk of allergic reactions	No match to known allergens; low risk of allergic reactions	No match to known allergens; low risk of allergic reactions
Reference	[97]	[98]	[99]	[100]	[101]	[102]

(continued)

Table 4 (continued)

(b)

Application	Animal feed processing				
Manufacturing company	AB Vista Inc., UK	Kemin Industries Inc, USA	Biochem Zusatzstoffe Handels- und Produktionsges. mbH, Germany	GNC Bioferm, Canada	BASF, Germany
Brand name	Econase XT	Xygest™ HT	Axtra® XAP	Endofeed®	Natugrain® TS
Source	*Trichoderma reesei* CBS 140,027	*Komagataella phaffii* ATCC PTA-127,053	*Trichoderma reesei* PTA-5588*	*Aspergillus niger*	*Aspergillus niger**
Evaluation by EFSA (i) Safety	Safe for chickens, weaned piglets, and turkeys; cannot conclude on safety for laying hens or minor poultry species; all forms are respiratory sensitizers	Safe for all poultry, consumers; dermal and respiratory sensitizer but low inhalation risk	Uncertainty regarding strain safety; could not conclude on safety for target species, consumers or users	Safe for chickens, laying hens, pigs; extrapolated to minor species; potential skin/eye irritant and respiratory sensitizer	Dermal sensitizer; could not conclude on safety for use in chickens for fattening as a zootechnical additive
(ii) Toxicology	Lack of toxicology data for consumers	No toxicity concern on consumers; sensitization risk exists	Insufficient data to conclude toxicological potential	No consumer toxicology studies available	Safety for the user reassessed with new data; potential as a dermal sensitizer

(continued)

Table 4 (continued)

(iii) Efficacy on target species	Effective for chickens, piglets, pigs, turkeys; could not conclude efficacy for laying hens and minor poultry species	Effective in all poultry for digestibility enhancement at proposed feed levels; effective in laying hens and chickens for fattening	Effective for poultry species at recommended doses	Effective for chickens, laying hens, and pigs; can be extrapolated to minor species	Unclear efficacy in chickens for fattening, but reduced lesions in chickens' foot pads and hocks
(iv) Environmental safety	Safe for the environment	Safe for the environment	Uncertain due to viable cell concerns	Not explicitly mentioned	Not explicitly mentioned
Reference	[103]	[104]	[105]	[106]	[107]

Note *—Genetically modified; CEF—Panel on food contact materials, enzymes, flavourings and processing aids; CEP—Panel on food contact materials, enzymes and processing aids; FEEDAP—Panel on additives and products or substances used in animal feed; FEZ—Panel on food enzymes MoE, margin of exposure; NOAEL—No observed adverse effect level; QPS—Qualified presumption of safety; rDNA—recombinant DNA; TOS—Total organic solids

When using commercial xylanases, safety is of paramount importance. Most xylanases are classified as GRAS by regulatory authorities such as EFSA. These enzymes are predominantly derived from microbial and fungal sources, with strict quality control ensuring purity and avoiding contamination with harmful by-products. This could be the reason behind the limited research evaluating the toxicity or safety of xylanases [109, 110], except those concluded by EFSA as safe under intended use conditions. Nonetheless, it is still crucial to ensure the absence of toxicity, allergenicity or genotoxicity from enzyme residues in final food products although xylanases are designed to break down during processing and leave minimal residues in the finished products to guarantee consumer safety. Strict adherence to food safety standards remains critical in maintaining consumer confidence and ensuring the safe application of xylanases in commercial processes.

6 Conclusion, Future Perspectives and Research Directions

This book chapter comprehensively provided the insights of xylanase role in food processing. Through the stated applications, it was apparent that xylanase shows its effectiveness in enhancing food product quality and improving the efficiency of the process. Factors affecting enzyme activity, in particular pH, temperature, and the presence of activating agent or inhibitor, were elaborated to offer practical understandings for industrial implementation. In addition, various sources and production methods were discussed, including natural sources and microbial fermentation, which warrant a more sustainable and scalable xylanase supply. The safety considerations highlighted the importance of regulatory compliance. This information will allow researchers to fill the gap between functionalities and applications of xylanase, reinforcing its potential in the food industry.

Future research should focus on enhancing thermostability, pH tolerance and substrate specificity of xylanase through metabolic engineering and synthetic biology techniques to customize xylanases with superior stability and activity to meet industrial demands. Moreover, continuous research on the immobilization techniques of xylanases is not only useful to improve temperature and pH stability but also allows repeated use in industrial processes, thereby reducing production costs and increasing operational stability. Experiments on developing and formulating multi-enzyme blends containing xylanase are anticipated to boost its versatility beyond single applications. Finally, investigations on safety in terms of toxicity and the regulatory aspects of xylanase derived from genetically modified organisms will be critical for expanding its market potential. International collaborations in research and knowledge sharing are critical to advance biotechnological innovations and boost consumer awareness and market acceptance.

References

1. T. Collins, C. Gerday, G. Feller, Xylanases, xylanase families and extremophilic xylanases. FEMS Microbiol. Rev. **29**(1), 3–23 (2005). https://doi.org/10.1016/j.femsre.2004.06.005
2. E. Drula, M.L. Garron, S. Dogan et al., The carbohydrate-active enzyme database: functions and literature. Nucleic Acids Res. **50**(D1), D571–D577 (2022). https://doi.org/10.1093/nar/gkab1045
3. G.K. Gupta, M. Dixit, R.K. Kapoor et al., Xylanolytic enzymes in pulp and paper industry: new technologies and perspectives. Mol. Biotechnol. **64**(2), 130–143 (2022). https://doi.org/10.1007/s12033-021-00396-7
4. R. Chaudhary, T. Kuthiala, G. Singh et al., Current status of xylanase for biofuel production: a review on classification and characterization. Biomass Conversion and Biorefinery **13**, 1–19 (2021). https://doi.org/10.1007/s13399-021-01948-2
5. G. Hu, X. Hong, M. Zhu et al., Improving the quality of wheat flour bread by a thermophilic xylanase with ultra activity and stability reconstructed by ancestral sequence and computational-aided analysis. Molecules **29**(8), 1895 (2024). https://doi.org/10.3390/molecules29081895
6. X. Li, L. Zhang, Z. Jiang et al., A novel cold-active GH8 xylanase from cellulolytic myxobacterium and its application in food industry. Food Chem. **393**, 133463 (2022). https://doi.org/10.1016/j.foodchem.2022.133463
7. X. Liu, Q. Yan, Y. Xue et al., Biochemical characterization of a novel glycoside hydrolase family 11 xylanase from *Chaetomium* sp. suitable for bread making. Process Biochem. **117**, 1–9 (2022). https://doi.org/10.1016/j.procbio.2022.03.017
8. Y. Zhang, C. Liu, M. Yang et al., Characterization and application of a novel xylanase from *Halolactibacillus miurensis* in wholewheat bread making. Front. Bioeng. Biotechnol. **10**, 1018476 (2022). https://doi.org/10.3389/fbioe.2022.1018476
9. S. Wen, G. Wu, H. Wu, Biochemical characterization of a GH10 xylanase from the anaerobic rumen fungus *Anaeromyces robustus* and application in bread making. 3 Biotech. **11**(9), 406 (2021). https://doi.org/10.1007/s13205-021-02956-9
10. A.T.P. Passarinho, R.Z. Ventorim, G.P. Maitan-Alfenas, et al., Engineered GH11 xylanases from *Orpinomyces* sp. PC-2 improve techno-functional properties of bread dough. J. Sci. Food Agric. **99**(2), 741–747 (2019). https://doi.org/10.1002/jsfa.9242
11. C.C. de Queiroz Brito Cunha, A.R. Gama, Cintra, et al., Improvement of bread making quality by supplementation with a recombinant xylanase produced by *Pichia pastoris*. Plos One **13**(2), e0192996 (2018). https://doi.org/10.1371/journal.pone.0192996
12. S. Yegin, B. Altinel, K. Tuluk, A novel extremophilic xylanase produced on wheat bran from *Aureobasidium pullulans* NRRL Y-2311-1: effects on dough rheology and bread quality. Food Hydrocolloids **81**, 389–397 (2018). https://doi.org/10.1016/j.foodhyd.2018.03.012
13. E.A. Carvalho, L.M. dos Santos Góes, A.P.T. Uetanabaro et al., Thermoresistant xylanases from *Trichoderma stromaticum*: application in bread making and manufacturing xylo-oligosaccharides. Food Chem. **221**, 1499–1506 (2017). https://doi.org/10.1016/j.foodchem.2016.10.144
14. V. Kumar, T. Satyanarayana, Production of thermo-alkali-stable xylanase by a novel polyextremophilic *Bacillus halodurans* TSEV1 in cane molasses medium and its applicability in making whole wheat bread. Bioprocess Biosyst. Eng. **37**, 1043–1053 (2014). https://doi.org/10.1007/s00449-013-1075-3
15. Q.F. Li, Y. Liu, X. Yan et al., Enzymatic properties of recombinant thermophilic xylanase from *Pseudothermotoga thermarum* and its application in beer brewing. Food Biosci. **44**(10), 181–187 (2023). https://doi.org/10.7506/spkx1002-6630-20220503-024
16. X.Q. Liu, Z.Q. Jiang, J. Yu et al., High-level expression of an engineered thermostable xylanase b from *Thermotoga maritima* and its application in beer brewing. Food Sci. **43**(6), 66–73 (2022). https://doi.org/10.7506/spkx1002-6630-20210504-021

17. J. Yu, X. Liu, L. Guan et al., High–level expression and enzymatic properties of a novel thermostable xylanase with high arabinoxylan degradation ability from *Chaetomium* sp. suitable for beer mashing. Int. J. Biol. Macromol. **168**, 223–232 (2021). https://doi.org/10.1016/j.ijbiomac.2020.12.040
18. D. Zhu, X. Liu, X. Xie, et al., Characteristics of a XIP-resistant xylanase from *Neocallimastix* sp. GMLF 1 and its advantage in barley malt saccharification. Int. J. Food Sci. Technol. **55**(5), 2152–2160 (2020). https://doi.org/10.1111/ijfs.14341
19. X. Wang, H. Luo, W. Yu et al., A thermostable *Gloeophyllum trabeum* xylanase with potential for the brewing industry. Food Chem. **199**, 516–523 (2016). https://doi.org/10.1016/j.foodchem.2015.12.028
20. Y. Du, P. Shi, H. Huang et al., Characterization of three novel thermophilic xylanases from *Humicola insolens* Y1 with application potentials in the brewing industry. Biores. Technol. **130**, 161–167 (2013). https://doi.org/10.1016/j.biortech.2012.12.067
21. H. Tai, Q. Guo, J. Zhao et al., A thermostable xylanase hydrolyzes several polysaccharides from *Bacillus altitudinis* JYY-02 showing promise for industrial applications. Carbohyd. Res. **538**, 109080 (2024). https://doi.org/10.1016/j.carres.2024.109080
22. S.A. Fasiku, M.A. Bello, O.A. Odeniyi, Production of xylanase by *Aspergillus niger* GIO and *Bacillus megaterium* through solid-state fermentation. Access Microbiol. **5**(6), 000506-v5 (2023). https://doi.org/10.1099/acmi.0.000506.v5
23. J. Chauhan, A. Bishoyi, R. Gohel, et al., Purification and characterization of acidophilic xylanase from *Bacillus* sp. with potential application as fruit juice clarifier. Emirates J. Food Agric. (EJFA), **35**(7) (2023). https://doi.org/10.9755/ejfa.2023.v35.i7.3112
24. M. Algan, Y. Sürmeli, G. Şanlı-Mohamed, A novel thermostable xylanase from *Geobacillus vulcani* GS90: production, biochemical characterization, and its comparative application in fruit juice enrichment. J. Food Biochem. **45**(5), e13716 (2021). https://doi.org/10.1111/jfbc.13716
25. C. Li, A. Kumar, X. Luo et al., Highly alkali-stable and cellulase-free xylanases from *Fusarium* sp. 21 and their application in clarification of orange juice. Int. J. Biol. Macromol. **155**, 572–580 (2020). https://doi.org/10.1016/j.ijbiomac.2020.03.249
26. G. Adiguzel, O. Faiz, M. Sisecioglu et al., A novel endo-β-1,4-xylanase from *Pediococcus acidilactici* GC25; purification, characterization and application in clarification of fruit juices. Int. J. Biol. Macromol. **129**, 571–578 (2019). https://doi.org/10.1016/j.ijbiomac.2019.02.054
27. P.O. da Silva, N.C. de Alencar Guimarães, J.D.M. Serpa et al., Application of an endo-xylanase from *Aspergillus japonicus* in the fruit juice clarification and fruit peel waste hydrolysis. Biocatal. Agric. Biotechnol. **21**, 101312 (2019). https://doi.org/10.1016/j.bcab.2019.101312
28. R. Ravindran, G.A. Williams, A.K. Jaiswal, Spent coffee waste as a potential media component for xylanase production and potential application in juice enrichment. Foods **8**(11), 585 (2019). https://doi.org/10.3390/foods8110585
29. D.M. Salas-Veizaga, L.R. Rocabado-Villegas, J.A. Linares-Pastén et al., A novel glycoside hydrolase 43-like enzyme from *Clostridium boliviensis* is an endo-xylanase and a candidate for xylooligosaccharide production from different xylan substrates. Appl. Environ. Microbiol. **90**(4), e02223-23 (2024). https://doi.org/10.1128/aem.02223-23
30. F. Zheng, J. Chen, J. Wang et al., Transformation of corncob into high-value xylooligosaccharides using glycoside hydrolase families 10 and 11 xylanases from *Trichoderma asperellum* ND-1. Biores. Technol. **394**, 130249 (2024). https://doi.org/10.1016/j.biortech.2023.130249
31. A.A.I. Ratnadewi, M.T. Rahma, N. Nurhayati et al., Production of xylooligosaccharide from cassava pulp's waste by endo-β-1, 4-D-xylanase and characterization of its prebiotic effect by fermentation of *Lactobacillus acidophilus*. Fermentation **8**(10), 488 (2022). https://doi.org/10.3390/fermentation8100488
32. J. Wang, S. Zhang, C. Li et al., Efficient secretion of xylanase in *Escherichia coli* for production of prebiotic xylooligosaccharides. Lwt **162**, 113481 (2022). https://doi.org/10.1016/j.lwt.2022.113481
33. G. Shivudu, K. Chandraraj, P. Selvam, Production of xylooligosaccharides from xylan catalyzed by endo-1, 4-β-D-xylanase-immobilized nanoscale carbon, silica and zirconia matrices. Mol. Catal. **484**, 110745 (2020). https://doi.org/10.1016/j.mcat.2019.110745

34. N. Bhardwaj, B. Kumar, K. Agarwal et al., Purification and characterization of a thermo-acid/alkali stable xylanases from *Aspergillus oryzae* LC1 and its application in xylo-oligosaccharides production from lignocellulosic agricultural wastes. Int. J. Biol. Macromol. **122**, 1191–1202 (2019). https://doi.org/10.1016/j.ijbiomac.2018.09.070
35. C. Ding, M. Li, Y. Hu, High-activity production of xylanase by *Pichia stipitis*: purification, characterization, kinetic evaluation and xylooligosaccharides production. Int. J. Biol. Macromol. **117**, 72–77 (2018). https://doi.org/10.1016/j.ijbiomac.2018.05.128
36. X. Liu, Y. Liu, Z. Jiang et al., Biochemical characterization of a novel xylanase from *Paenibacillus barengoltzii* and its application in xylooligosaccharides production from corncobs. Food Chem. **264**, 310–318 (2018). https://doi.org/10.1016/j.foodchem.2018.05.023
37. R. Ma, Y. Bai, H. Huang et al., Utility of thermostable xylanases of *Mycothermus thermophilus* in generating prebiotic xylooligosaccharides. J. Agric. Food Chem. **65**(6), 1139–1145 (2017). https://doi.org/10.1021/acs.jafc.6b05183
38. M. Nieto-Domínguez, L.I. de Eugenio, M.J. York-Durán et al., Prebiotic effect of xylooligosaccharides produced from birchwood xylan by a novel fungal GH11 xylanase. Food Chem. **232**, 105–113 (2017). https://doi.org/10.1016/j.foodchem.2017.03.149
39. F. Kallel, D. Driss, F. Bouaziz et al., Production of xylooligosaccharides from garlic straw xylan by purified xylanase from *Bacillus mojavensis* UEB-FK and their in vitro evaluation as prebiotics. Food Bioprod. Process. **94**, 536–546 (2015). https://doi.org/10.1016/j.fbp.2014.07.012
40. I.S. Sanguine, G.F. Cavalheiro, N.F.L. Garcia et al., Xylanases of *Trichoderma koningii* and *Trichoderma pseudokoningii*: production, characterization and application as additives in the digestibility of forage for cattle. Biocatal. Agric. Biotechnol. **44**, 102482 (2022). https://doi.org/10.1016/j.bcab.2022.102482
41. A.C. da Costa, G.F. Cavalheiro, E.R. de Queiroz Vieira et al., Catalytic properties of xylanases produced by *Trichoderma piluliferum* and *Trichoderma viride* and their application as additives in bovine feeding. Biocatal. Agric. Biotechnol. **19**, 101161 (2019). https://doi.org/10.1016/j.bcab.2019.101161
42. S. Tundo, G. Mandalà, L. Sella et al., Xylanase inhibitors: defense players in plant immunity with implications in agro-industrial processing. Int. J. Mol. Sci. **23**(23), 14994 (2022). https://doi.org/10.3390/ijms232314994
43. L.L. Newstead, K. Varjonen, T. Nuttall et al., Staphylococcal-produced bacteriocins and antimicrobial peptides: their potential as alternative treatments for Staphylococcus aureus infections. Antibiotics **9**(2), 40 (2020). https://doi.org/10.3390/antibiotics9020040
44. A. Palaniappan, U. Antony, M.N. Emmambux, Current status of xylooligosaccharides: production, characterization, health benefits and food application. Trends Food Sci. Technol. **111**, 506–519 (2021). https://doi.org/10.1016/j.tifs.2021.02.047
45. S. Sureshkumar, J. Song, V. Sampath et al., Exogenous enzymes as zootechnical additives in monogastric animal feed: a review. Agriculture **13**(12), 2195 (2023). https://doi.org/10.3390/agriculture13122195
46. M.E. Duarte, F.X. Zhou, W.M. Dutra Jr. et al., Dietary supplementation of xylanase and protease on growth performance, digesta viscosity, nutrient digestibility, immune and oxidative stress status, and gut health of newly weaned pigs. Anim. Nutr. **5**(4), 351–358 (2019). https://doi.org/10.1016/j.aninu.2019.04.005
47. S. Zhang, C. Wang, Y. Sun et al., Xylanase and fermented polysaccharide of *Hericium caputmedusae* reduce pathogenic infection of broilers by improving antioxidant and anti-inflammatory properties. Oxid. Med. Cell. Longev. **2018**(1), 4296985 (2018). https://doi.org/10.1155/2018/4296985
48. F. Van Petegem, T. Collins, M.A. Meuwis, et al., The structure of a cold-adapted family 8 xylanase at 1.3 Å resolution: structural adaptations to cold and investigation of the active site. J. Biol. Chem. **278**(9), 7531–7539 (2003). https://doi.org/10.1074/jbc.M206862200
49. S.B. Ajeje, Y. Hu, G. Song et al., Thermostable cellulases/xylanases from thermophilic and hyperthermophilic microorganisms: current perspective. Front. Bioeng. Biotechnol. **9**, 794304 (2021). https://doi.org/10.3389/fbioe.2021.794304

50. M. Kumari, S. Padhi, S. Sharma, et al., Biotechnological potential of psychrophilic microorganisms as the source of cold-active enzymes in food processing applications. 3 Biotech, **11**(11), 479 (2021). https://doi.org/10.1007/s13205-021-03008-y
51. C.R. Santos, Z.B. Hoffmam, V.P. de Matos Martins et al., Molecular mechanisms associated with xylan degradation by *Xanthomonas* plant pathogens. J. Biol. Chem. **289**(46), 32186–32200 (2014). https://doi.org/10.1074/jbc.M114.605105
52. V.E.B. Hernández, C.J. Salas-Montantes, A.P. Barba-De la Rosa et al., Autodisplay of an endo-1,4-β-xylanase from *Clostridium cellulovorans* in *Escherichia coli* for xylans degradation. Enzyme Microb. Technol. **149**, 109834 (2021). https://doi.org/10.1016/j.enzmictec.2021.109834
53. S. Kumar, I. Haq, J. Prakash, et al., Purification, characterization and thermostability improvement of xylanase from *Bacillus amyloliquefaciens* and its application in pre-bleaching of kraft pulp. 3 Biotech, **7**, 1–12 (2017). https://doi.org/10.1007/s13205-017-0615-y
54. X. Zhang, H. Liu, L. Ju et al., Exploring the effect of Cu^{2+} on sludge hydrolysis and interaction mechanism between Cu^{2+} and xylanase by multispectral and thermodynamic methods. Water Air Soil Pollut. **228**, 1–11 (2017). https://doi.org/10.1007/s11270-017-3252-5
55. M. Hou, C. Liang, Y. Fei et al., Analysis of the effect of metal ions on the ability of xylanase to hydrolyze wheat bran by molecular dynamics simulations. Front. Bioeng. Biotechnol. **11**, 1142873 (2023). https://doi.org/10.3389/fbioe.2023.1142873
56. A. Abdella, S. Ramadan, R.A. Hamouda et al., *Paecilomyces variotii* xylanase production, purification and characterization with antioxidant xylo-oligosaccharides production. Sci. Rep. **11**(1), 16468 (2021). https://doi.org/10.1038/s41598-021-95965-w
57. M.D.L.T.D.M. Polizeli, A.C.S. Rizzatti, R. Monti et al., Xylanases from fungi: properties and industrial applications. Appl. Microbiol. Biotechnol. **67**, 577–591 (2005). https://doi.org/10.1007/s00253-005-1904-7
58. N. Bhardwaj, B. Kumar, P. Verma, A detailed overview of xylanases: an emerging biomolecule for current and future prospective. Bioresources and Bioprocessing **6**(1), 1–36 (2019). https://doi.org/10.1186/s40643-019-0276-2
59. F.I. Qeshmi, A. Homaei, P. Fernandes, et al., Xylanases from marine microorganisms: a brief overview on scope, sources, features and potential applications. Biochimica et Biophysica Acta (BBA)-Proteins and Proteomics, **1868**(2), 140312 (2020). https://doi.org/10.1016/j.bbapap.2019.140312
60. C. Li, J. Zhou, G. Du et al., Developing *Aspergillus niger* as a cell factory for food enzyme production. Biotechnol. Adv. **44**, 107630 (2020). https://doi.org/10.1016/j.biotechadv.2020.107630
61. A. Kunamneni, F.J. Plou, M. Alcalde, et al., *Trichoderma* enzymes for food industries, in *Biotechnology and Biology of Trichoderma*, ed by V.K. Gupta, M. Schmoll, A. Herrera-Estrella, I. Druzhinina, M.G. Tuohy (Elsevier, Newnes, 2014), pp. 339–344. https://doi.org/10.1016/B978-0-444-59576-8.00024-2
62. N.A. Zin, N.A. Badaluddin, Biological functions of *Trichoderma* spp. for agriculture applications. Ann. Agric. Sci. **65**(2), 168–178 (2020). https://doi.org/10.1016/j.aoas.2020.09.003
63. H.H.M. Fadel, M.G. Mahmoud, M.M.S. Asker et al., Characterization and evaluation of coconut aroma produced by *Trichoderma viride* EMCC-107 in solid state fermentation on sugarcane bagasse. Electron. J. Biotechnol. **18**(1), 5–9 (2015). https://doi.org/10.1016/j.ejbt.2014.10.006
64. A.M. Abdel-Azeem, S.A. El-Mansy, H.A.H. Abo Nahas, et al., Thermophilic *Chaetomium* in biotechnology, in *Recent Developments on Genus Chaetomium. Fungal Biology*, ed by A. Abdel-Azeem (Springer, Cham, 2020), pp. 421–439. https://doi.org/10.1007/978-3-030-31612-9_16
65. Y. Su, C. Liu, H. Fang et al., *Bacillus subtilis*: a universal cell factory for industry, agriculture, biomaterials and medicine. Microb. Cell Fact. **19**, 1–12 (2020). https://doi.org/10.1186/s12934-020-01436-8

66. I. Danilova, M. Sharipova, The practical potential of bacilli and their enzymes for industrial production. Front. Microbiol. **11**, 1782 (2020). https://doi.org/10.3389/fmicb.2020.01782
67. R. Aguilar Suárez, J. Stülke, J.M. Van Dijl, Less is more: toward a genome-reduced Bacillus cell factory for "difficult proteins." ACS Synth. Biol. **8**(1), 99–108 (2019). https://doi.org/10.1021/acssynbio.8b00342
68. B. Wang, F. van der Kloet, L.W. Hamoen, Induction of the CtsR regulon improves xylanase production in *Bacillus subtilis*. Microb. Cell Fact. **22**(1), 231 (2023). https://doi.org/10.1186/s12934-023-02239-3
69. Y. Yang, X. Fu, X. Zhao et al., Overexpression of a thermostable α-amylase through genome integration in *Bacillus subtilis*. Fermentation **9**(2), 139 (2023). https://doi.org/10.3390/fermentation9020139
70. M. Shariq, M. Sohail, Application of *Candida tropicalis* MK-160 for the production of xylanase and ethanol. J. King Saud University-Sci. **31**(4), 1189–1194 (2019). https://doi.org/10.1016/j.jksus.2018.04.009
71. C.G. Morais, L.M. Sena, M.R. Lopes et al., Production of ethanol and xylanolytic enzymes by yeasts inhabiting rotting wood isolated in sugarcane bagasse hydrolysate. Fungal Biol. **124**(7), 639–647 (2020). https://doi.org/10.1016/j.funbio.2020.03.005
72. J.L. Ravn, M.K. Engqvist, J. Larsbrink et al., CAZyme prediction in ascomycetous yeast genomes guides discovery of novel xylanolytic species with diverse capacities for hemicellulose hydrolysis. Biotechnol. Biofuels **14**, 1–18 (2021). https://doi.org/10.1186/s13068-021-01995-x
73. Z. Yang, Z. Zhang, Engineering strategies for enhanced production of protein and bio-products in *Pichia pastoris*: a review. Biotechnol. Adv. **36**(1), 182–195 (2018). https://doi.org/10.1016/j.biotechadv.2017.11.002
74. A. Puseenam, S. Tanapongpipat, N. Roongsawang, Co-expression of endoxylanase and endoglucanase in *Scheffersomyces stipitis* and its application in ethanol production. Appl. Biochem. Biotechnol. **177**, 1690–1700 (2015). https://doi.org/10.1007/s12010-015-1846-1
75. Y. Wu, X. Peng, D. Fan et al., Pathway reconstruction and metabolic engineering for the de novo and enhancing production of monacolin J in *Pichia pastoris*. Bioprocess Biosyst. Eng. **47**(11), 1789–1801 (2024). https://doi.org/10.1007/s00449-024-03069-2
76. R. Baghban, S. Farajnia, M. Rajabibazl et al., Yeast expression systems: overview and recent advances. Mol. Biotechnol. **61**, 365–384 (2019). https://doi.org/10.1007/s12033-019-00164-8
77. A. Eskandari, N.G. Nezhad, T.C. Leow et al., Current achievements, strategies, obstacles, and overcoming the challenges of the protein engineering in *Pichia pastoris* expression system. World J. Microbiol. Biotechnol. **40**(1), 39 (2024). https://doi.org/10.1007/s11274-023-03851-6
78. G.H. Hansen, M. Lübeck, J.C. Frisvad et al., Production of cellulolytic enzymes from ascomycetes: comparison of solid state and submerged fermentation. Process Biochem. **50**(9), 1327–1341 (2015). https://doi.org/10.1016/j.procbio.2015.05.017
79. U.S.P. Uday, P. Choudhury, T.K. Bandyopadhyay et al., Classification, mode of action and production strategy of xylanase and its application for biofuel production from water hyacinth. Int. J. Biol. Macromol. **82**, 1041–1054 (2016). https://doi.org/10.1016/j.ijbiomac.2015.10.086
80. F. Güler, F. Özçelik, Screening of xylanase producing *Bacillus* species and optimization of xylanase process parameters in submerged fermentation. Biocatal. Agric. Biotechnol. **51**, 102801 (2023). https://doi.org/10.1016/j.bcab.2023.102801
81. H.A. Shakir, A. Anwar, M. Irfan et al., Statistical optimization of xylanase from *Bacillus licheniformis* using banana peels in submerged fermentation. Iranian J. Sci. Technol. Trans. A: Sci. **44**, 981–991 (2020). https://doi.org/10.1007/s40995-020-00933-0
82. A. Ashok, K. Doriya, D.R.M. Rao et al., Design of solid state bioreactor for industrial applications: an overview to conventional bioreactors. Biocatal. Agric. Biotechnol. **9**, 11–18 (2017). https://doi.org/10.1016/j.bcab.2016.10.014
83. V.S. Kumar, V. Karthi, A. Ramkumar, et al., Isolation, screening, identification, and optimization of xylanase producing bacteria from forests of Western Ghats, India. Int. J. Agric. Environ. Biotechnol. **11**(1), 173–179 (2018). https://doi.org/10.30954/0974-1712.2018.00178.22

84. S. Kar, S. Sona Gauri, A. Das et al., Process optimization of xylanase production using cheap solid substrate by *Trichoderma reesei* SAF3 and study on the alteration of behavioral properties of enzyme obtained from SSF and SmF. Bioprocess Biosyst. Eng. **36**, 57–68 (2013). https://doi.org/10.1007/s00449-012-0761-x
85. A. Banu, S. Ingale, Xylanase production by *Bacillus pumilus* AB-1 under solid-state fermentation and its application. Bull. Environ. Sci. Res. **1**(1), 35–43 (2012)
86. H. Ho, Xylanase production by *Bacillus subtilis* using carbon source of inexpensive agricultural wastes in two different approaches of submerged fermentation (SmF) and solid state fermentation (SsF). J. Food Process. Technol. **6**(4), 1000437 (2015). https://doi.org/10.4172/2157-7110.1000437
87. I. Membrillo Venegas, J. Fuentes-Hernández, M. García-Rivero et al., Characteristics of *Aspergillus niger* xylanases produced on rice husk and wheat bran in submerged culture and solid-state fermentation for an applicability proposal. Int. J. Food Sci. Technol. **48**(9), 1798–1807 (2013). https://doi.org/10.1111/ijfs.12153
88. R.D. Kamble, A.R. Jadhav, Xylanase production under solid state and submerged fermentation conditions by bacterial strains. African J. Microbiol. Res. **6**(20), 4292–4297 (2012). https://doi.org/10.5897/AJMR11.823
89. D. Bedade, S. Pawar, Downstream processing of biotechnology products, in *Basic Biotechniques for Bioprocess and Bioentrepreneurship*, ed by A.K. Bhatt, R.K. Bhatia, T.C. Bhalla (Academic Press, 2023), pp. 377–390. https://doi.org/10.1016/B978-0-12-816109-8.00026-X
90. K. Shanu, S. Choudhary, S. Kumari, K. Anu, S. Devi, Downstream processing for bio-product recovery and purification, in *Recent Advances in Bioprocess Engineering and Bioreactor Design*, ed by S. Dhagat, S.E. Jujjavarapu, N.S. Sampath Kumar, C. Mahapatra (Springer Nature Singapore, Singapore, 2024), pp. 139–169. https://doi.org/10.1007/978-981-97-1451-3_7
91. P. Dhaver, T. Sithole, B. Pletschke et al., Enhanced production of a recombinant xylanase (XT6): optimization of production and purification, and scaled-up batch fermentation in a stirred tank bioreactor. Sci. Rep. **13**(1), 20895 (2023). https://doi.org/10.1038/s41598-023-48202-5
92. E. Grigorov, B. Kirov, M.B. Marinov et al., Review of microfluidic methods for cellular lysis. Micromachines **12**(5), 498 (2021). https://doi.org/10.3390/mi12050498
93. S. Tiwari, R. Singh, J. Yadav et al., Three-step purification and characterization of organic solvent-tolerant and alkali-thermo-tolerant xylanase from *Bacillus paramycoides* T4 [MN370035]. Catalysts **12**(7), 749 (2022). https://doi.org/10.3390/catal12070749
94. D. Garai, V. Kumar, Aqueous two phase extraction of alkaline fungal xylanase in PEG/phosphate system: optimization by Box-Behnken design approach. Biocatal. Agric. Biotechnol. **2**(2), 125–131 (2013). https://doi.org/10.1016/j.bcab.2013.03.003
95. A. Taddia, M. Rito-Palomares, K. Mayolo-Deloisa et al., Purification of xylanase from *Aspergillus niger* NRRL3 extract by an integrated strategy based on aqueous two-phase systems followed by ion exchange chromatography. Sep. Purif. Technol. **255**, 117699 (2021). https://doi.org/10.1016/j.seppur.2020.117699
96. M. Božinović, R. Vičević, N. Zekić et al., Intensification of endo-1,4-xylanase extraction by coupling microextractors and aqueous two-phase system. Processes **11**(2), 447 (2023). https://doi.org/10.3390/pr11020447
97. EFSA Panel on Food Contact Materials, Enzymes, Flavourings and Processing Aids (CEF), V. Silano, C. Bolognesi, et al., Safety evaluation of the food enzyme xylanase from a genetically modified *Bacillus subtilis* strain TD 160 (229). EFSA J. **16**(1), e05008 (2018). https://doi.org/10.2903/j.efsa.2018.5008
98. EFSA Panel on Food Contact Materials, Enzymes and Processing Aids (CEP), C. Lambré, J.M. Barat Baviera, et al., Safety evaluation of the food enzyme endo-1,4-β-xylanase from the genetically modified Bacillus subtilis strain XAN. EFSA J. **21**(7), e08017 (2023). https://doi.org/10.2903/j.efsa.2023.8017
99. EFSA Panel on Food Enzymes (FEZ), H. Zorn, J.M. Barat Baviera, et al., Safety evaluation of the food enzym endo-1,4-β-xylanase from the genetically modified *Bacillus subtilis* strain AR-153. EFSA J. 22(12), e9123 (2024). https://doi.org/10.2903/j.efsa.2024.9123

100. EFSA Panel on Food Contact Materials, Enzymes, Processing Aids (CEP), V. Silano, J.M. Barat Baviera, et al., Safety evaluation of the food enzyme endo-1,4-β-xylanase from a genetically modified *Bacillus licheniformis* (strain NZYM-CE). EFSA J. **17**(4), e05685 (2019). https://doi.org/10.2903/j.efsa.2019.5685
101. EFSA Panel on Food Contact Materials, Enzymes and Processing Aids (CEP), C. Lambré, J.M. Barat Baviera, et al., Safety evaluation of the food enzyme endo-1,4-β-xylanase from the genetically modified *Trichoderma reesei* strain NZYM-ER. EFSA J. **20**(6), e07373 (2022). https://doi.org/10.2903/j.efsa.2022.7373
102. EFSA Panel on Food Enzymes (FEZ), H. Zorn, J.M.B. Baviera, et al., Revised dietary exposure assessment of the food enzyme endo 1,4-β-xylanase from the genetically modified *Aspergillus oryzae* strain NZYM-FB. EFSA J. **22**(12), e9163 (2024). https://doi.org/10.2903/j.efsa.2024.9163
103. EFSA Panel on Additives and Products or Substances used in Animal Feed (FEEDAP), V. Bampidis, G. Azimonti et al., Safety and efficacy of a feed additive consisting of endo-1,4-β-xylanase (ECONASE® XT) produced by *Trichoderma reesei* CBS 140027 as a feed additive for piglets (weaned), pigs for fattening, chickens for fattening, chickens reared for laying, laying hens, turkeys for fattening, turkeys reared for breeding and minor poultry species (Roal Oy). EFSA J. **19**(4), e06536 (2021). https://doi.org/10.2903/j.efsa.2021.6536
104. EFSA Panel on Additives and Products or Substances used in Animal Feed (FEEDAP), V. Bampidis, G. Azimonti et al., Efficacy of a feed additive consisting of endo-1,4-beta-xylanase produced by *Komagataella phaffii* ATCC PTA-127053 (Xygest™ HT) for all poultry species (Kemin Europa NV). EFSA J. **21**(6), e08047 (2023). https://doi.org/10.2903/j.efsa.2023.8047
105. EFSA Panel on Additives and Products or Substances used in Animal Feed (FEEDAP), V. Bampidis, G. Azimonti et al., Safety and efficacy of Axtra® XAP 104 TPT (endo-1,4-xylanase, protease and alpha-amylase) as a feed additive for chickens for fattening, laying hens and minor poultry species. EFSA J. **18**(6), e06165 (2020). https://doi.org/10.2903/j.efsa.2020.6165
106. EFSA Panel on Additives and Products or Substances used in Animal Feed (FEEDAP), Scientific Opinion on the safety and efficacy of Endofeed® DC (endo-1, 3(4)-beta-glucanase and endo-1,4-beta-xylanase) as a feed additive for chickens for fattening, laying hens, pigs for fattening and minor poultry and porcine species. EFSA J. **11**(8), 3322 (2013). https://doi.org/10.2903/j.efsa.2013.3322
107. EFSA Panel on Additives and Products or Substances used in Animal Feed (FEEDAP), Safety and efficacy of Natugrain® TS (endo-1,4-β-xylanase and endo-1,4-β-glucanase) for chickens for fattening. EFSA J. **14**(1), 4347 (2016). https://doi.org/10.2903/j.efsa.2016.4347
108. B. Gorenz, V. Iseri, J. Rubach et al., Xylanase supplementation of pelleted wheat-based diets increases growth efficiency and apparent metabolizable energy and decreases viscosity of intestinal contents in broilers. Poult. Sci. **101**(12), 102220 (2022). https://doi.org/10.1016/j.psj.2022.102220
109. L. Harbak, H.V. Thygesen, Safety evaluation of a xylanase expressed in *Bacillus subtilis*. Food Chem. Toxicol. **40**(1), 1–8 (2002). https://doi.org/10.1016/S0278-6915(01)00092-8
110. R. Van Dorn, D. Shanahan, V. Ciofalo, Safety evaluation of xylanase 50316 enzyme preparation (also known as VR007), expressed in *Pseudomonas fluorescens*, intended for use in animal feed. Regul. Toxicol. Pharmacol. **97**, 48–56 (2018). https://doi.org/10.1016/j.yrtph.2018.05.016

100. EFSA Panel on Food Contact Materials, Enzymes, Flavourings and Processing Aids (CEF), Silano, M., [illegible] safety evaluation of the food enzyme [illegible] from a genetically modified [illegible] EFSA J. [illegible] https://doi.org/10.2903/j.efsa.[illegible]
101. EFSA Panel on Food Contact Materials, Enzymes and Processing Aids (CEP), Lambré, C., Barat Baviera, J.M., et al.: Safety evaluation of the food enzyme endo-1,4-β-xylanase from the genetically modified [illegible] strain [illegible] EFSA J. [illegible] (2022). https://doi.org/10.2903/j.efsa.2022.[illegible]
102. EFSA Panel on Food Enzymes (FEZ), Zorn, H., Barat Baviera, J.M., et al.: Revised dietary exposure assessment of the food enzyme endo-1,4-β-xylanase from the genetically modified [illegible] strain [illegible] EFSA J. 22(12), e9163 (2024). https://doi.org/10.2903/j.efsa.2024.9163
103. EFSA Panel on Additives and Products or Substances used in Animal Feed (FEEDAP), [illegible] Safety and efficacy of [illegible]

Chapter 12
Xylanase in Biofuel Production

Hala Bakr El-Nassan

Abstract Xylanases are hydrolytic enzymes secreted by different strains of bacteria and fungi. They are responsible for the hydrolysis of xylan into monomeric xylose. Xylanases have wide industrial applications in baking industry, wood industry and laundry detergents production. One of the most important applications of xylanases is second-generation biofuel production where they serve as hydrolytic enzymes for lignocellulosic biomass in the saccharification step to produce sugars for the subsequent fermentation. The second generation biofuel production especially biohydrogen, bioethanol and biobutanol using xylanases received great attention in the last two decades. This chapter spotlights the efforts exerted in the last two decades for the biofuels production using xylanase enzymes.

Keywords Biocthanol · Biobutanol · Biohydrogen · Biofuel · Xylanases · Saccharification · Fermentation

1 Introduction

Lignocellulose is the most abundant and renewable biomass source in nature. Cellulose, hemicellulose and lignin are the main components of lignocellulose and they are formed by the plant to protect it from osmotic and mechanical stress. Dry weight of biomass contains about 40% cellulose, 33% hemicelluloses and 23% lignin. Cellulose is formed of β-*d*-glucose units connected by β-1,4 glycosidic bonds. Cellulose occurs in nature as micro-fibrils due to the strong H-bonds within chains of cellulose. This micro-fibrils form results in insoluble crystalline structure that resists degradation. In addition, cellulose is integrated into a hemicellulose matrix. The latter is formed of heteropolymer of pentose monosaccharides (D-arabinose and D-xylose) and hexose monosaccharides (D-glucose, D-galactose and D-mannose,). Xylan is the most abundant hemicellulose in nature and accounts for 30–35% of the plants total dry

H. B. El-Nassan (✉)
Department of Pharmaceutical Organic Chemistry, Faculty of Pharmacy, Cairo University, Cairo, Egypt
e-mail: hala.el-nassan@pharma.cu.edu.eg

A. Sattar Jatoi (ed.), *Xylanase from Microorganisms*,
https://doi.org/10.1007/978-981-95-0824-2_12

weight. Higher percentage of xylan is present in hard woods than soft woods. Xylan is composed of *β-D*-xylopyranosyl residues connected by *β*-1,4-glycosidic bonds (polymer of xylose). There are four types of xylans according to their structure; arabinoxylans, glucuronoxylans, glucuronoarabinoxylans and galactoglucuronoarabinoxylans. Lignin has polymeric structure formed mainly of polyphenol. Lignin binds to cellulose and hemicellulose through covalent and non-covalent interactions [1, 2].

Xylanase enzymes represent a group of hydrolases that catalyze the hydrolysis of xylan polymer to form xylose monosaccharaides. They act by hydrolyzing the glycosidic linkage (*β*-1,4) of xyloside sugars. Xylanases are formed of *endo*-*β*-1,4-xylanase, *β*-1,4-xylosidase, acetylxylan esterase, phenolic acid esterase, *α*-glucuronidase, and *α*-L-arabinofuranosidase. Xylanase has different synonyms including pentosanase, *β*-1,4-xylanase, endo-1,4-*β*-*d*-xylanase, *β*-xylanase, and *β*-1,4-*d*-xylan-xylanohydrolase [1, 3–6].

The *endo*-*β*-1,4-xylanase and the *β*-xylosidase are the two main xylanase enzymes that catalyze the cleavage of the glycosidic linkage (*β*-1,4) and the release of the xylose sugar from xylan, respectively. Other xylanase enzymes catalyze the removal of the other side chain groups from xylan. The overall action of xylanase results in depolymerization of xylan into monomeric pentose units of xylose (Fig. 1) [6–8].

Xylanases have wide applications in industry including their use in baking and pastry industry, and in the production of laundry detergents as well as their use in bio-bleaching of wood pulps [1, 2, 5].

Xylanase enzymes have another significant application in the production of second-generation biofuels. This chapter focuses on the utilization of xylanase enzymes in the production of different biofuels and the efforts exerted during the last two decades to improve the production yield of the xylanase enzymes and to enhance their characteristics to produce biofuels.

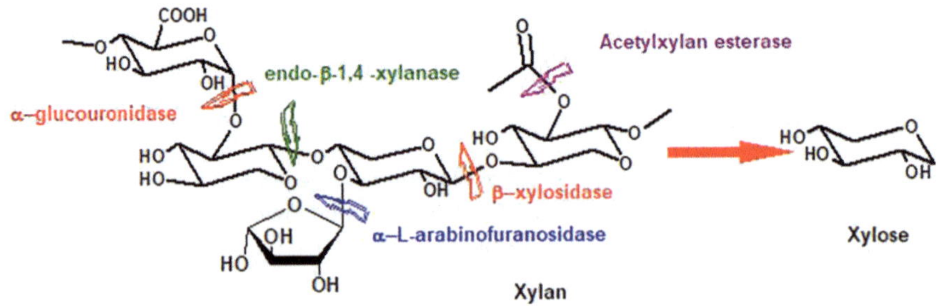

Fig. 1 The main cleavage points of xylanase enzyme in xylan structure to give xylose

2 Xylanase Classification

Xylanases as glycoside hydrolase (GH) are classified according to the similarity in hydrophobic clusters and structural elements into multiple families such as families 5, 7, 8, 9, 10, 11, 12, 16, 26, 30, 43, 44, 51 and 62 (http://www.cazy.org). Furthermore, these families are classified as GH with one catalytic domain (families 5, 7, 8, 10, 11 and 43) and as GH with two catalytic domains (families 16, 51 and 62) [2, 6, 9, 10].

Xylanases can be designated as GH10 and GH11 family on the basis of similarity in the hydrophobic cluster analysis of the catalytic domains and homology in the amino acid sequences [2, 11, 12]. The GH10 family (also called Family F) has a high molecular weight and acidic pI between 8 and 9.5. Family F consists of a cellulose-binding domain linked to a catalytic domain via a peptide linker. On contrary, the GH11 family (Family G) has a low molecular weight and lower pI values (alkaline pI) and can be subdivided into acidic and basic enzymes based on their pI. The GH11 family members have higher substrate selectivity, smaller size, and a broader range of pH and temperature optimum values which make them more suitable in many industrial applications [2, 9–14]. The mode of action, biochemical properties and substrates specificities depend on the type and source of microorganism that produces the enzyme [1, 15].

3 Xylanase Resources

The xylanase enzyme can be extracted from various sources including bacteria, fungi, protozoans, insects, snails, mollusks, marine algae and some type of immature seeds [1, 2, 15]. The characters and production yield of the enzyme vary with each source. However, microorganisms are preferred for commercial and industrial production of xylanases due to their rapid growth in small space, ability to consume waste biomass as cheap source of nutrients, low production cost and ease of handling [2]. Fungi; especially filamentous fungi; are considered the best source for the commercial production of xylanase enzyme [8]. Most of the bacterial xylanases belong to family F; while most of the fungal xylanases belong to family G [16].

Xylanases are secreted by different strains of the *Bacillus sp., Pseudomonas sp. and Streptomyces sp.* These types of xylanases are effective at pH 5–9, and temperature 35ºC to 60 ºC. Since bacterial xylanases are stable and active at elevated temperature and in alkaline medium, they are widely used in industrial applications [1, 8, 15, 17]. Examples of the studied bacteria include *Bacillus cereus, Bacillus pumilus, Bacillus circulans, Bacillus subtilis, Bacillus amyloliquefaciens, Bacillus licheniformis, Streptomyces roseiscleroticus, Streptomyces cuspidosporus, Clostridium absonum* and *Thermoactinomyces thalophilus* [1, 8, 15, 18].

Fungal xylanases are obtained from *Aspergillus sp., Penicillium sp. and Fusarium sp.* The mesophilic fungi especially *Aspergillus* and *Trichoderma* are the most potent xylanase producers [16, 19]. Generally, the fungal xylanases have higher activity and are produced in higher yields than bacterial xylanases. However, they are active at pH 4–6 and at temperature below 50 °C. Their low thermostability and intolerance to alkaline medium preclude their use in some industrial applications that need high temperature and alkaline pH such as pulp and paper industry [1, 8, 15]. Examples of the studied fungi include *Aspergillus niger, A. terreus, Trichoderma reesei, T. viride, T. harzianum, Thermomyces lanuginosus, Fusarium oxysporum, Schizophyllum commune and Talaromyces emersonii.* Other examples include the white-rot fungi such as *Penicillium echinulatum, P. chrysosporium, C. versicolor* and *C. subvermispora* [1, 2, 8, 15, 17, 20].

The optimum conditions for xylanase activity include high temperatures (40–60 °C) and neutral pH (for bacterial xylanase) or slightly acidic pH (for fungal xylanase). Some wild strains of microorganisms can produce thermostable xylanases [21, 22]. Examples of microorganisms that produce thermostable xylanases include *Thermoascus aurantiacus, Fusarium proliferatum, Dictyoglomus sp., Thermoactinomyces thalophilus, Alicyclobacillus acidocaldarius* and *Clostridium abusonum CFR-702* [23]. Furthermore, genetically modified microorganisms are developed to produce xylanases that are stable and active at temperature (5 to 105 °C), extreme pH (2–11) and high salt concentrations (up to 30%) [2, 6, 13, 16, 23].

4 Strategies Applied for Xylanase Production

The selection of the xylanase production method is dependent on the type and strain of microorganism, the physicochemical growth parameters, the type of substrate and the fermentation strategy. Two strategies are followed for the microbial production of xylanases; solid state fermentation (SSF) and submerged fermentation (SmF).

The solid-state fermentation (SSF) uses a solid matrix that serves as both a physical support and a nutrition source for the microorganism. The SSF technology is more suitable for the growth of fungi which can grow at low water activity unlike bacteria which need higher volume of water. Therefore, SSF is used for xylanase production more frequently than SmF. The SSF technology uses agriculture biomass and lignocellulose biomass as substrate for xylanase production from fungi. Challenges of SSF technology include the design of bioreactor, the use of different gradients of temperature, moisture and substrate concentrations. More research is required to improve the design of bioreactors and to optimize the growth parameters [15, 19].

The submerged fermentation (SmF) depends on the use of liquid fermentation media and the substrate is soluble in water. SmF doesnot need any equipment for enzyme production and affords more purified enzymes compared to SSF. However, SmF needs large space and high water activity [15, 20].

A specific substrate is needed for each bacterial or fungal strain and the cost of xylanase production is highly dependent on the type of the selected substrate. The solid substrates (seeds, roots, fecal matter or animal bones) are the best substrates used to grow bacteria and fungi. Other factors that also affect the xylanase production include the temperature, humidity, pH, agitation/aeration and incubation time. With the aid of biotechnology, the production yield of xylanase was increased to fulfil the demand for the mass production of xylanase for commercial and industrial use [11, 17, 19, 24].

The purification of xylanase enzyme can be achieved using conventional methods such as ultrafiltration, ion exchange chromatography and ammonium sulfate precipitation. However, these methods are uneconomical, time-consuming and afford low production yield of the enzyme [14, 15]. The liquid–liquid fractionation (aqueous two-phase system; ATPS) is a more effective technique that provides higher yield of xylanase enzyme and is more energy-efficient and economical method [25–27].

5 Importance of Biofuels

Biofuels are any type of fuel derived from organic material known as biomass. Biofuels include biodiesel, biogasoline such as bioalcohols (bioethanol, and biobutanol), biogas and biohydrogen. Biofuels represent promising alternatives to fossil fuels with lesser effect on environmental pollution as they reduce the carbon dioxide and sulfur emission. Thus, they reduce the greenhouse gases emission and donot contribute to the global warming effect. Additionally, biofuels are obtained from sustainable sources such as biomasses and wastewater. Moreover, the production of biofuel from biomass can increase the national income and aid in rural development of many countries [28–30].

The production of fuels and chemicals from biomass received great attention in last two decades to solve the fossil fuel depletion crisis and the environmental pollution crisis. The focus was on the use of agricultural wastes which are rich in polysaccharides. The plants contain high percentage of cellulose, hemicellulose and lignin that are used to protect their cells. Enzymatic hydrolysis of these compounds affords monosaccharides (sugar monomers) which can be further fermented to generate biofuels like ethanol and butanol [31].

6 Biofuels Classification

Biofuels are divided into four generations according to the production source and the production technology [29, 30, 32].

The first generation biofuels are obtained by the fermentation of edible biomass. This includes starch based feedstock like wheat, potato and corn; sugar based feedstock like sugar beet and sugarcane; and oil crops like palm oil, corn oil and soy bean oil [2]. The conventional technologies applied for biofuel production include chemical, thermo-chemical and biological conversion technologies which are used effectively to generate biodiesel, bioethanol and biobutanol. However, the production of biofuels using these resources on large scale needs cultivation of wide areas by these crops which constitutes a competition with food production for water and land.

This can increase the cost of crops and contributes to resources depletion. In addition, these biomass are used for animal feeds and their use as biofuel source increases the prices of animal foods. Therefore, the use of this source is not economical [28, 31].

The second generation biofuels are obtained mainly from lignocellulosic biomass such as woody crops and waste plant residues, grasses, waste and non-edible vegetables. This source can provide energy and meanwhile consumes a large amount of wastes and thus can solve energy, environmental and economic crisis. Moreover, they are produced from non-edible waste and thus are not competing with food crops and donot require cultivation land for their production. The greenhouse gases GHG emission from this type of biofuels is less than that produced from first generation biofuels. However, the production cost for the second generation biofuels is much higher than that of first generation biofuels as they require pretreatment steps and more complicated technology. The production of enzymes that catalyze the biofuel formation and the extraction of biofuels from these biomass are the two main limitations of this class as will be discussed later in this chapter [28, 32, 33].

The third generation of biofuel is obtained from certain types of phototrophic microalgae with natural high oil content. The algae convert solar energy into chemical energy and store it in the form of oils. The oils are extracted from algae and fermented to obtain biodiesel. The terpenes produced as secondary metabolites from algae can be used as substituents to petroleum fuels. This method affords the high energy biodiesel from low cost and renewable source. The microalgae have high growth rate (5–6 days), produce very large amount of oils and donot need large area for cultivation. They can be also cultivated in wastewater to reduce the production cost. However, the biofuels obtained by microalgae are more volatile compared to other generations biofuels [30, 34, 35].

The fourth generation biofuels (also known as photosynthetic biofuels) are obtained using microorganisms (like cyanobacteria and algae) that are genetically engineered to produce biofuels from optimized feedstock. These genetically modified microorganisms can capture high volume of carbon dioxide and use wastewater as a source of water and nutrients. Thus, the amount consumed of carbon dioxide is lower than the amount emitted and therefore they are carbon negative sources. The most important biofuels produced by this technology include bioethanol, biodiesel and biohydrogen. This method is sustainable, ecofriendly and reduces the carbon dioxide pollution. However, the production cost is still a limitation for large scale application of this technology [28, 32, 36].

7 Production of Second Generation Bioethanol

Bioethanol is the most common biofuel and can be used as a blend with petrol up to 85% content (E85; 85% ethanol and 15% petrol). E85 requires certain modifications in the engine and is thus used only in Flexible Fuel Vehicles (FFVs). USA and Brazil are the two major bioethanol producers from corn starch and sugarcane juice, respectively. Bioethanol is environmental friendly fuel that produces less amount of CO_2

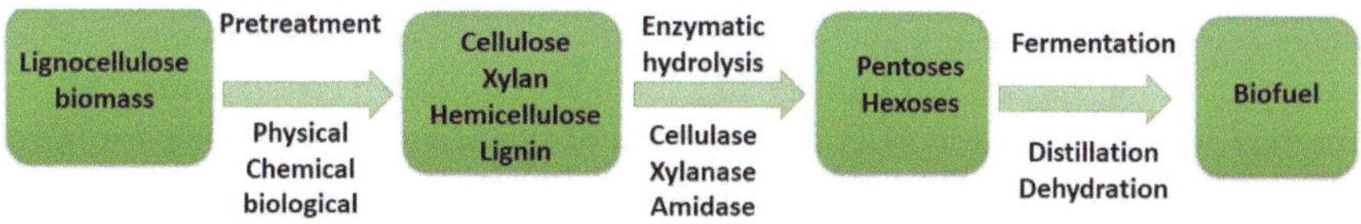

Fig. 2 The main steps used for biofuel generation from lignocellulose biomass

emission by 78%. However, bioethanol is hygroscopic and its blending with gasoline must be done right before use. Therefore, the transport, storage and distribution of bioethanol is a problem and this adds additional cost on the final product. Moreover, bioethanol is corrosive and its vapor pressure is lower than gasoline [37].

Bioethanol can be obtained as a first generation biofuel from sugar and starch edible biomass or as a second generation biofuel from lignocellulose biomass. The first generation bioethanol is obtained by hydrolysis of the biomass by amylases (saccharification) to yield sugars that are then fermented by yeast or bacteria. This is followed by distillation and dehydration of the bioethanol [37]. The bioethanol production from lignocellulose biomass (second generation bioethanol) involves the following key stages: pretreatment, enzyme hydrolysis or saccharification, fermentation, distillation and dehydration (Fig. 2).

The pretreatment step is needed to break up the fibers of lignocellulose and thereby facilitates the hydrolysis step and increases the sugar yield. The pretreatment step is carried out by mechanical, chemical or biological route. The mechanical route uses physical milling or grinding to decrease the size of the lignocellulose fibers but these techniques are energy consuming and thus not economical. The chemical route involves the use of acid or base with subsequent formation of organic acids, furfural and 5-hydroxymethylfurfural as sugar degradation products. This inhibits the microbial growth in the next fermentation step. To get rid of these inhibitors, a detoxification process is conducted using over-liming, activated carbon or molecular sieves. Other detoxification methods include extraction with organic solvents and the use of ion exchange resins. Again, this add additional cost and time to the process of bioethanol production [37]. The biological pretreatment route, on contrary, utilizes microorganisms (like white, brown or soft rot fungi) and is conducted under milder conditions. The biological route is environmental friendly, energy-efficient pathway and doesnot generate any toxic products [37].

During the enzymatic hydrolysis step, the polysaccharides in the biomass hydrolysate are converted into simple monosaccharaides (mainly glucose and xylose) that are suitable for bioethanol production in the next fermentation step. The enzymes used for the hydrolysis of lignocellulose biomass are mainly xylanase and cellulase which exert a synergistic effect and increase the yield of the monosaccharides [38, 39]. Microorganisms are frequently used to produce the enzymes necessary for this hydrolysis [40]. Cellulase is used to convert cellulose into glucose; while xylanase converts xylan into simple xylose monomers which yield bioethanol by fermentation [4, 33, 41, 42]. The combined production of xylanase and cellulase enzymes by the same microorganism or co-cultured microorganisms were studied by many research teams and proved to be an effective way of fermentation and production of biofuel. Examples of the studied microorganisms include *Penicillium echinulatum* [20] and *Bacillus amyloliquefaciens* [18]. The addition of surfactants like tween 80 or tween 20 enhanced the enzymatic activity of xylanase [38, 43].

The fermentation step is performed using different types of bacteria and fungi. The ideal microorganisms should be capable of fermentation of hexoses and pentoses. Biotechnology and genetic modifications are used to develop bacterial strains that can ferment two types of sugars simultaneously [38].

Three techniques are employed for bioethanol production. The first is a two-step method and is called separate hydrolysis and fermentation (SHF) process. The SHF is not an economical process as separate vessels are needed for the saccharification and fermentation steps and the process is time-consuming [43].

The second is a simultaneous saccharification and fermentation (SSF) method. The SSF is more cost-effective as it reduces the number of vessels used for production. Nevertheless, adjusting the suitable conditions for simultaneous enzymatic hydrolysis and fermentation remains a challenge. The enzymatic hydrolysis process requires high temperature (50–55 °C) which is not suitable for the mesophilic yeasts responsible for fermentation (35 °C). Therefore, natural or genetically modified thermophilic microorganisms such as *Saccharomyces, Candida, Pichia stipitis, Alkalothermophilic Thermomonospora* and *Wickerhamomyces* are preferred in SSF process [37, 43].

The third method is consolidated bioprocessing (CBP) in which a single strain is used for both saccharification and fermentation. The microorganism produces hydrolyzing enzymes and ferments the produced sugar to generate bioethanol [37, 41, 43, 44]. The co-culture system is frequently used in CBP where enzyme producing microorganism and biofuel producing microorganism are used together in the same reactor. CBP can thus reduce the overall cost of bioethanol production as multiple steps are conducted in a single reactor and the cost of enzyme separation and purification is omitted [44] (Table 1).

8 Production of Second Generation Biobutanol

Biobutanol has significant potential as a renewable fuel that resembles gasoline in its characteristics. Biobutanol has low hydrophilicity and high caloric content which make it superior biofuel over bioethanol [49]. Biobutanol can be blended with gasoline in any ratio and used in the engine systems without any modification, unlike bioethanol which can be mixed with gasoline with a maximum ratio of 15% [50]. Biobutanol is also less corrosive and less hygroscopic than ethanol. Moreover, the low vapor pressure of biobutanol enables its safe handling and transportation [51–54]. In spite of these characteristics, its production on a large scale is still expensive mainly due to the elevated cost of biomass substrates used [55]. Other obstacles include the low production yield due to formation of side products and the hard recovery steps owing to the low yield and high boiling point of biobutanol. The normal distillation

Table 1 Examples of bioethanol production from lignocellulose biomass using xylanase for enzymatic hydrolysis

Xylanase source	Biomass used	Bioethanol yield	References
Alkalothermophilic Thermomonospora and *Debaryomyces hansenii*	Oat-spelt xylan	9.1 g/L	[43]
Alkalothermophilic Thermomonospora and *Debaryomyces hansenii*	Wheat bran hemicellulose	9.5 g/L	[43]
Geobacillus thermoglucosidasius and Geobacillus sp. DUSELR13	Corn stover	3.53 g/L	[21]
Geobacillus thermoglucosidasius and Geobacillus sp. DUSELR13	Prairie cordgrass	3.72 g/L	[21]
Saccharomyces cerevisiae and Escherichia coli KO11	Low-moisture anhydrous ammonia pretreated napiegrass	4.82 g/L	[45]
		21.4 g/L	[46]
Candida tropicalis MK-160	Sugarcane-bagasse and wheat bran	5.45% (v/v)	[40]
Recombinant Saccharomyces cerevisiae	Birchwood xylan	0.31 g/g	[47]
Clostridium pasteurianum GL11	Birchwood xylan	5.12 g/L	[48]

processes are not effective in the recovery of biobutanol and therefore more advanced and economical processes are needed such as adsorption, reverse osmosis, gas stripping, liquid–liquid extraction and using heat pump (vapor recompression)-assisted azeotropic dividing-wall column [54, 56–59].

Biobutanol is produced by certain types of anaerobic bacteria through acetone-butanol-ethanol (ABE) fermentation. The most common species used for this process is *Clostridia sp.* The main strains of *Clostridia* reported in the literature for ABE fermentation include *C. saccharoperbutylacetonicum, C. acetobutylicum, C. beijerinckii, C. pasteurianum* and *C. saccharobutylicum* [52, 54, 60–64]. The solventogenic *Clostridia sp.* can efficiently use both hexoses and pentose for biofuel production. However, glucose is preferentially consumed before xylose. Some modified strains like *Clostridium sp. strain BOH3* can consume both glucose and xylose simultaneously to generate biobutanol [65]. The microorganisms produce butanol as a major product along with smaller amounts of acetone and ethanol (the mass ratio of butanol: acetone: ethanol is 6:3:1) [50, 52, 60, 66–70]. Some *Clostridia sp.* are also able to synthesize xylanase enzyme and hence can hydrolyze xylan and hemicellulose [14, 22, 63, 71]. Alternatively, xylanase together with other enzymes like cellulase, *β*-glucanase and mannase may be used initially to hydrolyze the lignocellulose biomass then fermentation of the resulting sugar can be effected by bacteria [39, 49, 72].

The production of biobutanol from lignocellulose biomass includes three steps: pretreatment, hydrolysis and fermentation (Fig. 2). Similar to bioethanol production, three main strategies are currently applied for biobutanol production:

- Performing the three steps separately (known as separate hydrolysis and fermentation (SHF)) which is not an economic process [73, 74].
- Merging the hydrolysis and fermentation steps (simultaneous saccharification and fermentation (SSF)) followed by separate recovery step [75].
- Performing the three steps in one process referred to as consolidated bioprocessing (CBP). Herein, the microorganism produces the hydrolytic enzymes which hydrolyze the biomass to simple sugars followed by fermentation of these sugars to afford biofuels [55, 63, 70, 71, 76–78]. Tables 2 and 3 include examples of production of biobutanol using xylanase

Overliming and activated carbon are used after the prehydrolysis of lignocellulose biomass step to adsorb toxic compounds (like aliphatic acids, phenols and furan) that can interfere with the microbial fermentation. This step is known as detoxification step [61, 62, 73, 75, 79]. Other effective detoxicification methods include flocculation, nanofiltration and adsorption to ion exchange resin [54, 58, 80].

The ABE fermentation involves two steps: acidogenesis and solventogenesis. In the acidogenesis phase, organic acids are produced from sugar fermentation and when their concentartions increase, the medium pH decreases and the solventogenesis phase starts resulting in ABE production [50]. The process is influenced by the amount and type of the fermentable sugar as well as the pH of the medium [54].

Table 2 Examples of biobutanol production from lignocellulose biomass using xylanase for enzymatic hydrolysis

Xylanase source	Biomass used	Biobutanol Yield	References
Kluyvera OM3 and *Clostridium* BOH3	Palm oil fiber	1.2 g/L	[22]
Hypocrealixii SS1and Trichoderma atroviridae	Sunflower crops	18.94 g/L	[89]
Thermoanaerobacterium sp. M5	Beechwood xylan	1.17 g/L	[78]
Thermoanaerobacterium sp. M5 and C. acetobutylicum NJ4	Beechwood xylan	8.34 g/L	[78]
Clostridium acetobutylicum	Hemicellulose	4.03 g/L	[63]
Clostridium acetobutylicum	Xylan	5.43 g/L	[63]
Clostridium sp. strain NJP7	Hemicelluloses	25.58 g/L	[71]
Clostridium strain BOH3	Sugarcane bagasse hydrolysate	14.80 g/L	[87]
Clostridium strain BOH3	Beechwood xylan	12.15 g/L	[88]
Clostridium strain BOH3	Rice bran and sesame oil cake	13.50 ± 0.12 g/L	[90]
Clostridium pasteurianum GL11	Birchwood xylan	1.48 g/L	[48]

Table 3 Examples of ABE production from lignocellulose biomass using xylanase for enzymatic hydrolysis

Bacterial strain used for fermentation	Biomass used	ABE yield	References
Clostridium acetobutylicum P260	Corn fiber arabinoxylan	24.67 g/L	[55]
C. acetobutylicum ATCC 824	Hemicellulose hydrolysate of pulping woods	10.8 g/L for poplar (hardwood) and 13.2 g/L for southern pine (softwood)	[75]
Clostridium acetobutylicum DSM 1731	Barley straw pretreated by dil. H_2SO_4	10.8 g/L ABE containing 7.9 g/L butanol	[83]
C. acetobutylicum YM1	Acacia mangium	2.87 g/L	[50]
Clostridium beijerinckii P260	Sweet Sorghum Bagasse	7.91 to 32.38 g/L	[76]
	Corn stover	20.79 g/L	[91]
	Barley straw hydrolysate (untreated)	7.09 g/L	[92]
	Barley straw hydrolysate (Lime treated)	26.64 g/L	[92]
Clostridium acetobutylicum ATCC 824	Steam-exploded corn stover	12.38 ± 0.23 g/L	[79]
Clostridium saccharoperbutylacetonicum N1-4	Wastewater algae	9.74 g/L	[72]
Clostridium beijerinckii BA 101	Fiber-enriched distillers dried grains with solubles	5.35 g/L containing 3.55 g/L butanol	[93]

The pretreatment step of lignocellulose biomass is done usually using dilute acids which degrade the cross-linked matrix of lignin or hemicelluloses. During the subsequent hydrolysis step, enzymes like xylanase and cellulase can easily work on xylan or cellulose. However, in some cases remains of lignin or hemicelluloses are still present and adsorb the enzymes onto their surfaces thereby disrupting their hydrolytic actions on xylan or cellulose [54, 81, 82]. This issue can be overcome by adding surfactant like polyethylene glycol PEG and Tween together with the enzyme in the hydrolysis step. The added surfactant stabilizes the enzymes and can adsorb to lignin and xylan surface and inhibits or reduces the enzyme binding to these surfaces and thus the enzymes can effectively work [54, 83–86]. The use of surfactants with xylanase enzyme increases the sugar yield significantly and results in higher yield of biobutanol. In one comparative study, it was found that the simultaneous use of xylanase and PEG 4000 on pretreated barley straw resulted in a significant improvement in glucose and xylose yields compared to the yields from the enzyme or the surfactant alone. This was accompanied by an increase in ABE yield from 93.8 to 135.0 g/kg [83].

Enhancement of xylanase production and activity was achieved by optimization of the culture medium via addition of iron (II) and (III) and ammonium sulfate. The enhanced xylanase activity boosted the butanol and hydrogen production by strain *BOH3* to reach concentrations comparable to those produced directly from xylose fermentation [87, 88].

The thermophilic *Thermoanaerobacterium* sp. M5 strain was recently identified as butanol producer from xylan via butanol–ethanol (BE) fermentation pathway rather than ABE fermentation. This strain can efficiently secrete xylanase and *β*-xylosidase that degrade xylan. The biobutanol yield was increased from 1.17 g/L to 8.34 g/L upon microbial co-cultivation with the highly butanol producing strain *C. acetobutylicum NJ4.* The biobutanol production was achieved via the CBP process under thermophilic conditions [78].

The wild-type *Clostridium* sp. strain NJP7 was reported to produce biobutanol through acetone–isopropanol–butanol fermentation pathway and afforded 2.06 g/L butanol from xylan fermentation [71].

9 Production of Second Generation Biohydrogen

Biohydrogen provides a clean and sustainable energy alternative. It does not contain carbon or sulfur; so no GHGs are emitted by its combustion. Moreover, biohydrogen has an energy content of 143 GJ/ton which is higher than other gaseous fuels. Biohydrogen can be also used in fuel cells to produce electricity [94–96]. However, its production cost is relatively high owing to the high cost of substrates. Hydrogen can be produced by the conventional methods (thermochemical and electrochemical) or the more advantageous biological method. The process of biohydrogen production can be performed via dark fermentation or photosynthesis. The dark fermentation is more environmental friendly and various substrates can be used to generate hydrogen in higher yields than photosynthesis. Upon using lignocellulose as a substrate, a pretreatment step is needed to break up the cellulose fibers and hydrolyze the sugars. Furthermore, the production of biohydrogen from lignocellulose biomass is limited by the number of microorganisms that can ferment glucose and xylose to produce hydrogen. Examples of these strains include *Clostridium strain BOH3* [87, 88], *Thermoanaerobacterium thermosaccharolyticum KKU-ED1* [97] and *T. thermosaccharolyticum W16* [98]. The ABE fermentation yields both biohydrogen and biobutanol. Therefore, the main microorganism strains used for biohydrogen production belong to *Clostridium* strain and in order to favor the hydrogen production pathway, the carbon source was reduced (<10 g/L) and the pH was regulated around pH7. The *clostridium* strains used for the biohydrogen production include both mesophilic strains such as *Clostridium populeti FZ10* [99] and thermophilic strains like *C. thermocellum* [100].

Table 4 Examples of biohydrogen production from lignocellulose biomass using xylanase for enzymatic hydrolysis

Bacterial strain used for fermentation	Biomass used	Biohydrogen yield	Reference
Thermoanaerobacterium thermosaccharolyticum W16	Corn stover hydrolysate	108.5 mM/L	[98]
Clostridium strain BOH3	Sugarcane bagasse hydrolysate	2.65 ± 0.15 L/L	[87]
	Beechwood xylan	3.29 L/L	[88]
	Rice bran and sesame oil cake	4.41 L/L	[90]
	Oat spelt xylan	1.93 mmol	[102]
Clostridium sp. Strain X53	Oat spelt xylan	1184 mL/L	[101]
Acinetobacter junii F6-02 and Clostridium butyricum CGS5	Hydrolysate of xylan	1288 mL/L	[103]
	Hydrolysate of pretreated rice straw	1006 mL/L	[103]
Thermoanaerobacterium thermosaccharolyticum KKU-ED1	Sugarcane bagasse hydrolysate	86.58 ± 6.18 mL/L	[97]
Clostridium populeti FZ10	Microcrystalline cellulose	177.5 ± 4.8 mL/g	[99]
Thermoanaerobacterium thermosaccharolyticum and Clostridium sp	Sugarcane bagasse	2.60 mmol/g	[104]
Clostridium thermocellum ATCC 27,405	Dried distillers grain	1.27 mmol/g	[100]
T. thermosaccharolyticum M18	Corn cob, corn stalk, and wheat straw	3.23–3.48 mmol/g	[105]

Xylanase enzymes are combined with cellulase to hydrolyze lignocellulose in the laboratory studies. However, some strains such as *Clostridium strain BOH3* [87, 88], *Clostridium populeti FZ10* [99] and *Clostridium sp. Strain X53* [101] are capable of producing xylanase enzyme and produce hydrogen simultaneously. Table 4 summarizes the biohydrogen production from lignocellulose using xylanase enzyme.

10 Challenges and Limitations of Biofuel Production from Lignocellulose Biomass

Many challenges face the biofuel production from lignocellulosic biomass. The cost of the biomass represents the major challenge in biofuel production. Optimization of the operation process to elevate the sugar content of the hydrolysate and to enhance the biofuel yield and recovery are all possible solutions to decrease the production cost. The use of xylanase enzyme aids to increase the percentage of the sugar content as indicated by many research teams. The application of genetic engineering to develop microorganisms with high bioethanol or biobutanol production ability is another effective strategy to increase the biofuel yield. The use of microorganisms capable of fermenting both hexoses and pentoses can also assist in resolving the problem of low production yield of biofuel. The use of advanced and economical methods of recovery like liquid–liquid extraction and gas stripping can also increase the production yield.

The lignocellulose hydrolysis by xylanases is restricted by multiple factors including the hard texture of the lignocellulose that prevents the accessibility of the enzymes to xylan polymer. This can be overcome by the pretreatment that break up the fibrous structure of lignocellulose. Other limitations include the small pH range and low thermal stability of the enzyme. Both can be overcome by the use of thermostable xylanases. The elevated enzyme production cost can be reduced by the use of cheap substrates, high enzyme producing strains and solid-state fermentation (SSF) technique for enzyme production.

The application of consolidated bioprocessing can also decrease the expenses of the biofuel production process as it decreases the cost of providing commercial enzymes for saccharification process and reduces the number of stages and culture media required for fermentation. The microorganism produces cellulase, xylanase or amylase in situ and the produced enzymes hydrolyze lignocellulose that can be fermented in the same reactor to produce biofuel. However, this step requires the use of microorganisms that produce enzymes in high yield. Additionally, the production of thermostable enzymes is more favorable as they can tolerate the alkaline and high temperature conditions used for fermentation. Some strains like *Clostridia* are solventogenic and meanwhile enzyme producing. Further research is still needed to develop xylanases with high stability to extreme conditions and high enzymatic activity for industrial applications. The isolation of xylanase-producing wild bacterial and fungi strains that can withstand different ranges of temperatures and pH and the optimization of their growth conditions become essential. Alternatively, genetic modification of the known microbial strains with thermophilic genes can also lead to production of thermophilic xylanases.

11 Conclusion

The drastic climatic changes and the possibility of fossil fuel depletion initiate the urgent need for more energy sources that are sustainable and environment friendly. Biofuels represent a promising solution for these problems as they can be obtained from renewable sources and meanwhile do not increase the greenhouse gases emission. Second generation biofuels are obtained by fermentation of lignocellulosic biomass using different strains of bacteria and fungi.

The bioethanol or biobutanol production steps involve pretreatment to break the lignocellulose fiber into smaller size polymers; enzymatic hydrolysis to convert polysaccharides into glucose and xylose; fermentation of these sugars to give biofuel; and finally recovery and drying of biofuel. The pretreatment process is usually achieved using acids or bases and results in the production of toxic inhibitors that inhibit the microorganism growth and decrease the biofuel yield. The detoxification of these inhibitors can be carried out using overliming, activated carbon or adsorption on ion exchange resin. The enzyme hydrolysis of lignocellulose is achieved using mainly two enzymes cellulase and xylanases. These enzymes are either added to the fermentation system or produced in situ by the fermentation microorganisms. Xylanase enzyme efficiently catalyzes the conversion of the polymeric xylan structure into the simple xylose sugar that can produce biofuel via fermentation processes. However, the application of these processes on large scale is limited by the low production of xylanase enzyme and the limited activity on lignocellulosic biomass. This problem can be overcome through the application of recombinant DNA techniques to enhance the enzyme production in large amount. Filamentous fungi are widely used for the commercial production of xylanases. The data presented in this chapter spotlighted the importance of xylanase enzymes in the process of biofuel production from lignocellulosic biomass and indicated that more effort are still needed to increase the yield of xylanases produced by microorganisms and to enhance the thermostability of the produced enzymes.

References

1. A. Burlacu, F. Israel-Roming, C.P. Cornea, Microbial xylanase: a review. Sci. Bull. Ser. F. Biotechnol. **20**, 335–342 (2016)
2. A. Walia, S. Guleria, P. Mehta, A. Chauhan, J. Parkash, Microbial xylanases and their industrial application in pulp and paper biobleaching: a review. 3 Biotech. **7,** 11 (2017). https://doi.org/10.1007/s13205-016-0584-6
3. D. Dodd, I.K.O. Cann, Enzymatic deconstruction of xylan for biofuel production. GCB Bioenergy. **1**, 2–17 (2009). https://doi.org/10.1111/j.1757-1707.2009.01004.x
4. P.V. Harris, F. Xu, N.E. Kreel, C. Kang, S. Fukuyama, New enzyme insights drive advances in commercial ethanol production. Curr. Opin. Chem. Biol. **19**, 162–170 (2014). https://doi.org/10.1016/j.cbpa.2014.02.015
5. V. Juturu, J.C. Wu, Microbial xylanases: engineering, production and industrial applications. Biotechnol. Adv. **30**, 1219–1227 (2012). https://doi.org/10.1016/j.biotechadv.2011.11.006
6. M. Sharma, A. Kumar, Xylanases: an overview. Br. Biotechnol. J. **3**, 1–28 (2013). https://doi.org/10.9734/BBJ/2013/1784
7. Y. Yang, N. Zhu, J. Yang, Y. Lin, J. Liu, R. Wang, F. Wang, H. Yuan, A novel bifunctional acetyl xylan esterase/arabinofuranosidase from Penicillium chrysogenum P33 enhances enzymatic hydrolysis of lignocellulose. Microb. Cell Fact. **16**, 1–12 (2017). https://doi.org/10.1186/s12934-017-0777-7
8. A. Mandal, Review on microbial xylanases and their applications. Int. J. Life Sci. **4**, 178–187 (2015)
9. U.S.P. Uday, P. Choudhury, T.K. Bandyopadhyay, B. Bhunia, Classification, mode of action and production strategy of xylanase and its application for biofuel production from water

hyacinth. Int. J. Biol. Macromol. **82**, 1041–1054 (2016). https://doi.org/10.1016/j.ijbiomac.2015.10.086
10. R. Chaudhary, T. Kuthiala, G. Singh, S. Rarotra, A. Kaur, S.K. Arya, P. Kumar, Current status of xylanase for biofuel production: a review on classification and characterization. Biomass Convers. Biorefinery. **13**, 8773–8791 (2023). https://doi.org/10.1007/s13399-021-01948-2
11. S.H. Lee, V. Lim, C.K. Lee, Newly isolate highly potential xylanase producer strain from various environmental sources. Biocatal. Agric. Biotechnol. **16**, 669–676 (2018). https://doi.org/10.1016/j.bcab.2018.09.024
12. Y.A. Denisenko, A.V. Gusakov, A.M. Rozhkova, I.N. Zorov, A.V. Bashirova, V.Y. Matys, V.A. Nemashkalov, A.P. Sinitsyn, Protein engineering of GH10 family xylanases for gaining a resistance to cereal proteinaceous inhibitors. Biocatal. Agric. Biotechnol. **17**, 690–695 (2019). https://doi.org/10.1016/j.bcab.2019.01.042
13. I.J. Kim, S.R. Kim, K.H. Kim, U.T. Bornscheuer, K.H. Nam, Characterization and structural analysis of the endo-1,4-β-xylanase GH11 from the hemicellulose-degrading Thermoanaerobacterium saccharolyticum useful for lignocellulose saccharification. Sci. Rep. **13**, 17332 (2023). https://doi.org/10.1038/s41598-023-44495-8
14. C.H. Ng, J. He, K.L. Yang, Purification and characterization of a GH11 xylanase from biobutanol-producing clostridium beijerinckii G117. Appl. Biochem. Biotechnol. **175**, 2832–2844 (2015). https://doi.org/10.1007/s12010-014-1470-5
15. T. Abena, A. Simachew, A review on xylanase sources, classification, mode of action, fermentation processes, and applications as a promising biocatalyst. Biotechnologia **105**, 273–285 (2024). https://doi.org/10.5114/bta.2024.141806
16. L. Liu, J. Cheng, H. Chen, X. Li, S. Wang, A. Song, M. Wang, B. Wang, J. Shen, Directed evolution of a mesophilic fungal xylanase by fusion of a thermophilic bacterial carbohydrate-binding module. Process Biochem. **46**, 395–398 (2011). https://doi.org/10.1016/j.procbio.2010.07.026
17. C. Kuancha, J. Apiraksakorn, Cultural condition improvement for xylanase production by Bacillus subtilis GN156. KKU Res. J. **17**, 933–938 (2012)
18. A. Amore, B. Parameswaran, R. Kumar, L. Birolo, R. Vinciguerra, L. Marcolongo, E. Ionata, F. La Cara, A. Pandey, V. Faraco, Application of a new xylanase activity from Bacillus amyloliquefaciens XR44A in brewer's spent grain saccharification. J. Chem. Technol. Biotechnol. **90**, 573–581 (2015). https://doi.org/10.1002/jctb.4589
19. K. Vimalashanmugam, T. Viruthagiri, Production of xylanase enzyme by Aspergillus terreus under SSF using response surface methodology. Int. J. ChemTech Res. **5**, 2365–2374 (2013)
20. C.E. Todero Ritter, M. Camassola, D. Zampieri, M.M. Silveira, A.J.P. Dillon, Cellulase and xylanase production by penicillium echinulatum in submerged media containing cellulose amended with sorbitol. Enzyme Res. **2013**, 240219(2013). https://doi.org/10.1155/2013/240219
21. M. Bibra, V.R. Kunreddy, R.K. Sani, Thermostable xylanase production by geobacillus sp. Strain duselr13, and its application in ethanol production with lignocellulosic biomass, Microorganisms. **6**, 93 (2018). https://doi.org/10.3390/microorganisms6030093
22. F. Xin, J. He, Characterization of a thermostable xylanase from a newly isolated Kluyvera species and its application for biobutanol production. Bioresour. Technol. **135**, 309–315 (2013). https://doi.org/10.1016/j.biortech.2012.10.002
23. G.D. Haki, S.K. Rakshit, Developments in industrially important thermostable enzymes: a review. Bioresour. Technol. **89**, 17–34 (2003). https://doi.org/10.1016/S0960-8524(03)00033-6
24. V. Kumar, A.K. Dangi, P. Shukla, Engineering thermostable microbial xylanases toward its industrial applications. Mol. Biotechnol. **60**, 226–235 (2018). https://doi.org/10.1007/s12033-018-0059-6
25. A. Glyk, T. Scheper, S. Beutel, PEG–salt aqueous two-phase systems: an attractive and versatile liquid–liquid extraction technology for the downstream processing of proteins and enzymes. Appl. Microbiol. Biotechnol. **99**, 6599–6616 (2015). https://doi.org/10.1007/s00253-015-6779-7

26. M. Iqbal, Y. Tao, S. Xie, Y. Zhu, D. Chen, X. Wang, L. Huang, D. Peng, A. Sattar, M.A.B. Shabbir, H.I. Hussain, S. Ahmed, Z. Yuan, Aqueous two-phase system (ATPS): an overview and advances in its applications. Biol. Proced. Online. **18**, 18 (2016). https://doi.org/10.1186/s12575-016-0048-8
27. F. Rahimpour, R. Hatti-Kaul, G. Mamo, Response surface methodology and artificial neural network modelling of an aqueous two-phase system for purification of a recombinant alkaline active xylanase. Process Biochem. **51**, 452–462 (2016). https://doi.org/10.1016/j.procbio.2015.12.018
28. R. Luque, L. Herrero-Davila, J.M. Campelo, J.H. Clark, J.M. Hidalgo, D. Luna, J.M. Marinas, A.A. Romero, Biofuels: a technological perspective. Energy Environ. Sci. **1**, 542–564 (2008). https://doi.org/10.1039/b807094f
29. H.K. Jeswani, A. Chilvers, A. Azapagic, Environmental sustainability of biofuels: a review: environmental sustainability of biofuels. Proc. R. Soc. A Math. Phys. Eng. Sci. **476**, 20200351 (2020). https://doi.org/10.1098/rspa.2020.0351
30. K. Malik, S.C. Capareda, B.R. Kamboj, S. Malik, K. Singh, S. Arya, D.K. Bishnoi, Biofuels production: a review on sustainable alternatives to traditional fuels and energy sources. Fuels. **5**, 157–175 (2024). https://doi.org/10.3390/fuels5020010
31. A. Bušić, S. Kundas, G. Morzak, H. Belskaya, N. Mardetko, M.I. Šantek, D. Komes, S. Novak, B. Šantek, Recent trends in biodiesel and biogas production. Food Technol. Biotechnol. **56**, 152–173 (2018). https://doi.org/10.17113/ftb.56.02.18.5547
32. N.S. Mat Aron, K.S. Khoo, K.W. Chew, P.L. Show, W.H. Chen, T.H.P. Nguyen, Sustainability of the four generations of biofuels—a review. Int. J. Energy Res. **44**, 9266–9282 (2020). https://doi.org/10.1002/er.5557
33. A. Kuila, V. Sharma, V.K. Garlapati, A. Singh, L. Roy, R. Banerjee, Present status on enzymatic hydrolysis of lignocellulosic biomass for bioethanol production. Adv. Biofeedstocks Biofuels 85–95 (2017). https://doi.org/10.1002/9781119117322.ch4
34. Y. Chisti, Biodiesel from microalgae. Biotechnol. Adv. **25**, 294–306 (2007). https://doi.org/10.1016/j.biotechadv.2007.02.001
35. J. Singh, S. Gu, Commercialization potential of microalgae for biofuels production. Renew. Sustain. Energy Rev. **14**, 2596–2610 (2010). https://doi.org/10.1016/j.rser.2010.06.014
36. B. Abdullah, S.A.F. ad Syed Muhammad, Z. Shokravi, S. Ismail, K.A. Kassim, A.N. Mahmood, M.M.A. Aziz, Fourth generation biofuel: a review on risks and mitigation strategies. Renew. Sustain. Energy Rev. **107**, 37–50 (2019). https://doi.org/10.1016/j.rser.2019.02.018
37. K. Robak, M. Balcerek, Review of second generation bioethanol production from residual biomass. Food Technol. Biotechnol. **56**, 174–187 (2018). https://doi.org/10.17113/ftb.56.02.18.5428
38. M.G. Tabka, I. Herpoël-Gimbert, F. Monod, M. Asther, J.C. Sigoillot, Enzymatic saccharification of wheat straw for bioethanol production by a combined cellulase xylanase and feruloyl esterase treatment. Enzyme Microb. Technol. **39**, 897–902 (2006). https://doi.org/10.1016/j.enzmictec.2006.01.021
39. S. Garg, Xylanase: applications in biofuel production. Curr. Metabolomics. **4**, 23–37 (2015). https://doi.org/10.2174/2213235x03666150915211224
40. M. Shariq, M. Sohail, Application of Candida tropicalis MK-160 for the production of xylanase and ethanol. J. King Saud Univ. - Sci. **31**, 1189–1194 (2019). https://doi.org/10.1016/j.jksus.2018.04.009
41. S.M. Scully, J. Orlygsson, Recent advances in second generation ethanol production by thermophilic bacteria. Energies **8**, 1–30 (2015). https://doi.org/10.3390/en8010001
42. A. Thakur, A. Sharma, K.C. Khaire, V.S. Moholkar, P. Pathak, N.K. Bhardwaj, A. Goyal, Two-Step saccharification of the xylan portion of sugarcane waste by recombinant xylanolytic enzymes for enhanced xylose production. ACS Omega **6**, 11772–11782 (2021). https://doi.org/10.1021/acsomega.1c01262
43. V. Menon, G. Prakash, A. Prabhune, M. Rao, Biocatalytic approach for the utilization of hemicellulose for ethanol production from agricultural residue using thermostable xylanase and

thermotolerant yeast. Bioresour. Technol. **101**, 5366–5373 (2010). https://doi.org/10.1016/j.biortech.2010.01.150
44. S. Horisawa, A. Inoue, Y. Yamanaka, Direct ethanol production from lignocellulosic materials by mixed culture of wood rot fungi schizophyllum commune. Bjerkandera adusta, and fomitopsis palustris, Fermentation. **5**, 21 (2019). https://doi.org/10.3390/fermentation5010021
45. M. Yasuda, K. Takeo, H. Nagai, T. Uto, T. Yui, T. Matsumoto, Y. Ishii, K. Ohta, Enhancement of ethanol production from napiergrass (pennisetum purpureum schumach) by a low-moisture anhydrous ammonia pretreatment. J. Sustain. Bioenergy Syst. **3**, 179–185 (2013). https://doi.org/10.4236/jsbs.2013.33025
46. M. Yasuda, H. Nagai, K. Takeo, Y. Ishii, K. Ohta, Bio-ethanol production through simultaneous saccharification and co-fermentation (SSCF) of a low-moisture anhydrous ammonia (LMAA)-pretreated napiegrass (Pennisetum purpureum Schumach). Springerplus **3**, 333 (2014). https://doi.org/10.1186/2193-1801-3-333
47. J. Sun, F. Wen, T. Si, J.H. Xu, H. Zhao, Direct conversion of xylan to ethanol by recombinant Saccharomyces cerevisiae strains displaying an engineered minihemicellulosome. Appl. Environ. Microbiol. **78**, 3837–3845 (2012). https://doi.org/10.1128/AEM.07679-11
48. F. Xin, C. Wang, W. Dong, W. Zhang, H. Wu, J. Ma, M. Jiang, Comprehensive investigations of biobutanol production by a non-acetone and 1,3-propanediol generating Clostridium strain from glycerol and polysaccharides. Biotechnol. Biofuels **9**, 220 (2016). https://doi.org/10.1186/s13068-016-0641-8
49. O.O. Tigunova, D.S. Kamenskyh, T.V. Tkachenko, V.A. Yevdokymenko, V.I. Kashkovskiy, D.B. Rakhmetov, Y.B. Blume, S.M. Shulga, Biobutanol production from plant biomass. Open Agric. J. **14**, 187–197 (2020). https://doi.org/10.2174/1874331502014010187
50. R. Jalil, M.S. Kalil, N.A. Rahman, A.I. Al-Tabib, A.A. Hamid, M.F. Ahmad, Effect of initial fermentation medium on bioacetone, biobutanol, and bioethanol (BioABE) production from fermentable sugars of acacia mangium using clostridium acetobutylicum YM1. BioResources **15**, 6912–6927 (2020). https://doi.org/10.15376/biores.15.3.6912-6927
51. H. Kuszewski, Physical and chemical properties of 1-butanol-diesel fuel blends. Energy Fuels **32**, 11619–11631 (2018). https://doi.org/10.1021/acs.energyfuels.8b02912
52. B. Kolesinska, J. Fraczyk, M. Binczarski, M. Modelska, J. Berlowska, P. Dziugan, H. Antolak, Z.J. Kaminski, I.A. Witonska, D. Kregiel, Butanol synthesis routes for biofuel production: trends and perspectives. Materials (Basel). **12**, 350 (2019). https://doi.org/10.3390/ma12030350
53. H. Liu, X. Wang, D. Zhang, F. Dong, X. Liu, Y. Yang, H. Huang, Y. Wang, Q. Wang, Z. Zheng, Investigation on blending effects of gasoline fuel with n-butanol, DMF, and ethanol on the fuel consumption and harmful emissions in a GDI vehicle. Energies **12**, 1845 (2019). https://doi.org/10.3390/en12101845
54. J. Sarkar, S. Bhattacharyya, Application of graphene and graphene-based materials in clean energy-related devices Minghui. Arch. Thermodyn. **33**, 23–40 (2012). https://doi.org/10.1002/er.1598
55. N. Qureshi, X.L. Li, S. Hughes, B.C. Saha, M.A. Cotta, Butanol production from corn fiber xylan using Clostridium acetobutylicum. Biotechnol. Prog. **22**, 673–680 (2006). https://doi.org/10.1021/bp050360w
56. I. Patraşcu, C.S. Bîldea, A.A. Kiss, Eco-efficient downstream processing of biobutanol by enhanced process intensification and integration. ACS Sustain. Chem. Eng. **6**, 5452–5461 (2018). https://doi.org/10.1021/acssuschemeng.8b00320
57. N. Qureshi, S. Hughes, I.S. Maddox, M.A. Cotta, Energy-efficient recovery of butanol from model solutions and fermentation broth by adsorption. Bioprocess Biosyst. Eng. **27**, 215–222 (2005). https://doi.org/10.1007/s00449-005-0402-8
58. C. Lu, J. Dong, S.T. Yang, Butanol production from wood pulping hydrolysate in an integrated fermentation-gas stripping process. Bioresour. Technol. **143**, 467–475 (2013). https://doi.org/10.1016/j.biortech.2013.06.012

59. C. Birgen, P. Dürre, H.A. Preisig, A. Wentzel, Butanol production from lignocellulosic biomass: revisiting fermentation performance indicators with exploratory data analysis. Biotechnol. Biofuels **12**, 167 (2019). https://doi.org/10.1186/s13068-019-1508-6
60. F. Xin, W. Yan, J. Zhou, H. Wu, W. Dong, J. Ma, W. Zhang, M. Jiang, Exploitation of novel wild type solventogenic strains for butanol production. Biotechnol. Biofuels **11**, 252 (2018). https://doi.org/10.1186/s13068-018-1252-3
61. R.L. Kudahettige-Nilsson, J. Helmerius, R.T. Nilsson, M. Sjöblom, D.B. Hodge, U. Rova, Biobutanol production by Clostridium acetobutylicum using xylose recovered from birch Kraft black liquor. Bioresour. Technol. **176**, 71–79 (2015). https://doi.org/10.1016/j.biortech.2014.11.012
62. Z. Sun, S. Liu, Production of n-butanol from concentrated sugar maple hemicellulosic hydrolysate by Clostridia acetobutylicum ATCC824. Biomass Bioenerg. **39**, 39–47 (2012). https://doi.org/10.1016/j.biombioe.2010.07.026
63. Z. Wang, X. Cao, N. Li, Z. Yang, M. Lei, Y. Zhao, L. Wang, Z. Li, D. Liu, H. Niu, H. Ying, Production of butanol directly from hemicellulose through secretory expression of a xylanase in clostridium acetobutylicum. Energy Fuels **34**, 3376–3382 (2020). https://doi.org/10.1021/acs.energyfuels.9b04489
64. K. Kumar, L. Barbora, V.S. Moholkar, Genomic insights into clostridia in bioenergy production: comparison of metabolic capabilities and evolutionary relationships. Biotechnol. Bioeng. **121**, 1297–1312 (2024). https://doi.org/10.1002/bit.28610
65. F. Xin, Y.R. Wu, J. He, Simultaneous fermentation of glucose and xylose to butanol by Clostridium sp. strain BOH3, Appl. Environ. Microbiol. **80**, 4771–4778 (2014). https://doi.org/10.1128/AEM.00337-14
66. L. Lépiz-Aguilar, C.E. Rodríguez-Rodríguez, M.L. Arias, G. Lutz, Acetone-butanol-ethanol (ABE) production in fermentation of enzymatically hydrolyzed cassava flour by Clostridium beijerinckii BA101 and solvent separation. J. Microbiol. Biotechnol. **23**, 1092–1098 (2013). https://doi.org/10.4014/jmb.1301.01021
67. K. Kumar, S.M. Jadhav, V.S. Moholkar, Acetone-Butanol-Ethanol (ABE) fermentation with clostridial co-cultures for enhanced biobutanol production. Process. Saf. Environ. Prot. **185**, 277–285 (2024). https://doi.org/10.1016/j.psep.2024.03.027
68. C. Bellido, M. Loureiro Pinto, M. Coca, G. González-Benito, M.T. García-Cubero, Acetone-butanol-ethanol (ABE) production by Clostridium beijerinckii from wheat straw hydrolysates: efficient use of penta and hexa carbohydrates. Bioresour. Technol. **167**, 198–205 (2014). https://doi.org/10.1016/j.biortech.2014.06.020
69. M. Hijosa-Valsero, J. Garita-Cambronero, A.I. Paniagua-García, R. Díez-Antolínez, A global approach to obtain biobutanol from corn stover. Renew. Energy **148**, 223–233 (2020). https://doi.org/10.1016/j.renene.2019.12.026
70. F. Xin, W. Dong, W. Zhang, J. Ma, M. Jiang, Biobutanol production from crystalline cellulose through consolidated bioprocessing. Trends Biotechnol. **37**, 167–180 (2019). https://doi.org/10.1016/j.tibtech.2018.08.007
71. F. Xin, T. Chen, Y. Jiang, W. Dong, W. Zhang, M. Zhang, H. Wu, J. Ma, M. Jiang, Strategies for improved isopropanol-butanol production by a Clostridium strain from glucose and hemicellulose through consolidated bioprocessing. Biotechnol. Biofuels **10**, 118 (2017). https://doi.org/10.1186/s13068-017-0805-1
72. J.T. Ellis, N.N. Hengge, R.C. Sims, C.D. Miller, Acetone, butanol, and ethanol production from wastewater algae. Bioresour. Technol. **111**, 491–495 (2012). https://doi.org/10.1016/j.biortech.2012.02.002
73. K. Liu, H.K. Atiyeh, O. Pardo-Planas, T.C. Ezeji, V. Ujor, J.C. Overton, K. Berning, M.R. Wilkins, R.S. Tanner, Butanol production from hydrothermolysis-pretreated switchgrass: quantification of inhibitors and detoxification of hydrolyzate. Bioresour. Technol. **189**, 292–301 (2015). https://doi.org/10.1016/j.biortech.2015.04.018
74. C. Sasaki, Y. Kushiki, C. Asada, Y. Nakamura, Acetone-butanol-ethanol production by separate hydrolysis and fermentation (SHF) and simultaneous saccharification and fermentation (SSF) methods using acorns and wood chips of Quercus acutissima as a carbon source. Ind. Crops Prod. **62**, 286–292 (2014). https://doi.org/10.1016/j.indcrop.2014.08.049

75. W. Guan, G. Xu, J. Duan, S. Shi, Acetone-Butanol-Ethanol production from fermentation of hot-water-extracted hemicellulose hydrolysate of pulping woods. Ind. Eng. Chem. Res. **57**, 775–783 (2018). https://doi.org/10.1021/acs.iecr.7b03953
76. N. Qureshi, B.C. Saha, S. Liu, T.C. Ezeji, N.N. Nichols, Cellulosic butanol biorefinery: production of biobutanol from high solid loadings of sweet sorghum bagasse—simultaneous saccharification, fermentation, and product recovery. Fermentation. **7**, 310 (2021). https://doi.org/10.3390/fermentation7040310
77. Z. Qin, G.J. Duns, T. Pan, F. Xin, Consolidated processing of biobutanol production from food wastes by solventogenic Clostridium sp. strain HN4, Bioresour. Technol. **264**, 148–153 (2018). https://doi.org/10.1016/j.biortech.2018.05.076
78. Y. Jiang, D. Guo, J. Lu, P. Dürre, W. Dong, W. Yan, W. Zhang, J. Ma, M. Jiang, F. Xin, Consolidated bioprocessing of butanol production from xylan by a thermophilic and butanologenic Thermoanaerobacterium sp. M5, Biotechnol. Biofuels. **11**, 89 (2018). https://doi.org/10.1186/s13068-018-1092-1
79. L. Wang, H. Chen, Increased fermentability of enzymatically hydrolyzed steam-exploded corn stover for butanol production by removal of fermentation inhibitors. Process Biochem. **46**, 604–607 (2011). https://doi.org/10.1016/j.procbio.2010.09.027
80. F. Mechmech, H. Chadjaa, M. Rahni, M. Marinova, N. Ben Akacha, M. Gargouri, Improvement of butanol production from a hardwood hemicelluloses hydrolysate by combined sugar concentration and phenols removal. Bioresour. Technol. **192**, 287–295 (2015). https://doi.org/10.1016/j.biortech.2015.05.012
81. J. Zhang, M. Tang, L. Viikari, Xylans inhibit enzymatic hydrolysis of lignocellulosic materials by cellulases. Bioresour. Technol. **121**, 8–12 (2012). https://doi.org/10.1016/j.biortech.2012.07.010
82. X. Ge, Z. Sun, D. Xin, J. Zhang, Enhanced xylanase performance in the hydrolysis of lignocellulosic materials by surfactants and non-catalytic protein. Appl. Biochem. Biotechnol. **172**, 2106–2118 (2014). https://doi.org/10.1007/s12010-013-0673-5
83. M. Yang, J. Zhang, S. Kuittinen, J. Vepsäläinen, P. Soininen, M. Keinänen, A. Pappinen, Enhanced sugar production from pretreated barley straw by additive xylanase and surfactants in enzymatic hydrolysis for acetone-butanol-ethanol fermentation. Bioresour. Technol. **189**, 131–137 (2015). https://doi.org/10.1016/j.biortech.2015.04.008
84. J. Börjesson, R. Peterson, F. Tjerneld, Enhanced enzymatic conversion of softwood lignocellulose by poly(ethylene glycol) addition. Enzyme Microb. Technol. **40**, 754–762 (2007). https://doi.org/10.1016/j.enzmictec.2006.06.006
85. T. Eriksson, J. Börjesson, F. Tjerneld, Mechanism of surfactant effect in enzymatic hydrolysis of lignocellulose. Enzyme Microb. Technol. **31**, 353–364 (2002). https://doi.org/10.1016/S0141-0229(02)00134-5
86. J. Ouyang, Z. Dong, X. Song, X. Lee, M. Chen, Q. Yong, Improved enzymatic hydrolysis of microcrystalline cellulose (Avicel PH101) by polyethylene glycol addition. Bioresour. Technol. **101**, 6685–6691 (2010). https://doi.org/10.1016/j.biortech.2010.03.085
87. G. Rajagopalan, J. He, K.L. Yang, Direct fermentation of xylan by clostridium strain BOH3 for the production of butanol and hydrogen using optimized culture medium. Bioresour. Technol. **154**, 38–43 (2014). https://doi.org/10.1016/j.biortech.2013.11.094
88. G. Rajagopalan, Effect of nitrogen salts on butanol and hydrogen production from xylan by using clostridium strain BOH3. J. Appl. Biotechnol. Bioeng. **1**, 44–51 (2016). https://doi.org/10.15406/jabb.2016.01.00008
89. P. Sakthiselvan, R. Madhumathi, N. Partha, Eco friendly bio-butanol from sunflower oil sludge with production of xylanase. Eng. Agric. Environ. Food. **8**, 212–221 (2015). https://doi.org/10.1016/j.eaef.2015.01.008
90. G. Rajagopalan, J. He, K.L. Yang, One-pot fermentation of agricultural residues to produce butanol and hydrogen by Clostridium strain BOH3. Renew. Energy **85**, 1127–1134 (2016). https://doi.org/10.1016/j.renene.2015.07.051
91. N. Qureshi, V. Singh, S. Liu, T.C. Ezeji, B.C. Saha, M.A. Cotta, Process integration for simultaneous saccharification, fermentation, and recovery (SSFR): production of butanol from

corn stover using Clostridium beijerinckii P260. Bioresour. Technol. **154**, 222–228 (2014). https://doi.org/10.1016/j.biortech.2013.11.080
92. N. Qureshi, B.C. Saha, B. Dien, R.E. Hector, M.A. Cotta, Production of butanol (a biofuel) from agricultural residues: part I—use of barley straw hydrolysate. Biomass Bioenerg. **34**, 559–565 (2010). https://doi.org/10.1016/j.biombioe.2009.12.024
93. X. Wang, Y. Wang, B. Wang, H. Blaschek, H. Feng, Z. Li, Biobutanol production from fiber-enhanced DDGS pretreated with electrolyzed water. Renew. Energy **52**, 16–22 (2013). https://doi.org/10.1016/j.renene.2012.10.011
94. K.Y. Show, D.J. Lee, J.H. Tay, C.Y. Lin, J.S. Chang, Biohydrogen production: current perspectives and the way forward. Int. J. Hydrogen Energy **37**, 15616–15631 (2012). https://doi.org/10.1016/j.ijhydene.2012.04.109
95. W.J. Martinez-Burgos, E. de Souza Candeo, A.B. Pedroni Medeiros, J. Cesar de Carvalho, V. Oliveira de Andrade Tanobe, C.R. Soccol, E.B. Sydney, Hydrogen: current advances and patented technologies of its renewable production. J. Clean. Prod. **286**, 124970 (2021). https://doi.org/10.1016/j.jclepro.2020.124970
96. X.Y. Wu, Y. Luo, F. Hess, W. Lipiński, Editorial: sustainable hydrogen for energy, fuel and commodity applications. Front. Energy Res. **9**, 698669 (2021). https://doi.org/10.3389/fenrg.2021.698669
97. A.F. Saripan, A. Reungsang, Biohydrogen production by Thermoanaerobacterium thermosaccharolyticum KKU-ED1: culture conditions optimization using xylan as the substrate. Int. J. Hydrogen Energy **38**, 6167–6173 (2013). https://doi.org/10.1016/j.ijhydene.2012.12.130
98. N.Q. Ren, G.L. Cao, W.Q. Guo, A.J. Wang, Y.H. Zhu, B. feng Liu, J.F. Xu, Biological hydrogen production from corn stover by moderately thermophile Thermoanaerobacterium thermosaccharolyticum W16. Int. J. Hydrogen Energy. **35**, 2708–2712 (2010). https://doi.org/10.1016/j.ijhydene.2009.04.044
99. J. Zhang, B. Jiang, H. Zhang, S. Qian, T. Wei, Z. Zhang, L. Song, X. Yang, Fermentative hydrogen production from lignocellulose by mesophilic clostridium populeti FZ10 newly isolated from microcrystalline cellulose-acclimated compost. Appl. Sci. **12**, 9562 (2022). https://doi.org/10.3390/app12199562
100. L. Magnusson, R. Islam, R. Sparling, D. Levin, N. Cicek, Direct hydrogen production from cellulosic waste materials with a single-step dark fermentation process. Int. J. Hydrogen Energy **33**, 5398–5403 (2008). https://doi.org/10.1016/j.ijhydene.2008.06.018
101. F. Taguchi, K. Hasegawa, T. Saito-Taki, K. Hara, Simultaneous production of xylanase and hydrogen using xylan in batch culture of Clostridium sp. strain X53. J. Ferment. Bioeng. **81**, 178–180 (1996). https://doi.org/10.1016/0922-338X(96)87600-8
102. S.E. Bramono, Y.S. Lam, S.L. Ong, J. He, A mesophilic Clostridium species that produces butanol from monosaccharides and hydrogen from polysaccharides. Bioresour. Technol. **102**, 9558–9563 (2011). https://doi.org/10.1016/j.biortech.2011.07.077
103. Y.C. Lo, W.C. Lu, C.Y. Chen, J.S. Chang, Dark fermentative hydrogen production from enzymatic hydrolysate of xylan and pretreated rice straw by Clostridium butyricum CGS5. Bioresour. Technol. **101**, 5885–5891 (2010). https://doi.org/10.1016/j.biortech.2010.02.085
104. A.F. Saripana, A. Reungsang, Thermophilic fermentative biohydrogen production from xylan by anaerobic mixed cultures in elephant dung. Int. J. Green Energy **12**, 900–907 (2015). https://doi.org/10.1080/15435075.2014.887567
105. G.L. Cao, L. Zhao, A.J. Wang, Z.Y. Wang, N.Q. Ren, Single-step bioconversion of lignocellulose to hydrogen using novel moderately thermophilic bacteria. Biotechnol. Biofuels **7**, 82 (2014). https://doi.org/10.1186/1754-6834-7-82

Chapter 13
Xylanase in Pulp and Paper Industry

Muhammad Ammar Khan, Zainab Rasheed, Sher Ali, Muhammad Saeed ud Din, Muhammad Waseem, Tazeen Asghar, and Muhammad Yahya Hammad

Abstract The pulp and paper (P&P) industry faces significant sustainability challenges due to its reliance on chlorine-based bleaching agents, which generate toxic organochlorine compounds like adsorbable organic halides (AOX). These chemicals conflict with global sustainability goals, particularly SDG 6 and 12. This chapter explores xylanase enzymes as a sustainable alternative to conventional bleaching methods. Xylanases degrade xylan, a hemicellulose component, hence facilitates lignin removal and reducing chlorine consumption. Their application enhances paper quality, reduces energy use, and minimizes environmental pollution. However, challenges such as enzyme stability, production costs, and gaps in molecular understanding hinder widespread adoption. Advances in genetic engineering, thermostable enzyme development, and process optimization are critical for scaling xylanase use. By addressing these challenges, the P&P industry can transition toward greener practices, aligning with global sustainability targets.

Keywords Xylanase enzymes · Pulp and paper industry · Enzymatic bleaching · Sustainable bleaching · AOX reduction · Enzyme stability · Industrial biotechnology

---

M. A. Khan (✉) · M. Waseem · T. Asghar
Department of Food Science and Technology, Faculty of Agriculture and Environment, The Islamia University of Bahawalpur, Bahawalpur, Pakistan
e-mail: ammar@iub.edu.pk

Z. Rasheed
Food Science and Engineering, Food College, Sichuan Agricultural University, Ya'An, China

S. Ali
Institute of Food Science and Technology, Sindh Agriculture University Tando Jam, Hyderabad, Sindh, Pakistan

M. S. Din
Department of Food Sciences and Technology, School of Food and Agricultural Sciences, University of Management and Technology, Lahore, Pakistan

M. Y. Hammad
Faculty of Industrial Management, Universiti Malaysia Pahang Al-Sultan Abdullah, Pekan, Malaysia

A. Sattar Jatoi (ed.), *Xylanase from Microorganisms*,
https://doi.org/10.1007/978-981-95-0824-2_13

1 Introduction

The P&P Industry (P&P industry) is a global sector that transforms raw fibrous plant materials into pulp and finished paper products. The primary raw materials include primarily wood from softwood and hardwood trees, as well as non-wood sources including residues obtained from agricultural sources (e.g., wheat straw, rice straw, jute). Global paper production was about 400 billion kg in year 2015 [1] and it has been continuous increasing with about 8% coming from lignocellulosic residues and remaining from softwood or hardwood [2].

The Sustainable Development Goals (SDGs) are a universal set of 17 targets established by the United Nations to address global challenges [3]. The P&P industry faces significant sustainability challenges, including excessive energy and water use, as well as chemical spillage in effluents—issues that directly conflict with SDG 6 (Clean Water and Sanitation, particularly Target 6.3, which calls for reducing pollution and hazardous chemical discharges) and SDG 12 (Responsible Consumption and Production, especially Target 12.4, which emphasizes the management of chemicals and waste in environment-friendly ways).

Bleaching is a critical step in paper manufacturing that comprises the elimination of lignin and other impurities from the pulp to achieve the desired brightness and quality [4]. Traditionally, the industry has relied on chlorine-based chemicals to oxidize and break down lignin. These include molecular chlorine (Cl_2) chlorine dioxide (ClO_2) [5], sodium hypochlorite (NaOCl) [6], and calcium hypochlorite [$Ca(OCl)_2$ [7]. The chlorine-based bleaching process in P&P industry results in the creation of organochlorine compounds due to reactions between chlorine and residual lignin available in wood fiber [8]. The discharge of chlorinated wastewater of P&P industry into water bodies causes the eradication of aquatic species, water pollution, and bio-accumulation of these compounds in fish bodies. Moreover, some of these compounds are such as Adsorbable Organic Halides (AOX) are toxic, mutagenic, and persistent [9]. In response, the industry has increasingly embraced non-chlorine bleaching agents—such as hydrogen peroxide (H_2O_2), ozone (O_3), peracetic acid (PAA), and oxygen (O_2) delignification methods—which offer enhanced sustainability by reducing harmful effluent loads, as well as reduced usage of energy and water.

In the pulp and paper industry, various chlorine-based chemicals have been traditionally employed for bleaching purposes. Initially, molecular chlorine (Cl_2) was introduced due to its cost-effectiveness and high reactivity with lignin [10]. However, environmental concerns, particularly the formation of toxic compounds like dioxins and furans, have led to a decline in its use. Chlorine dioxide (ClO_2) emerged as a preferred alternative, offering selective oxidation of lignin with reduced environmental impact. Other chlorine-containing compounds, including NaOCl and $Ca(OCl)_2$, have also been utilized for their oxidative properties in lignin removal. Nonetheless, the industry has been shifting towards more environmentally friendly bleaching agents to minimize the release of harmful chlorinated organic compounds

[11]. The form of carbon processed to create a porous structure is known as activated carbon. It can bind (adsorb) many molecules mainly due to a high surface area. AOXs have halogens adsorbed onto activated carbon. These compounds are often associated with industrial processes, such as water treatment, bleaching, and chemical manufacturing, and can pose environmental and genetic risks [12] due to their persistence and toxicity to aquatic life. Utilization of chlorine-based bleaching agents in P&P industry results in emission of AOX, mainly by reaction of chlorinated chemicals with lignin and other organic substances present in the pulp. Moreover, significant amount of black liquor is carried over when pulp washing is improper, while upon exposure to chlorine-based bleaches, this black liquor also forms AOX [13]. Some other chemicals such as neutral wet strength agents and cationic additives used during papermaking also contribute to total AOX formation. Hence substitution of bleaching agents is a dire call.

In the pulp and paper industry, several non-chlorine bleaching agents are utilized to achieve desired pulp brightness while minimizing environmental impact. Notable non-chlorine bleaching agents include H_2O_2, O_3, PAA, and O_2 delignification. Widely used for bleaching mechanical and chemical pulps, H_2O_2 is effective in achieving high brightness levels [14]. It operates under alkaline conditions and is particularly beneficial for bleaching recycled fibers, as it aids in ink removal during the re-pulping stage [15]. Additionally, H_2O_2 produces fewer colored effluents, reducing the pollution load on surface waters. O_3 is a powerful oxidizing agent employed in the bleaching process to reduce lignin content in pulp [16]. Its application can lead to high levels of pulp brightness with minimal environmental impact, as it decomposes into oxygen after the reaction. PAA is used in pulp bleaching due to its strong oxidative properties. It effectively breaks down lignin and other chromophores, contributing to the brightness of the pulp [17]. O_2 delignification is a process applied before the bleaching stages to reduce the lignin content in pulp, thereby reducing the demand for bleaching chemicals in subsequent steps [18].

P&P industry is transitioning from traditional bleaching toward more sustainable mode of production in the scenario of SDGs. The chapter has: (1) defined key concepts and terminology related to the P&P industry, bleaching processes, and sustainability; (2) contextualized the scale of global P&P industry, including an overview of traditional chlorine-based and emerging non-chlorine bleaching technologies; (3) examined the role of xylanases—detailing their mode of action, production methods, and industrial applications; and (4) identified the benefits and limitations of enzyme-based pretreatment relative to conventional chemical bleaching. P&P industry professionals seeking to implement cleaner bleaching strategies, researchers and academics working on enzyme technology and process optimization, and policymakers interested in promoting eco-friendly P&P industry could take benefit of this book chapter (Fig. 1).

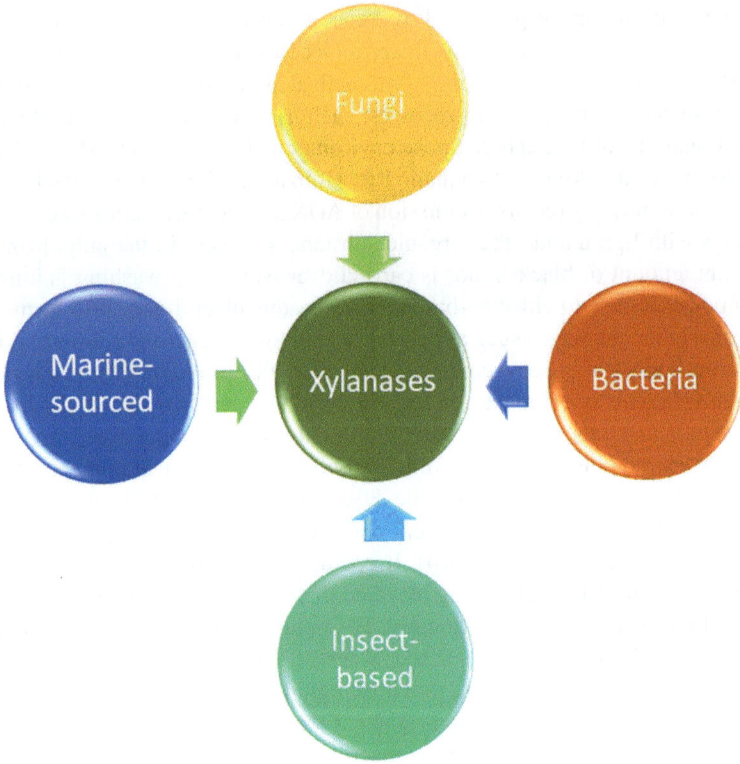

Fig. 1 Sources of xylanases

2 Use of Xylanases as an Enzymatic Bleaching Agent in P&P Industry

As part of the drive toward greener technologies, enzymatic bleaching has emerged as an innovative alternative to conventional chemical methods. Xylanases are a group of enzymes that degrade xylan, a major component of plant cell walls, into xylose. Xylanases represent a significant advancement in sustainable P&P processing, offering improvements across multiple dimensions of production. Xylanases, which catalyze the breakdown of xylan (key hemicellulosic constituent of plant cell wall) into into xylooligosaccharides, are particularly valuable because they help break the bond between xylan and lignin [19]. This pre-bleaching or bleach boosting action increases fiber permeability and facilitates subsequent chemical treatments, thereby reducing the required dose of chlorine-based agents. While other enzymes such as cellulases, pectinases, and laccases are also used in the P&P industry for tasks like deinking and fiber modification, cellulase-free xylanases are especially important in bleaching applications [20]. Enzyme pretreatment can significantly reduce consumption of chemicals in bleaching and AOX formation. Moreover, adopting greener technologies such as xylanase-based bleaching can reduce reliance on chlorine-based chemicals, thereby lowering water contamination and energy consumption, thereby contributing to the achievement of SDG targets 6.3 and 12.4.

The use of chemical bleaching methods in P&P industry has environmental and economic drawbacks. Traditional chlorine-based treatments discharge organochlorine compounds and AOX into wastewater. Moreover, traditional chemical bleaching methods cause lignin reprecipitation on fiber surfaces, leading to brown-colored pulp. In contrast, the application of xylanases offers a promising biotechnological intervention. Moreover, traditional bleaching processes demand extensive energy, water, and chemical inputs while often compromising the quality of the final product through issues like lignin reprecipitation [21]. These enzymes enhance paper quality by improving fiber flexibility, bonding, and overall mechanical properties while simultaneously reducing energy requirements in the refining process. These enzymes not only curtail the utilization of harsh chemicals in the wastewater by reducing up to 10–15% (for hardwood) and 20–25% (for softwood) chlorine consumption during bleaching [22] but also diminish the formation of AOX [23]. From an economic perspective, cost-effective production methods and the potential for chemical reduction create a compelling value proposition for manufacturers. Xylanase facilitates delignification without harming cellulose. Particularly, Kraft pulp bleaching involves extensive use of chemicals for lignin removal [24]. In contrast, xylanases enhance the delignification process [25], and gives higher brightness in both conventional and Kraft pulp. Xylanases remove chromophoric groups and enhance lignin removal [26]. Furthermore, the recycling of waste paper pulp is hindered by the inefficiency of chemical deinking processes. Chemical deinking of recycled paper generates pollutants [27]. However, xylanases improve the efficiency of processes mainly by hydrolyzing bonds between fibrils and ink particles with minimal ecological fallout [28]. This improves the quality of recycled paper.

Although some challenges remain—primarily centered on enzyme stability, production costs, and achieving high expression levels—the continuous development of thermostable and pH-tolerant xylanases is gradually mitigating these obstacles. This progressive shift not only fosters a reduction in chemical dependency but also promotes a significantly lower environmental footprint and enhanced process efficiency. These advancements collectively set the stage for a transformative move away from conventional chemical bleaching towards greener, enzyme-based technologies (Fig. 2).

3 Mechanism of Action of Xylanases

During paper production, the chemical process used to separate wood into its fibers is known as Kraft pulping [29]. Wood is composed primarily of cellulose, hemicellulose, and lignin. Lignin is a complex organic polymer that provides structural support,

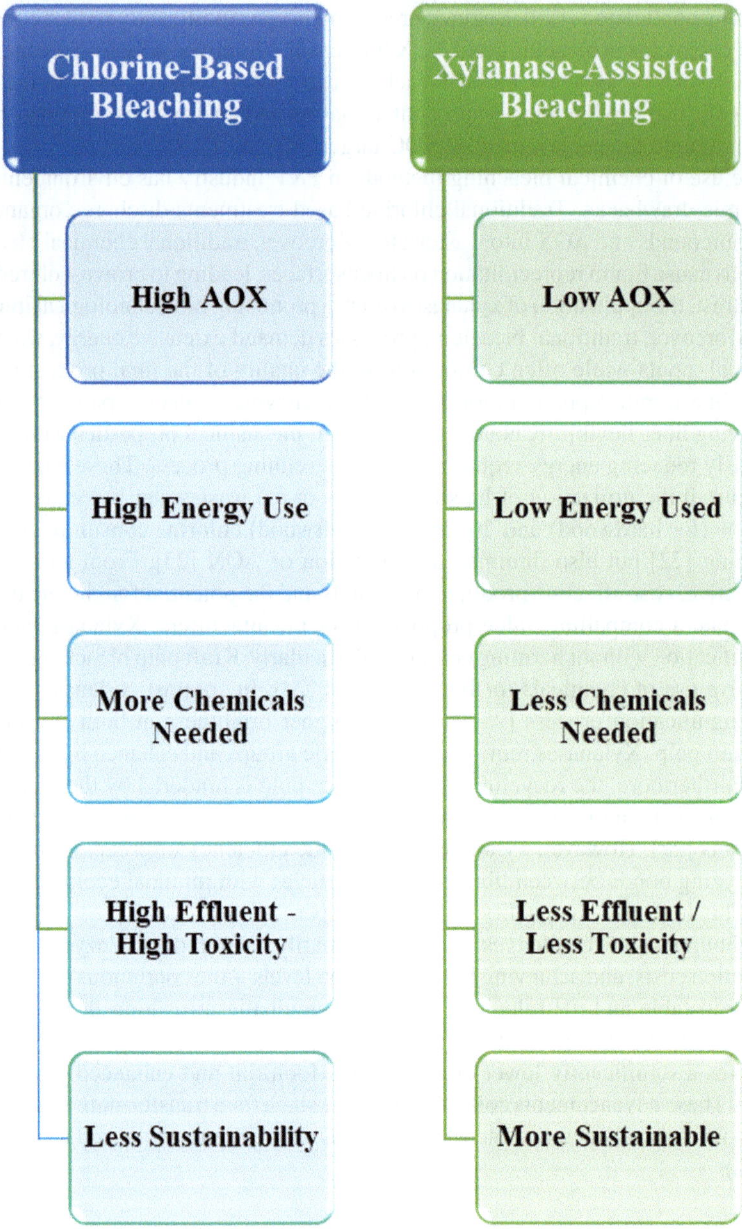

Fig. 2 Traditional versus enzymatic bleaching processes

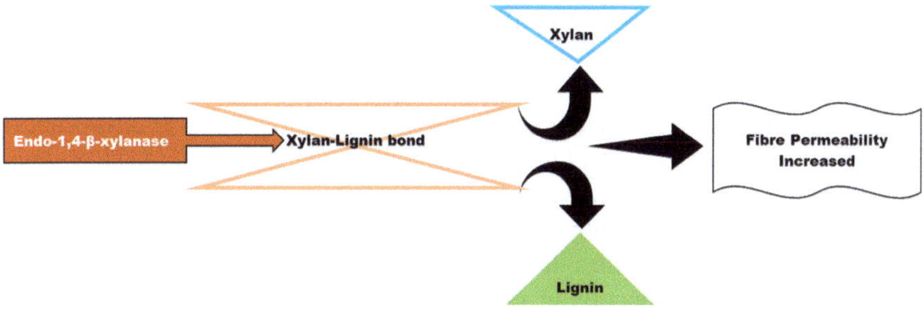

Fig. 3 Mechanism of xylanase action in lignin removal

while xylan—a natural component derived from hemicellulose—is also present in the wood. Bleaching is essential for whitening the pulp, ensuring that the final paper is free of unwanted color or impurities. A common bleaching agent used during this process is ClO_2. During Kraft pulping, short chains of xylan adhere to the surface of the fibers and bind to some of the lignin. This absorbs water and swells, forming a protective barrier that hinders bleaching agents from fully penetrating the fibers and removing the remaining lignin. Consequently, higher quantities of bleaching agents are required during the first stage of bleaching to remove lignin effectively. However, xylanase treatment as a pre-bleaching step prepares the fibers for subsequent chemical bleaching. It breaks the bonds between xylan and lignin, facilitating the removal of xylan from the fibers [30]. Once xylan is removed, lignin is no longer strongly attached to the fiber surface, and it increases the permeability of fibers, allowing bleaching chemicals to penetrate thoroughly. As a result, the process becomes more efficient during subsequent bleaching stages, reduces the reliance on chlorine-based chemicals, and paves the way for sustainable industrial practices (Fig. 3).

4 Production of Xylanases

Traditionally, xylanases are sourced from a variety of microorganisms. Fungi and bacteria represent the principal commercial sources of xylanases due to their efficiency in enzyme production, ease of cultivation, and scalability for industrial applications. Fungal xylanases are less effective as their optimal pH is too low for direct brown stock pulp treatment, whereas bacterial xylanases meet all requirements for pulp bleaching. Xylanase can be effectively obtained from various fungal species such as *Aspergillus niger* [31] and *Trichoderma reesei* [32] through various methods. Xylanases can also be obtained from yeast, although yeast-derived xylanases are less commonly studied compared to those from other fungi, with certain yeast species like *Sugiyamaella lignohabitans* [33], *Pichia kudriavzevii* [34], and *Cryptococcus laurentii* [35] showing potential for xylanase production. Wood-feeding termite guts are a promising reservoir for novel xylanases-producing and xylose-fermenting yeasts, potentially benefiting the biorefinery industry [36]. Bacterial xylanases are preferred over fungal ones due to their alkaline-thermostable properties, higher optimal pH and temperature, and suitability for industrial use. Bacillus strains, for instance, produce extracellular, cellulase-free xylanases stable at high temperatures and alkaline pH. Among bacteria, *Bacillus sonorensis* [37], *Bacillus australimaris* [38], *Bacillus firmus* [39], *Bacillus safensis* [40], *Bacillus subtilis* [41], *Bacillus paramycoides* [42], and *Bacillus tequilensis* [43] are among the Bacillus species that produce xylanases commercially. Moreover, *Caldicoprobacter algeriensis* produce thermophilic and thermostable xylanase [44]. Several other bacterial species are also used for commercial production of xylanases, which have potential application for bio bleaching in P&P industry.

Biotechnological advances continue to drive innovations in xylanase applications, promising improved industrial processes and product quality. The ongoing research in this field is essential for developing more efficient, stable, and environmentally friendly enzymatic solutions that can address the growing demands of sustainable industrial processes. Genetically modified organism (GMO) xylanases are produced using engineered microbial strains to improve functionality and yield. Genetically modified microbial strains are engineered for enhanced enzyme production efficiency. *Bacillus subtilis* Strain AR-153 is noted for its efficient enzyme production capabilities [45]. *Trichoderma reesei* Strain RF5703 has been engineered for safe and effective xylanase production [46]. Biotechnological advancements significantly enhance the properties and applications of xylanases. According to the European Food Safety Authority (EFSA), GMO xylanases do not pose safety concerns under intended conditions of use, as they do not contain the production organism or recombinant DNA. Genetic modifications help improve the production efficiency of xylanases, enabling higher yields and more stable enzyme activity. Moreover, engineering xylanases at the gene and protein levels enhances their thermal stability and catalytic efficiency, making them more robust in industrial applications. Specific manipulation of xylanase genes helps achieve desired enzymatic properties for developing enzymes suited for various industrial applications for the creation of custom enzymes with specific activity profiles. Gene manipulation for xylanase production can involve cloning xylanase genes into various hosts like *E. coli* [47] and *Pichia pastoris* [48]. It involves overexpressing transcription factors such as *Xyr1* [49], optimizing codon usage [50], introducing mutations for heat resistance, and employing chromosomal integration and gene amplification techniques. Building on gene manipulation, advanced protein engineering techniques are used to create xylanases that can function effectively under adverse conditions, improving thermal stability and catalytic efficiency [51]. These engineered proteins demonstrate enhanced performance in industrial settings. Gene and protein engineering technologies have significantly improved production efficiency, enhancing thermal stability and catalytic properties [52]. These improvements facilitate more economical enzyme production at industrial scales. The thermostable variants are particularly valuable in processes requiring high-temperature conditions. Engineering thermostable xylanases emphasizes strategies for creating gene diversity and high-throughput screening methods to identify robust variants. Techniques from synthetic biology are increasingly being integrated into the engineering of xylanases,

Table 1 Comparison of xylanase sources and their industrial suitability

Citations	Source	Optimal pH	Thermostability	Key advantages
[41–45]	*Bacillus subtilis*	6–9	High (up to 80 °C)	Cellulase-free, alkaline-tolerant
[31, 59]	*Aspergillus niger*	3–5	Moderate (50–60 °C)	High yield, cost-effective
[41, 56]	Marine *Bacillus* sp.	7–10	Halotolerant	Stable in high salinity
[36, 58]	Termite gut yeasts	5–7	Moderate	Novel enzymes, synergistic degradation

expanding their potential applications [53]. This emerging field combines biological components in novel ways to create enzymes with unprecedented properties.

Marine xylanases are produced by microorganisms that thrive in a wide variety of environmental conditions, making them suitable for various industrial processes. Marine-derived *Trichoderma pleuroticola* 08K001 strains are prolific sources for xylanase production, potentially useful in various industrial processes [54]. Moreover, The xylanase enzyme isolated from bacterial strains HR05 and HR06, which coexist with alga Ulva flexuosa, could be potential candidates for biotechnology and various industries, such as pulp and paper production [41]. In addition, marine-derived Streptomyces viridochromogenes strain M11 can be used as a alkaline tolerant xylanase producer, and ribosome engineering can improve xylanase production in Streptomyces [55]. Marine bacterium Bacillus subtilis cho40 produces a novel halotolerant xylanase that can play a crucial role in bioethanol production from marine seaweeds [56]. Furthermore, marine-derived fungus Aspergillus cf. tubingensis LAMAI 31 is another genetic resource for xylanase production [57].

Specific insect species produce xylanases as part of their digestive processes. Particularly wood-feeding termites, produce xylanases through symbiotic microorganisms in their guts, which have potential for commercial production [36]. Novel xylanase enzymes with unusual primary sequences and domains were discovered in insect guts, catalyzing the hydrolysis of various xylose substrates and producing unique hydrolysis products [58]. Table 1 summarizes microbial sources (fungi, bacteria, marine/insect-derived) discussed in the section, highlighting their pH ranges, thermostability, and advantages. It complements textual details about production methods and microbial strains.

5 Applicability

The environmental benefits of xylanase use are substantial, with significantly improved effluent biodegradability and reduced harmful chemical usage compared to traditional bleaching methods. In addition, xylanase treatment significantly enhances the physical properties of paper through multiple mechanisms. The enzymatic action improves fiber flexibility by altering the xylan structure and enhancing wetting properties. This increased flexibility leads to better conformability of the fibers, resulting in denser paper sheets with a higher bonded area. The alteration of xylan in the fiber network simultaneously improves the overall structural integrity of paper, creating a stronger and more durable paper structure. Xylanase pretreatment yields substantial improvements in paper strength properties, including increase in tensile index, tear index, and burst index. Its pretreatment improves pulp viscosity resulting in enhanced mechanical strength properties of the finished paper. These improvements confirm the ability of enzyme to enhance the fundamental mechanical properties of paper products [60]. Beyond paper quality improvements, xylanases enhance the susceptibility of pulp to refining, thereby reduces the energy requirement for this process while maintaining paper static strength properties. This dual benefit of energy conservation and quality maintenance represents a significant advantage for manufacturers seeking both sustainability and performance [61]. Traditional chlorine-based bleaching agents generate harmful organochlorine compounds that persist in the environment. Xylanase pretreatment significantly reduces the need for these chemicals. Pre-bleaching with xylanase reduces subsequent bleach chemical requirements, directly decreasing organochlorine emissions. Moreover, by minimizing chlorinated compound formation, xylanase treatment reduces the overall toxicity of paper production effluents [62]. Particularly, xylanase-treated pulp produces effluents yield lower levels of AOX. In addition, xylanase treatment offers significant environmental advantages over conventional bleaching processes, particularly regarding effluent quality. Effluents from xylanase-treated pulp are twice as biodegradable as those from standard pulp bleaching methods, representing a substantial environmental improvement [63]. Furthermore, optimizing xylanase application in pre-bleaching processes yields higher brightness. Table 2 consolidates key trade-offs discussed in the text.

Table 2 Benefits and limitations of xylanase application

Citations	Benefits	Limitations
[22, 64]	Reduces chlorine use by 10–25%	High production costs
[23, 65]	Lowers AOX emissions by 30–50%	Sensitivity to pH/temperature
[60, 66]	Enhances paper strength (tensile index)	Requires complementary lignin treatments
[63, 67]	Improves effluent biodegradability	Scalability challenges in fermentation

6 Challenges

6.1 *Technical Challenges*

Xylan has a complex and diverse structure, making its degradation challenging. The enzymatic breakdown of xylan requires multiple enzymes with specific actions and properties. While xylanases offer eco-friendly advantages in pulp bleaching, several factors can limit their effectiveness compared to non-enzymatic agents. Particularly, xylanase efficacy is sensitive to specific conditions such as pH, temperature, and retention time. Moreover, residual substances from previous processing stages can hinder xylanase performance, necessitating thorough pulp washing to achieve desired results. Xylanases primarily target hemicellulose and have limited direct action on lignin, often requiring subsequent chemical treatments to achieve complete bleaching. However, cellulase contamination that risks damaging cellulose fibers also creates xylanase bleaching challenges. The implementation of xylanase in P&P industrial settings faces several technical challenges. Xylanase must maintain stability and activity under the harsh conditions typical in pulp processing, including high alkaline pH and elevated temperatures. Engineering suitable xylanase variants remains an ongoing challenge [65]. In addition, incorporating xylanase treatment into existing production lines requires careful optimization to manage factors such as pulp carryover, which can significantly impact enzyme efficiency [68]. Furthermore, differences in wood species and pulp quality across operations necessitate tailored enzymatic approaches, complicating standardization efforts. Finally, ensuring that the production and application of xylanase are themselves environmentally sustainable requires careful lifecycle assessment and optimization [63].

6.2 *Cost Implications*

The production cost of xylanases and related enzymes is high, which hinders their widespread commercialization in various industries. Xylanase production is costly due to the need for high enzyme expression levels. Genetic engineering is essential for large-scale, economical enzyme production. However, in the P&P industry, traditional methods involve harsh chemicals (e.g., chlorine derivatives), leading to higher costs, environmental harm, and energy consumption. Xylanases offer an eco-friendly alternative for bleaching and deinking. The economic viability of xylanase in industrial applications stems from efficient production methods. The production and stabilization of xylanase enzymes can be more expensive than traditional chemical bleaching agents, impacting their economic feasibility, which limits the integration of xylanases into existing bleaching processes, where non-enzymatic agents may offer more straightforward and cost-effective solutions. While xylanase offers long-term cost benefits, initial implementation presents economic hurdles. The initial capital investment and ongoing costs for enzyme procurement and process modifications can deter adoption, particularly for smaller operations. Moreover, developing new xylanase variants specifically suited to P&P industry conditions also require comprehensive research and investment. In this regard, microbial sources have been optimized for cost-effective xylanase production, particularly when scaled to industrial levels [64]. This cost-efficiency is further enhanced by low-cost xylan extraction from agro-residue materials, providing economical raw materials for enzyme production [59], as well as by streamlining industrial production processes for bacterial xylanase.

6.3 Gaps in Research

Despite significant advancements over the decades, several unresolved issues and gaps persist in the application of xylanases within the P&P industry. These gaps hinder the full potential of xylanases as environmentally friendly alternatives to traditional chemical methods in pulping and bleaching processes. Xylanases offer promising solutions for reducing chemical usage, energy consumption, and environmental impact in paper production, yet their industrial implementation faces several challenges that require further research and development.

6.3.1 Enzyme Stability

One of the primary challenges is the stability of xylanases under the extreme conditions typical of pulping and bleaching processes. These processes often involve high temperatures, varying pH levels, and the presence of chemical reagents, all of which can denature xylanases and reduce their effectiveness. Xylanases often have thermal instability and are dependent on specific pH conditions, limiting their effectiveness in industrial applications. Efforts to enhance the thermal stability and catalytic properties of xylanases through genetic modifications have shown promise. Techniques such as DNA shuffling, error-prone PCR, and the identification of specific residues affecting stability are being explored. Extracting xylanase from thermophilic fungi and bacteria has been another approach to obtain more stable enzymes. Xylanases from *Thermomyces lanuginosus* [69] and *Bacillus stearothermophilus* [70] have demonstrated activity at temperatures up to 60–70 °C, making them suitable candidates for industrial applications. However, large-scale production of these enzymes remains challenging.

6.3.2 Purity and Efficacy

For xylanases to be effective in industrial applications, they need to be of high purity. Impurities can inhibit enzyme activity and lead to inconsistent results. The purification of xylanases remains a costly and complex process. Although several cost-effective purification methods have been proposed, achieving the necessary purity levels at an industrial scale is challenging. Current methods include ammonium sulfate precipitation, aqueous two-phase systems (ATPS), and ultrafiltration. ATPS using polyethylene glycol and potassium phosphate achieved 85% recovery with a 12-fold increase in specific activity, but scaling this process remains difficult [71]. The polyethylene glycol-phosphate aqueous two-phase system (ATPS) effectively purified thermotolerant xylanase from *Cladophora hutchinsiae*, retaining 93.21% of its activity and a 7.18 purification factor [72] The cost of purification is too high. Combining precipitation techniques with membrane filtration could reduce purification costs, but further optimization is needed for commercial viability.

6.3.3 Limited Understanding of Molecular Mechanisms

Understanding the molecular mechanisms of xylanases is crucial for optimizing their use and improving their performance in industrial applications. This knowledge enables targeted modifications to enhance enzyme properties relevant to P&P processing. There is still insufficient knowledge about how xylanases interact with various substrates at the molecular level, particularly in the complex matrix of wood pulp. Studies have focused on the structure and function of xylanases, but the relationship between specific structural features and performance in industrial conditions remains unclear [73]. The interaction between xylanases and other enzymes (such as cellulases and laccases) in pulp treatment is also not fully understood. Combination with other enzymes such as laccases generally increases the delignification efficiency of xylanases, but the molecular basis for this synergy needs further elucidation [66] Moreover, computational modeling such as homology modeling and molecular docking may further provide insights into xylanase-substrate interactions. These may predict binding affinities and identified potential modification sites, but experimental validation of these models is still needed [74].

6.4 Production Challenges

The production of xylanases for industrial use typically involves fermentation processes. Efficiency and cost-effectiveness are critical for large-scale production to make xylanase treatment economically competitive with traditional chemical methods. Current fermentation methods may not be efficient or cost-effective enough for large-scale applications, with production costs remaining a significant barrier to wider adoption. Innovations in Solid State Fermentation have shown potential for cost-effective xylanase production. However, scaling up SSF processes presents challenges in maintaining uniform conditions and efficient product recovery [67] While recombinant expression in bacterial hosts has improved xylanase yields, the costs associated with induction, purification, and maintaining sterile conditions remain high. Moreover, continuous fermentation systems could significantly reduce

Table 3 Key challenges and research gaps in xylanase use

Citations	Challenge	Proposed solutions
[65, 69]	Enzyme instability at high pH	Engineering thermostable variants
[71, 72]	High purification costs	Low-cost methods (aqueous two-phase systems)
[25, 66]	Limited lignin degradation	Synergistic use with laccases
[64, 67]	Scalability of production	Solid-state fermentation optimization

production costs but face technical challenges related to long-term stability and contamination control.

6.5 Environmental Impact

The use of xylanases in the P&P industry is promoted as an environmentally friendly alternative to chemical methods, potentially reducing water consumption, chemical usage, and harmful effluents. However, more comprehensive research is needed to assess the environmental benefits and potential drawbacks of using xylanases throughout the entire production chain. In this regard, existing Life Cycle Analysis (LCA) studies have shown promising results. Enzyme-assisted pulping could reduce water consumption and chemical oxygen demand (COD) in effluents compared to conventional chemical pulping. However, these analyses often do not account for the environmental impact of enzyme production itself. In this regard, a more holistic approach is needed to evaluate the entire value chain, from enzyme production to final paper product [75] Finally, the lack of standardized methodologies for assessing the environmental impact of enzyme applications in P&P processing makes it difficult to compare different studies and draw definitive conclusions about sustainability benefits. Table 3 directly maps the technical challenges and gaps in research.

7 Future Directions

The economic advantages of xylanase are reflected in market projections, with the global xylanase market expected to grow from USD 1.2 billion in 2023 to USD 2.5 billion by 2032 [76]. This substantial growth trajectory is driven by the cost benefits of this enzyme and the increasing demand for sustainable processing solutions in the pulp and paper industry. However, in order to achieve these targets, developing xylanases with enhanced stability at temperatures above 80 °C and pH tolerance between 4 and 10 through targeted genetic modifications and exploration of extremophilic sources need to be achieved. Moreover, establishing cost-effective purification methods that can achieve >90% purity while reducing purification costs by at least 50% will make the usage of xylanase for the industrial applications more economically viable. In addition, deepening the understanding of molecular mechanisms through advanced structural studies and computational modeling, particularly focusing on substrate specificity and synergistic effects with other enzymes will further help in achieving specific conditions-based microbes and genes, which will help achievement of more toleration of enzymes to inactivity at various processing conditions. Furthermore, process optimization for scaling up solid-state fermentation by improving recombinant expression systems will be needed to enhance the current yield levels for ensuring stable continuous production systems. For optimized enzyme production, to maximize yield and enhance enzyme activity, statistical modeling tools need to be used, such as response surface methodology [77] Finally, conducting comprehensive environmental impact assessments that consider the entire value chain, with standardized methodologies will enable meaningful comparisons between enzymatic and chemical processes. Achieving all these targets with active research in this field will not only enhance the efficiency, cost-effectiveness, and sustainability of xylanase applications in the P&P industry, but also potentially transform traditional paper manufacturing into a more environmentally friendly process with reduced chemical dependency.

8 Conclusion

The integration of xylanase enzymes into pulp bleaching processes offers a transformative solution to the environmental and economic drawbacks of traditional chlorine-based methods. By selectively degrading xylan, these enzymes enhance lignin removal, reduce chlorine consumption, and lower AOX emissions, thereby improving effluent biodegradability and paper quality. Despite their promise, technical barriers—such as sensitivity to industrial conditions, high production costs, and incomplete understanding of enzyme–substrate interactions—limit their full-scale application. Future advancements in protein engineering, cost effective purification methods, and synergistic enzyme systems are essential to overcome these challenges. Collaborative efforts among researchers, industry stakeholders, and policymakers will be pivotal in driving the adoption of xylanase-based technologies. As the global demand for sustainable paper production grows, xylanases represent a cornerstone for achieving cleaner, energy-efficient industrial processes that align with the United Nations Sustainable Development Goals.

References

1. E. Nzediegwu, *Valorization of Paper Waste into Platform Chemicals* (McGill University, Canada, 2022)
2. S.K. Gülsoy, A literature review on the wood fiber morphology of fruit trees. Res. Rev. Agric. Forestry Aquacul. Sci. **I**, 103 (2021)

3. H. Hassani, X. Huang, S. MacFeely, M.R. Entezarian, Big data and the united nations sustainable development goals (UN SDGs) at a glance. Big Data and Cognitive Computing **5**, 28 (2021)
4. G.K. Gupta, G. Pathak, P. Shukla, R.K. Kapoor, *Microbial Bioprocesses* (Elsevier, 2023), pp. 299–318
5. X. Wang et al., Structural evolution and reaction kinetics of lignin-carbohydrate complex oxidized by chlorine dioxide. Chem. Eng. J. **496**, 154018 (2024)
6. K. Nakamura, N. Furuta, J. Miyazaki, N. Nakamura, Sodium hypochlorite–based binderless technology for solid wood adhesion: application of a commercial bleaching agent for adhesion and observation of surface characteristics. J. Wood Sci. **70**, 53 (2024)
7. V. Kumar, P. Verma, A critical review on environmental risk and toxic hazards of refractory pollutants discharged in chlorolignin waste of pulp and paper mills and their remediation approaches for environmental safety. Environ. Res. **236**, 116728 (2023)
8. D. da Silva Vilar, et al., Lignin-modifying enzymes: a green and environmental responsive technology for organic compound degradation. J. Chem. Technol. Biotechnol. **97**, 327–342 (2022)
9. M. Sonkar, V. Kumar, D. Dutt, A novel sequence batch treatment of wastewater using Bacillus sp. IITRDVM-5 mixing with paper mill and sewage sludge powders. Environ. Technol. Innovation **21**, 101288 (2021)
10. F. Di Turo, The chemistry behind paper restoration: diagnostic techniques and cutting-edge innovation. ChemTexts **11**, 7 (2025)
11. M.M. Alam, A. Greco, Z. Rajabimashhadi, C.E. Corcione, Efficient and environmentally friendly techniques for extracting lignin from lignocellulose biomass and subsequent uses: a review. Clean. Mater. 100253 (2024)
12. N. Kumar, G. Parvathi, S. Lenita, K.S. Pavithra, Chinnadurai muralidharan 11 role of microbial oxidative enzymes. Microbial Oxidative Enzymes: Biotechnol. Appl. 221 (2023)
13. X. An et al., Evaluation of bioremediation and detoxification potentiality for papermaking black liquor by a new isolated thermophilic and alkali-tolerant Serratia sp. AXJ-M. J. Hazardous Mater. **406**, 124285 (2021)
14. A. Caglak, H. Sari-Erkan, G. Onkal Engin, Chemical oxygen demand and tannin/lignin removal from paper mill wastewater by electrocoagulation combined with peroxide and hypochlorite treatments. Environ. Technol. **45**, 3076–3094 (2024)
15. T. Chen, H. Wang, L. Chen, 2021 4th International conference on information systems and computer aided education, 16–21
16. X. An et al., Ozone pretreatment facilitating cellulase hydrolysis of unbleached bamboo pulp for improved fiber flexibility. Ind. Crops Prod. **178**, 114577 (2022)
17. R. Ma et al., Role of peracetic acid on the disruption of lignin packing structure and its consequence on lignin depolymerisation. Green Chem. **23**, 8468–8479 (2021)
18. C.V. Esteves, Evaluation of oxygen delignified fibers with high water absorbency, as a greener alternative to fully bleached fibers for tissue paper. Nord. Pulp Pap. Res. J. **39**, 527–539 (2024)
19. G. Kaur, P. Kaur, J. Kaur, D. Singla, M.S. Taggar, Xylanase, xylooligosaccharide and xylitol production from lignocellulosic biomass: Exploring biovalorization of xylan from a sustainable biorefinery perspective. Ind. Crops Prod. **215**, 118610 (2024)
20. G.K. Gupta, M. Dixit, R.K. Kapoor, P. Shukla, Xylanolytic enzymes in pulp and paper industry: new technologies and perspectives. Mol. Biotechnol. **64**, 130–143 (2022)
21. Á. Vaz et al., Effect of enzymatic treatment on Eucalyptus globulus vessels passivation. Sci. Rep. **13**, 2832 (2023)
22. A. Dukare et al., Microbial xylanase aided biobleaching effect on multiple components of lignocelluloses biomass based pulp and paper: a review. Nord. Pulp Pap. Res. J. **38**, 459–480 (2023)
23. J.M. Matos, D.V. Evtuguin, A.P. Sousa, M.G.V. Carvalho, Effect of xylanase-assisted treatment of oxygen-delignified eucalypt kraft pulp on ECF bleaching. Forests **14**, 396 (2023)
24. E.R. Senthilkumar, G. Henriksson, M.E. Lindström, T. Vikström, O. Sevastyanova, Effects of chemical environment on softwood kraft pulp: exploring beyond conventional washing methods. Nord. Pulp Pap. Res. J. **40**, 83–93 (2025)

25. L. Sharma, D. Bera, S.K. Pandey, Biotechnological approaches for increasing the. Emerg. Trends Environ. Biotechnol. (2022)
26. S.K. Ghatora, B.S. Chadha, A. Badhan, H. Saini, M. Bhat, Identification and characterization of diverse xylanases from thermophilic and thermotolerant fungi. BioResources **1**, 18–33 (2006)
27. V. Kumar, P. Pathak, N.S.K. Harsh, N.K. Bhardwaj, Biodeinking: an eco-friendly alternative for chemicals based recycled fiber processing. Phys. Sci. Rev. **8**, 1941–1965 (2023)
28. G. Fan et al., Impact of the disulfide bond on hydrolytic characteristics of a xylanase from Talaromyces thermophiles F1208. Int. J. Biol. Macromol. **164**, 1748–1757 (2020)
29. X. Guo et al., Efficient separation of functional xylooligosaccharide, cellulose and lignin from poplar via thermal acetic acid/sodium acetate hydrolysis and subsequent kraft pulping. Ind. Crops Prod. **153**, 112575 (2020)
30. D. Kaur, A. Joshi, V. Sharma, N. Batra, A.K. Sharma, An insight into microbial sources, classification, and industrial applications of xylanases: a rapid review. Biotechnol. Appl. Biochem. **70**, 1489–1503 (2023)
31. F.S. Ire, I.J. Chima, V. Ezebuiro, Enhanced xylanase production from UV-mutated Aspergillus niger grown on corn cob and sawdust. Biocatal. Agric. Biotechnol. **31**, 101869 (2021)
32. A. Khalaji, M. Sedighi, F. Vahabzadeh, Optimization and kinetic evaluation of acetylxylan esterase and xylanase production by Trichoderma reesei using corn cob xylan. Environ. Process. **7**, 885–909 (2020)
33. K. Šuchová, A. Chyba, Z. Hegyi, M. Rebroš, V. Puchart, Yeast GH30 xylanase from Sugiyamaella lignohabitans is a glucuronoxylanase with auxiliary xylobiohydrolase activity. Molecules **27**, 751 (2022)
34. H.A. Salah, et al. Production and optimization of xylanase and α-amylase from non-Saccharomyces yeasts (Pichia membranifaciens). J. Pure Appl. Microbiol. **15** (2021)
35. D.M. Otero, A.R.C. Braga, S.J. Kalil, Diversification of nitrogen sources as a tool to improve endo-xylanase enzyme activity produced by Cryptococcus laurentii. Biocatal. Agric. Biotechnol. **32**, 101941 (2021)
36. S.S. Ali et al., Screening and characterizing of xylanolytic and xylose-fermenting yeasts isolated from the wood-feeding termite, Reticulitermes chinensis. PLoS One **12**, e0181141 (2017)
37. A. Kiribayeva et al., Cloning, expression, and characterization of a recombinant xylanase from Bacillus sonorensis T6. PLoS ONE **17**, e0265647 (2022)
38. K. Kumari, S. Nagar, S. Goyal, S. Maan, Hyper xylanase production and potential of xylooligosaccharides formation from a novel Bacillus australimaris KS2. Biocatal. Agric. Biotechnol. **54**, 102899 (2023)
39. L.C. Phukon et al., Production of xylanase under submerged fermentation from Bacillus firmus HS11 isolated from Sikkim Himalayan region. Indian J. Experime. Biol. (IJEB) **58**, 563–570 (2020)
40. S. Devi, D. Dwivedi, A. Bhatt (2022)
41. A. Pasalari, A. Homaei, Isolation and molecular identification of xylanase-producing bacteria from Ulva flexuosa of the Persian Gulf. Processes **10**, 1834 (2022)
42. S. Tiwari et al., Three-step purification and characterization of organic solvent-tolerant and alkali-thermo-tolerant xylanase from Bacillus paramycoides T4 [MN370035]. Catalysts **12**, 749 (2022)
43. K. Patel, P. Dudhagara, Optimization of xylanase production by Bacillus tequilensis strain UD-3 using economical agricultural substrate and its application in rice straw pulp bleaching. Biocatal. Agric. Biotechnol. **30**, 101846 (2020)
44. S. Mhiri et al., A thermophilic and thermostable xylanase from Caldicoprobacter algeriensis: recombinant expression, characterization and application in paper biobleaching. Int. J. Biol. Macromol. **164**, 808–817 (2020)
45. E.P.O.F. Enzymes, et al. Safety evaluation of the food enzym endo-1, 4-β-xylanase from the genetically modified Bacillus subtilis strain AR-153. EFSA J. **22**, e9123 (2024)
46. S. Yan, Y. Xu, X.-W. Yu, Rational engineering of xylanase hyper-producing system in Trichoderma reesei for efficient biomass degradation. Biotechnol. Biofuels **14**, 1–17 (2021)

47. S. Chang, Y. Guo, B. Wu, B. He, Extracellular expression of alkali tolerant xylanase from Bacillus subtilis Lucky9 in E. coli and application for xylooligosaccharides production from agro-industrial waste. Int. J. Biol. Macromol. **96**, 249–256 (2017)
48. J. Cai et al., Cloning and heterologous expression of a novel xylanase gene TAX1 from Trichoderma atroviride and its application in the deconstruction of corn stover. Appl. Biochem. Biotechnol. **193**, 3029–3044 (2021)
49. C. Derntl, R.L. Mach, A.R. Mach-Aigner, Fusion transcription factors for strong, constitutive expression of cellulases and xylanases in Trichoderma reesei. Biotechnol. Biofuels **12**, 1–18 (2019)
50. H. Hu, S. Dai, A. Wen, X. Bai, Efficient expression of xylanase by codon optimization and its effects on the growth performance and carcass characteristics of broiler. Animals **9**, 65 (2019)
51. Y. Sürmeli, G. Şanlı-Mohamed, Engineering of xylanases for the development of biotechnologically important characteristics. Biotechnol. Bioeng. **120**, 1171–1188 (2023)
52. D. Mu, et al. Advances in the understanding of the production, modification and applications of xylanases in the food industry. Enzyme and Microbial Technol. 110473 (2024)
53. C. Pandey, P. Sharma, N. Gupta, Engineering to enhance thermostability of xylanase: for the new era of biotechnology. J Appl Biol Biotechnol **11**, 41–54 (2023)
54. M.N. Korkmaz, S.C. Ozdemir, A. Uzel, Xylanase production from marine derived Trichoderma pleuroticola 08ÇK001 strain isolated from Mediterranean coastal sediments. J. Basic Microbiol. **57**, 839–851 (2017)
55. Z. Liu, X. Zhao, F. Bai, Production of xylanase by an alkaline-tolerant marine-derived Streptomyces viridochromogenes strain and improvement by ribosome engineering. Appl. Microbiol. Biotechnol. **97**, 4361–4368 (2013)
56. R. Khandeparker, P. Verma, D. Deobagkar, A novel halotolerant xylanase from marine isolate Bacillus subtilis cho40: gene cloning and sequencing. New Biotechnol. **28**, 814–821 (2011)
57. J.A. Dos Santos, et al., Marine-derived fungus Aspergillus cf. tubingensis LAMAI 31: a new genetic resource for xylanase production. AMB Express **6**, 1–10 (2016)
58. Y. Brennan et al., Unusual microbial xylanases from insect guts. Appl. Environ. Microbiol. **70**, 3609–3617 (2004)
59. K. Kumari et al., A fast, reliable, low-cost, and efficient xylan extraction for xylooligosaccharides production. Biofuels, Bioprod. Biorefin. **18**, 1355–1368 (2024)
60. N. Liu, et al. Pulp properties and fiber characteristics of xylanase-treated aspen APMP. BioResources **7** (2012)
61. K. Przybysz Buzała, P. Przybysz, H. Kalinowska, M. Derkowska, Effect of cellulases and xylanases on refining process and kraft pulp properties. PLoS ONE **11**, e0161575 (2016)
62. T. Abena, A. Simachew, A review on xylanase sources, classification, mode of action, fermentation processes, and applications as a promising biocatalyst. Biotechnologia **105**, 273 (2024)
63. N. Almeida et al., Use of a novel extremophilic xylanase for an environmentally friendly industrial bleaching of kraft pulps. Int. J. Mol. Sci. **23**, 13423 (2022)
64. M. Kapoor, L.M. Nair, R.C. Kuhad, Cost-effective xylanase production from free and immobilized Bacillus pumilus strain MK001 and its application in saccharification of Prosopis juliflora. Biochem. Eng. J. **38**, 88–97 (2008)
65. A. Walia, S. Guleria, P. Mehta, A. Chauhan, J. Parkash, Microbial xylanases and their industrial application in pulp and paper biobleaching: a review. 3 Biotech **7**, 1–12 (2017)
66. H. Al-Darkazali, V. Meevootisom, D. Isarangkul, S. Wiyakrutta, Gene expression and molecular characterization of a xylanase from chicken cecum metagenome. Int. J. Microbiol. **2017**, 4018398 (2017)
67. A. Shehzadi et al., Xylanase; production to characterization and substrate interactions. Int. J. Appl. Biol. Forensics **1**, 110–118 (2017)
68. J.M. Matos, D.V. Evtuguin, A.P.M. de Sousa, M.G.V. Carvalho, Xylanase treatment of eucalypt kraft pulps: effect of carryover. Appl. Microbiol. Biotechnol. **108**, 210 (2024)
69. T. Cesar, V. Mrša, Purification and properties of the xylanase produced by Thermomyces lanuginosus. Enzyme Microb. Technol. **19**, 289–296 (1996)

70. T. Nanmori, T. Watanabe, R. Shinke, A. Kohno, Y. Kawamura, Purification and properties of thermostable xylanase and beta-xylosidase produced by a newly isolated Bacillus stearothermophilus strain. J. Bacteriol. **172**, 6669–6672 (1990)
71. R.A. Deshmukh, S. Jagtap, M.K. Mandal, S.K. Mandal, Purification, biochemical characterization and structural modelling of alkali-stable β-1,4-xylan xylanohydrolase from Aspergillus fumigatus R1 isolated from soil. BMC Biotechnol. **16**, 11 (2016). https://doi.org/10.1186/s12896-016-0242-4
72. F. Oz Tuncay, U. Cakmak, Y. Kolcuoglu, Aqueous two-phase extraction and characterization of thermotolerant alkaliphilic Cladophora hutchinsiae xylanase: biochemical properties and potential applications in fruit juice clarification and fish feed supplementation. Preparative Biochem. Biotechnol. **54**, 553–563 (2024)
73. Z. Han, F. Shang-Guan, J. Yang, Molecular and biochemical characterization of a bimodular xylanase from Marinifilaceae bacterium strain SPP2. Front. Microbiol. **10**, 1507 (2019)
74. Homology modeling and molecular docking studies of xylanase enzyme from bacillus. Substilis. J. Adv. Zool. **45**, https://doi.org/10.53555/jaz.v45i2.4615 (2024)
75. H. El-Gendi, A.S. Badawy, E.K. Bakhiet, M. Rawway, S.G. Ali, Valorization of lignocellulosic wastes for sustainable xylanase production from locally isolated Bacillus subtilis exploited for xylooligosaccharides' production with potential antimicrobial activity. Arch. Microbiol. **205**, 315 (2023)
76. B. Anuradha. Xylanase market outlook. <https://dataintelo.com/report/global-xylanase-market> (2025)
77. A. Khusro, B.K. Kaliyan, N.A. Al-Dhabi, M.V. Arasu, P. Agastian, Statistical optimization of thermo-alkali stable xylanase production from Bacillus tequilensis strain ARMATI. Electron. J. Biotechnol. **22**, 16–25 (2016)

Chapter 14
Challenges in Industrial Xylanase Applications

Abdul Sattar Jatoi

Abstract Xylanases are hydrolytic enzymes with immense industrial potential; extensively used for industry such as pulp and paper, biofuels, textiles, food processing, and animal feed. Despite their recognized environmental and economic benefits, the widespread industrial application of xylanases remains limited by several significant challenges. This chapter provides a comprehensive exploration of the critical barriers to industrial xylanase utilization, including enzyme stability under extreme processing conditions, high production costs, substrate specificity limitations, and product inhibition. It also addresses the complexities of integrating xylanases into established industrial workflows and the environmental concerns associated with enzyme production and usage. Emerging strategies, such as advanced protein engineering, synthetic biology, and digital process optimization, are examined. As promising solutions to these problems, by presenting a more in-depth picture current limitation and highlighting innovative technological and scientific progress, this chapter intends to give a guideline for improving commercial viability and sustainability of xylanase-based industrial processes.

Keywords Xylanase · Enzyme stability · Industrial biotechnology · Process integration · Sustainable processing · Protein engineering

1 Introduction

Xylanases (endo-1, 4-xylanases; EC 3. 2. 1. 8) are essential enzymes responsible for the hydrolysis of the second most abundant polysaccharide in plant cell walls, xylan [1]. These enzymes belong to the hemicellulase family and can hydrolyze xylan into xylose and other xylo-oligosaccharides, which can then be used to produce biofuels, animal feed additives, food stabilizers and pulp-bleaching agents [2]. Because they

A. S. Jatoi (✉)
Department of Chemical Engineering, Dawood University of Engineering and Technology Karachi, Karachi, Pakistan
e-mail: abdul.sattar@duet.edu.pk

A. Sattar Jatoi (ed.), *Xylanase from Microorganisms*,
https://doi.org/10.1007/978-981-95-0824-2_14

are therefore excellent catalytic enzymes in mild and environmentally friendly conditions, xylanases have become an important element in the development of more sustainable industrial processes [3]. Rising environmental awareness in the world and increasing energy demand for technologies have stimulated interest in industrial enzymes, particularly those involved with degradation of biomass. Xylanases, since they are responsible for degrading hemicellulose into fermentable sugars, are an essential element of biorefineries and biobased processes [4]. They are used across a broad range of applications including pulp and paper (as bleachers), food and feed (as baking additives and feed enhancers), textiles (for fabric softening and finishing) and pharmaceuticals (for oligosaccharide synthesis) [5]. However, despite their widespread potential, the transition of xylanase from lab-scale discovery to full-scale industrial application is riddled with challenges. Foremost among these is the variability of xylan's structural complexity across different biomass sources, which necessitates highly specific and robust enzymatic action. Additionally, xylanase production must be scalable, cost-effective, and stable under the rigorous operational conditions typical of industrial settings—factors often not addressed during initial laboratory researching [6]. Due to many things the xylanases used in industry are inefficient. Factors, including enzyme instability under extreme pH and temperature, substrate inhibition, low catalytic efficiency on raw substrates, and difficulties in downstream processing [7]. Further complications arise from the biological source of xylanases. Most commercial the evolution of xylanases fungi and bacteria are similar in a lot of ways and limitations. While genetic engineering and recombinant DNA technologies have enabled the tailoring of xylanases with desirable traits, such modifications often bring regulatory, safety, and consumer acceptance concerns, particularly in food and pharmaceutical sectors [8].

The industrial application of xylanase also faces practical hurdles such as integration into pre-existing process lines, consistency in enzyme performance across feedstocks, and contamination control during enzyme production and installation and deployment; furthermore a major barrier to wider adoption is the cost associated with enzyme production and purification, especially when high purity or specific activity is required. In recent years, considerable progress has been made in enzyme engineering, process optimization, and strain improvement. Still, many of these advancements But it will remain at the laboratory/pilot scale, limited translation into commercial operations. The need to close that gap has led to an interest in interdisciplinarity combining biotechnology, systems biology, bioinformatics, and process engineering. This Chapter covers a comprehensive overview of the main challenges confronting the industrial application of xylanases. It highlights the bottlenecks encountered at various stages—from enzyme production and stability to regulatory issues and sustainability concerns. The subsequent sections will delve into specific challenges associated with xylanase sourcing, enzyme engineering, industrial compatibility, economic feasibility, and future research directions.

2 Source and Production Challenges

The industrial application of xylanases hinges significantly on the efficiency, scalability, and cost-effectiveness of enzyme production systems, yet sourcing and producing xylanase at commercial scale presents multifaceted challenges. Naturally occurring xylanase-producing microorganisms, including fungi like Aspergillus and Trichoderma and bacteria such as Bacillus and Streptomyces, usually make the enzyme in low yields and with variable purity, hence requiring extensive downstream processing Fungi have very high secretion capacities but bacteria have low [9]. Xylanases frequently offer greater thermos ability and pH tolerance, though they may be lacking efficient secretion systems. Native strains are very sensitive to and some require expensive inducers (xylan or xylose) for cultivation complicating production and increasing costs [10]. Fermentation strategies, particularly submerged (SmF) and solid-state fermentation (SSF), must be optimized for each strain and process with parameters e. g. pH, temperature aeration, and nutrient concentrations needing strict control to ensure consistent yields. However, this is far from straightforward when scaling these processes up from the lab to industry differences in oxygen transfer, mixing, and contamination risks in large fermenters [11]. The choice of substrate further complicates matters—while agricultural residues like wheat bran or corn stover offer low-cost. Alternatives to synthetic media, they introduce heterogeneity and may contain microbial growth inhibitors, necessitating pretreatment and adaptive microbial strains. To overcome these constraints, recombinant DNA technology is widely employed, with heterologous hosts such as Escherichia coli, Pichia pastoris, and Saccharomyces cerevisiae engineered to express xylanases with improved characteristics. Nevertheless, recombinant expression It has its own set of problems, like inclusion bodies (in E. coli) improper folding, codon usage issues, and differences in post-translational modifications [12]. While genetic engineering and directed evolution have been involved in improving enzyme yield, activity and stability (the use of) genetically modified organisms (GMOs) remains subject to rigorous regulation, particularly in food and pharmaceutical sectors, potentially limiting market acceptance. Traditional strain improvement methods such as random mutagenesis and selection have produced promising mutants, but they're laborious and expensive often lack predictability and long-term stability [13]. Modern tools like adaptive Laboratory evolution and omics technologies provide more targeted and efficient approaches to strain development, yet require substantial expertise and infrastructure [14]. Moreover, despite advances in upstream production, downstream processing is an issue; but crude extracts of enzymes might be sufficient for certain applications like pulp bleaching or biofuel production, high-purity preparations required in food or pharma demand additional purification steps, often involving expensive and complex chromatographic techniques. The presence of unwanted proteins, proteases, or inhibitory compounds in the fermentation broth further complicates purification. Overall, achieving industrial-scale xylanase production that is both economically viable and technically robust demands integrated solutions involving microbial strain development, process [15]. Engineering, and system-level optimization. Continued innovation in synthetic biology, bioprocess automation, and sustainable feedstock utilization will be we need to move beyond these obstacles to unlock the full potential of xylanases in diverse industrial sectors.

3 Enzyme Stability and Activity Under Industrial Conditions

One of the major problems encountered in industrial use of nanotechnology xylanases lies in ensuring their stability and catalytic activity under the harsh conditions typically encountered in commercial processes. Industrial usually environments in which enzymes must be active and stable for extended periods Periods of high temperatures, high pH values and with/without: various solvents, oxidizing agents, or metal ions. Unfortunately, many naturally occurring xylanases are sensitive to these conditions, leading to rapid denaturation and loss of enzymatic function, thereby reducing the efficiency and cost effectiveness of the process e. g. in the pulp and paper paper industry, xylanases must function under highly alkaline conditions and In the biofuels industry they're expected to tolerate a certain amount of acidity. Pretreatment residues and high thermal loads. Structural integrity of the... enzyme, including hydrogen bonds, hydrophobic interactions, and disulfide its structural integrity (usually under such stresses) is often compromised. Engineered variants with enhanced stability. Enzyme immobilization techniques, developed, such as binding to solid supports or encapsulation in gels. To increase operational sturdiness and to facilitate reuse, but these methods may sometimes reduce catalytic efficiency or introduce mass transfer limitations. Alternatively, protein engineering approaches, including site-directed mutagenesis and directed evolution, have been used to change amino acid residues responsible for structural flexibility, aiming to produce enzymes with greater rigidity and resistance to denaturing conditions. Despite notable successes in laboratory-scale studies, translating these engineered enzymes to in industrial environments—owing to issues like production costs, expression efficiency, and unpredictability in real process environments. Another aspect of this problem is the ability of the enzyme to maintain high levels of. Catalytic activity in the presence of process inhibitors or complex substrates. Many lignocellulosic feedstocks release inhibitors such as furfural, hydroxymethylfurfural (HMF), phenolic compounds, or organic acids during pretreatment all of which can inhibit xylanase activity. Conversely, lignin/osmotic barrier may prevent enzyme's access to xylan cellulose, necessitating synergistic action with other hemicellulases or lignin-degrading enzymes. Therefore, xylanases used industrially must not only be very stable but also have broad substrate specificity and high turnover rates to efficiently degrade heterogeneous biomass. Multifunctional or bifunctional enzymes that can degrade multiple polysaccharide components simultaneously are of growing interest, although their production and functional optimization remain technically demanding. Additionally, the pH and temperature optima of xylanases frequently not compatible with existing industrial processes, leading to additional energy costs or necessitate process redesign. For example, thermophilic xylanases suitable for use in bioethanol production at 60–70 °C may be incompatible with microbial fermentation stages that occur at lower temperatures.

4 Substrate Specificity and Efficiency

One of the biggest hurdles in using xylanases in industry is their substrate specificity and how well they work with a variety of lignocellulosic materials. Xylan, which is the main substrate for these enzymes, can vary quite a bit depending on where it comes from—different plants, types of tissues, and environmental conditions all play a role. This variability shows up in differences in the length of the backbone, how branched it is, and the presence of side chains like arabinose, glucuronic acid, acetyl groups, and ferulic acid. Because of this complexity, it's tough for a single xylanase to effectively break down all kinds of xylan, which can lead to less-than-ideal conversion rates in industrial applications. For instance, xylans sourced from hardwoods are quite different from those in grasses or cereals, and enzymes that work well on one type may not perform as effectively on another. This inconsistency poses a real challenge for industries that depend on a variety of feedstocks, especially those that change with the seasons, like biofuels and animal feed. Additionally, natural xylanases usually have a limited range of substrates and can be hindered by non-hydrolysable linkages or side chains, which further decreases efficiency. To tackle these challenges, researchers have been looking into enzyme cocktails that combine different xylanase isoforms with accessory enzymes like α-L-arabinofuranosidases, acetyl xylan esterases, and feruloyl esterases to help break down those tough xylan structures. While this approach can boost saccharification yields, it also adds complexity and cost to production, along with the need for careful process control to make sure the enzymes work well together. Another issue is that the structure of the substrate in industrial biomass is often tough to penetrate, being highly recalcitrant, cross-linked, or embedded in lignin, which limits how accessible the enzymes are. Common pretreatment methods like steam explosion, acid hydrolysis, or alkali treatment are used to improve substrate accessibility, but they can also create inhibitory byproducts.

5 Process Integration and Operational Compatibility

Effective use of xylanases in industry is difficult since their action is influenced by a range of process settings, including temperature, pH, how much time is needed for a reaction, the type of feedstock and post-enzymatic needs. In the pulp, paper, food and animal feed industries—for example, in bleaching, bioethanol production and making animal feed—xylanases must interact with chemical reactions, other enzymes and often, with microorganisms. To be truly compatible, xylanases must function well according to planned conditions, keep their activity over extended use, resist mechanical stress and not react to external additives or contaminants. Xylanases often help cellulases and lignin-active enzymes at close-range pH and temperature conditions in several biorefineries. Still, the ideal pH of xylanases generally differs from that of cellulases or fermentative bacteria, often leading to either waste during fermentation or the need to rebalance the process using costly energy-demanding means. Also, introducing xylanase into present production processes often means upgrading equipment, changing flow stages and confirming that the process follows all rules and standards. In the pulp and paper industry, chlorine-free bleaching methods must involve xylanase and fit seamlessly into lengthy bleaching systems that maintain the quality, color and fiber strength of the paper. Just as in other fields, xylanase treatment of lignocellulosic material during biofuel processing should be timed to follow hydrolysis and fermentation to prevent loss of sugars or toxicity to microbes. Batch-wise changes in biomass feedstock are a further challenge for integration, given that the changes in xylan content or chemical inhibitors may require adjusting enzyme performance and controls. When residual lignin, ash or different non-carbohydrate substances are found, they can take up or deactivate enzymes, causing their effective strength to drop and requiring more enzyme to achieve the same results—a costly situation. In industries where production is in continuous or semi-continuous mode, xylanases must be able to be used more than once. Using enzymes that cannot move offers a way to repeat the process, yet their enzymatic activity may lessen because of fouling, leaching or due to decreased effectiveness of mass transfer. In integration projects, the layout of the bioreactor also matters a lot; stirred tanks, beds made of particles or membrane reactors all offer special benefits and challenges based on the form of the substrate, speed of enzyme reactions and mixing requirements. Predicting how enzymes behave and deciding how reactors should be designed is now mostly done by using modeling and simulation, but it depends on complete kinetic and thermodynamic information which isn't often available for new or engineered xylanases. It is important to integrate design solutions so that they meet the rules for the environment, worker safety and cost management. Producers must use sustainability rules, including less water and energy use, a low chemical count and minor greenhouse gas production, with xylanase. In addition, making enzymes useful in industry is influenced by their purchase price, dependable performance when added in small amounts and the ability to cut down on expensive chemicals or use less energy than other methods. For this reason, proper comprehension of technical and economic aspects should take place during the development and upscaling of bioprocesses. Safety and purity must be carefully considered in food and feed applications because of the regulations that apply to xylanase. EFSA (European Food Safety Authority) and FDA (U.S. Food and Drug Administration) require detailed data on the strain used in production, its fermentation process, residual DNA, allergen potential and possible toxicity. Since getting this information takes time, it can slow down the use of genetically modified enzymes. As a result, a combined method is being used that chooses and engineers enzymes alongside the development of full processes and their constant monitoring online. Thanks to smart sensors,

AI in process control and digital twins, it is possible to optimize xylanase applications more flexibly, adding to their resilience. Partnerships among enzymologists, process engineers, chemists and experts in regulations are needed to develop the best strategy for xylanase use in every industrial field. Integrating xylanases into industry effectively means looking at the enzymes' characteristics, how they are produced, their costs and making sure they support the factory's main operations, environmental protection and the needs of the market.

6 Cost and Commercialization Constraints

High cost is the main factor that holds back the expansion of xylanase industry applications. Although there have been great advances in enzyme biotechnology, it is still very expensive for xylanase to be used in industries such as animal feed, biofuels and pulp and paper. The final cost of xylanase is made up of upstream charges for developing the strain, building and running the fermentation facility, obtaining raw materials and running downstream processing, along with charges for stabilizers, preservatives and delivery systems. Ensuring high yields and activity of enzymes translates to saving money, so biotech often uses specially designed microorganisms and works to make the bioprocessing steps as efficient as possible. Developing these strains requires professionals to use a lot of resources, spend significant amounts of time and acquire required approvals, mainly for GMOs. Fermentation media make up a large part of manufacturing costs and although agro-industrial residues are cheaper than synthetic media, they can add harm or contaminants that may lower enzyme yield or activity. The cost of enzyme production is further increased by extra steps such as filtration, concentration and purification, especially for enzymes needed for food or pharmaceutical applications. In addition, the capability of xylanase formulations to keep their properties during storage and transport is necessary for selling them. Temperature, humidity and oxidation make enzymes sensitive, so the logistical costs increase due to the use of cold chains or other encapsulation technology. In these markets, xylanases should excel technically and also be more economical or add more value than typical chemical routes. Xylanase treatment should lead to a lower need for chlorine without affecting the pulp quality or raising the time required for processing papers in the paper industry. For biofuels, xylanase should help make more sugar or ethanol, but should not slow things down or interfere with fermentation. As a result, the cost benefit of using enzymes must be evident to guarantee their use for companies with limited profits. The industry is also hesitant due to unproven security over long-term supplies, uncertain similarity from one batch to another and potential incompatibility with present manufacturing methods. Producing goods, fulfilling requirements and starting marketing all require big investments for small companies. Even though technology is completely ready, starting the commercial process may be stopped by intellectual property problems, not enough industry-specific testing or opposition from those using chemical methods. Global commercialization encounters various conditions set by different regulators. Procedures for getting regulatory approval differ from region to region, with unique requirements for certificate check, labeling and the levels of food, feed and environmental permitted use. Because markets are split, getting in becomes more challenging and results in bigger expenses for compliance. To add, as environmental concerns rise, industries are more interested in green technologies that don't involve substantial extra costs. Even though xylanases help decrease our impact on the environment, their production should also be green and sustainable as measured by low carbon production, use of renewable materials and little or no waste to fit both corporate and government requirements. Whether people support GMOs can affect how much products containing them are sold. Getting past these barriers requires enzyme producers, end users, academic institutions and regulators to work together in business models. Dealing with risks in the business can be lessened through public–private agreements, government help through subsidies and investment in showcase plants. Besides, constantly finding new ways in strain engineering, making better processes and improving enzyme formulations is necessary to cut prices and improve results. Today's technology, including systems that simplify work and provide transparency through blockchain during delivery, is making it simpler and safer to use precious enzymes. The success of xylanases in the market comes from considering both their technical results and ensuring they fit economic, regulatory and environmental criteria.

7 Regulatory and Safety Issues

In the food, feed, pharmaceutical and environmental fields, strict rules about safety and regulation make it very difficult to use xylanases on a large scale. The use of enzymes in these sectors has to fulfill strict safety checks that include considering if they are allergenic, their toxicity, how they affect the environment and where and how they are made. Using GMOs to improve enzymes is a major issue for regulators. Several agencies like the U.S. FDA, EFSA and many national groups expect detailed records of the genetic changes, including the source and host organisms, the carrier DNA, where the DNA is placed in the genome and how long the changes remain. Also, evaluations of leftover recombinant DNA, antibiotic resistance and residual products from fermentation are carried out and must be either not detected or present in safe amounts. At the same time, the purification part of the process must make certain that the enzyme is free from dangerous contaminants, so it won't cause health problems when used. In addition to making sure an enzyme is safe, regulators also check how it functions as expected in its practical environment. It is important that the xylanase in animal feed helps animals digest better and doesn't change their nutrient needs or introduce negative effects on their health. When food enzymes are used for baking or clarifying beverages, their taste, texture and safety must remain the same and customers should know if the enzyme is obtained from a genetically modified source. Around the world, there is disagreement about labeling GMO-derived enzymes, as some countries require notices while others permit some products to go unlabelled if their modifications took place during the process. In addition, occupational safety should be taken into account during work with enzymes such as xylanase, because frequent exposure to enzyme dust or aerosols may cause asthma in workers and because enzymes can also cause allergic reactions. So, companies need to use protective gear, install air purifiers and provide training for their staff to reduce their chances of exposure. Commercialization of xylanase is also affected by environmental rules, especially when dealing with wastewater enzyme discharge or when releasing genetically altered organisms into the environment. Though xylanases are both biodegradable and safe for the environment, producing and using these industrial enzymes must follow sustainability requirements, as people pay closer attention to carbon footprints, water use and waste management in industrial work. Thus, organizations generally have dialogues with regulatory authorities prior to marketing to make sure they stick to the rules and obtain approval process smoothly. On the other hand, this requires more time and funds from entities hoping to launch in places with such regulations. Besides, because rules differ from one region to another, it becomes difficult for businesses to access global markets and may mean they have to register products many times in parallel. Having international standards for enzyme regulations such as those set by the Codex Alimentarius Commission, could address these issues, but these efforts rarely move fast and are often shape by local and public factors. Responses from the public are important when considering the formulation of new rules. How people feel negatively about GMOs or synthetic biology-derived enzymes can impact both decisions related to policies and acceptance in the markets, mainly in Europe and parts of Asia. Thus, manufacturers might have to carry out public education activities, involve various parties and provide honest information to win trust and ensure the acceptable use of enzyme technologies. It is very important to involve regulatory and safety issues early in the process of building a new product. If companies act this way, they can make enzymes and set up their factories to fit the safety regulations that are expected, preventing any big issues at the approval stage. Better, faster development of safety assessment guidelines and enhanced consistency in how rules are applied can be achieved when academia joins with industry and regulatory groups. Looking at safety risks today, in silico methods and omics technologies are providing scientists with better and faster data. When it comes to these issues, an integrated process using science, knowledge of regulations and cooperation from stakeholders is necessary. If the industrial enzyme sector responds well to these difficulties, xylanases can help improve sustainability and performance in biotechnology across different industrial fields.

8 Integration with Industrial Processes

Adding xylanases to industrial steps is an important move toward using enzymes fully in pulp and paper, biofuels, textiles, the food and beverage industry and animal feed. Even so, bringing enzymes into a process isn't easy, since they must fit well with the enzyme needs, the process, the product aims and the plant's environment. Since industrial production is focused on being efficient, constant and following laws, enzymes used should easily fit into current plant systems and processes. In biobleaching which is important in the pulp and paper industry, xylanases are applied to decrease the use of strict chlorine chemicals. Yet, for this approach to succeed, the xylanase has to be stable at elevated temperatures (50–70 °C), alkali (8–10 pH) and with phenolic compounds from lignin that might act as blockers for it. Not only do you need to pick enzymes that can survive high temperatures and strong alkali, but you must also improve the process by optimizing how long the reactants mix with enzymes. In making biofuels, xylanases support breaking up hemicellulose in lignocellulosic biomass during the pretreatment and saccharification process, thereby increasing the release of useful sugars. Nevertheless, the integration takes into account differences in biomass kind, the method used to treat it and possible coexistence of cellulases and β-glucosidases. Making sure these enzymes work together effectively usually needs creating enzyme cocktails and checking the process pH, temperature and level of ingredients. Some of the biomass materials have chemicals called furfural and acetic acid which lower the activity of xylanase, so new types of resistant enzymes are required. In the baking industry, dough is made better by using xylanases which improve handling, shape and crumb quality. In these stages, integration mainly involves finding the best balance of enzyme concentration, time and temperature during brief dough mixing and proofing. Flour mixtures and interactions between enzymes and the substrates in heterogeneous dough can make it hard to predict what will happen. That is why it is necessary to keep formulations unchanged and to ensure quality control. Xylanases in brewing help get rid of arabinoxylans which might cause brewing to be too thick and hard to filter properly. To avoid spoiling the beer, places where integration comes into play are when you mix enzyme action with mashing and fermentation, where all temperatures, pH and the kinetics of enzymes play a role. In keeping with high-fiber diets for animals, xylanases are introduced to feed to boost their digestion by breaking down wheat and rye polysaccharides. To achieve integration, xylanases must handle extreme heat from pelleting and have the ability to work in a variety of conditions along the gastrointestinal tract. Often, enzymes are put into formulas that protect them until they reach their needed site. For all types of industries, the liquid, powder or granule format of the enzyme affects how smoothly it can be added to a process, checked for accuracy and stored safely. Another approach involves using enzymes repeatedly by fixing them to solid supports or including them in bioreactor systems to improve both function and affordability. But if improperly set up, the immobilization of catalysts can make mass transfer more increased and reduce the efficiency of the process. How well xylanases react with different chemicals is significant; they need to perform while being mixed with surfactants, salts, different solvents and other enzymes. Additionally, since large industrial facilities must process many batches of raw materials over time, the xylanase must perform consistently and can only do so through strict controls that can be monitored in real time. Using process simulation, AI optimization and digital twins, enzyme performance models are now used to make predictions about how best to use enzymes. Integration choices in food and feed industries are guided by rules and concerns for safety, along with managing how traceability, labeling and what consumers accept is handled. Sustainability purposes which involve cutting back on water, energy and chemical waste, add value to using xylanase and also put pressure on engineers to design new production processes. These challenges can be resolved by joining efforts from enzymologists, chemical engineers, process developers and regulatory specialists. In the end, integration of xylanases works best when one recognizes the activity of these enzymes and the dynamics of the process being used and then optimizes them from a broader standpoint to ensure they can be used on a commercial scale and meet intended product targets.

9 Environmental and Sustainability Concerns

Xylanases are now preferred by industry due to their eco-friendly nature, as firms worldwide want to limit their use of chemicals. Using xylanases as sustainable biocatalysts supports environmentally friendly industries, helps cut the use of harsh chemicals and results in less energy and water use. Using xylanase in pulp and paper biobleaching reduces the reliance on chlorine-based agents which are famous for manufacturing toxic dioxins and furans. There are negative ecological and health consequences of these pollutants, for example, disruption of hormone systems and forming cancer. When chemical bleaching is not used or used less, xylanases play a role in decreasing wastewater pollution and reduce the impact pulp mills make on the environment. Similarly, xylanases in textile treatment reduce the use of harmful chemicals, lower the risk of toxic discharges and make waste more biodegradable. Xylanases make it possible to change lignocellulosic biomass into fermentable sugars using milder methods compared to other approaches, reducing emissions and less energy is required. Even so, we need to evaluate the environmental advantages of xylanases during the whole process of production and application. Energy and resources are used in enzyme manufacture and large scales of fermentation often lead to waste and emissions. All three parts of microbial production—microbial strains, raw materials and methods used—impact the environment. As an example, employing genetically modified microorganisms (GMOs) can worry ecologists, since any unintentional release outside the lab might disrupt the balance in the microbial environment. A lot of the media for fermentation come from agricultural products or industrial leftovers and being able to source them sustainably is important to protect the environment. So, to properly examine whether xylanase applications are sustainable, a complete LCA is important because it reviews items like carbon emissions, water use, energy demand, how much land is used and the possibility of toxicity. Hemicelluloses are a major source for the substrates of xylanases in lignocellulosic biomass which is another important sustainability issue. The use of agricultural wastes, forestry scraps and other green materials for fuel encourages both circularity and less waste, yet it is necessary to consider that too much biomass taking can result in reduced soil health and loss of species. The application of xylanases in industrial processes ought to be guided by the green chemistry values of efficient use of atoms, less use of dangerous chemicals and less waste. The aim of enzyme-based processes should be to use a closed-loop system that uses by-products, reuses water and makes the most of available energy. Sustainable frameworks that achieve low-impact manufacturing such as those using Xylanases, help meet important environmental targets such as the UN Sustainable Development Goals (SDGs) for water, climate, production and nature. Despite everything, firms that use processes that save money but are environmentally harmful often find it difficult to move to xylanase-based methods because of costs and technology Work to increase the sustainability of xylanase applications should emphasize growing enzyme yields, maintaining their durability and improving how much enzyme is produced from a single batch of materials. Additionally, making enzymes that function in a wider range of process conditions saves the need for making changes and adding more materials. Immobilization and retrieval of enzymes support sustainability because they let enzymes be used many times and decrease waste. At the same time, governments may introduce green standards for purchases, give support to environmental efforts and increase regulations on factory discharges to encourage xylanase growth. Advocacy and requests for sustainably produced items by many people encourage more industries to use things like enzymatic solutions. At the same time, sustainability goals need to be easy to track, checked by outside bodies and linked to their sources to make sure everything is accurate and reliable. Making sure people in industry understand the advantages of xylanases helps to fight resistance to change and supports the growth of a green culture. Ultimately, xylanases are environmentally beneficial, although their real value is achieved through proper integration into sustainability, responsible enzyme production, appropriate regulations and what the market demands. Dealing with these matters in advance, xylanase-based methods serve to help industry cut down waste, operate more sustainably and satisfy world goals for sustainability.

10 Future Perspectives and Emerging Solutions

The next step for industrial xylanase applications is to bring in innovative and emerging solutions that help overcome present problems and offer more advantages, greater usage and better compatibility with sustainable industrial systems. Because pressure to decarbonize and improve the environment grows, xylanases will play an important role as biotechnology develops further with synthetic biology, work on proteins and studies centered on systems. Through directed evolution and rational design, enzyme engineering is being revolutionized, so that xylanases can be customized with better heat resistance, ability to operate within a wide pH range, selectivity in the substrate used and resistance to industrial inhibitors. Furthermore, machine learning and computational approaches now support the older strategies by identifying links between a protein's structure and its function. As a result, using metagenomic techniques has made it possible to identify numerous xylanase genes in difficult-to-culture microbes living in extreme or unexplored habitats such as hot springs, deep-sea vents or compost piles which could have strong uses in industry. Having access to omics technologies such as genomics, transcriptomics, proteomics and metabolomics is helping scientists find out how production hosts function in groups and improve their potential for producing enzymes economically. Applications of these approaches make microbial platforms more tolerant to difficult environments and allow them to use cheaper raw materials. At the same time, better enzyme protection in industry is being achieved with developments such as encapsulation, nanocarriers and responsive polymers which lead to better production and less waste. Progress with immobilization processes is allowing xylanases to be reused in continuous-flow reactors and lowering the expense of handling them afterwards. It ensures enzymes work longer and fits with the goals of zero-waste processes. More importantly, designing mixtures of different enzymes allows complex biomass to be broken down efficiently, reducing the time needed for processing and raising the yield of sugars in biofuels and biorefineries. Using xylanase in combination with other modern technologies such as membrane filtration, microfluidics and bioreactors equipped to monitor and control activities continuously, can help improve the use of enzymes and the conversion of substances at the same time. Thanks to artificial intelligence and digital twins, models and predictions can be used to make changes to xylanase production while limiting costly experiments. For businesses, modular, decentralized biomanufacturing of xylanases is helping local and small-scale operations process biomass at their sites, making transportation of materials less necessary and making it possible to apply circular economy ideas. New rules emerging internationally often recognize the importance of enzyme safety and sustainability, with a special focus on those from GRAS (Generally Recognized As Safe) microbes which helps improve their access in global markets. But dealing with intellectual property, safety in bioscience and public questions about genetically engineered enzymes will call for continuous interaction with key groups like policymakers, consumers and stakeholders. It will be important to educate, be open and use ethical guidelines to encourage both the community and researchers to accept and use innovation responsibly. Additionally, joining efforts between industries and academic institutions and using open innovation technologies are helping to quickly introduce and use advanced xylanase products. Projects that encourage working across fields, using shared infrastructure and basing experiments on data are likely to bring products to market sooner and reduce development costs. More and more money from the government and companies is going towards green enzyme technologies, thanks in part to new incentives like carbon credits, certifications and measuring impact. It appears that there are many more opportunities for xylanases in modern applications than in traditional areas. Enzymes are now being studied for some unique uses such as in biodegradable packaging, agricultural soil improvement, making bioplastics, recycled textiles and medicines that prevent the sticking of bacterial plaques (known as biofilms). All in all, future prospects for industrial xylanase use depend on the relationship among biotechnology, digital innovation and sustainability science. Thanks to new tools and collaboration models, the enzyme sector is poised to provide efficient, flexible and eco-friendly solutions that respond to today's industrial dilemmas and support a stable bio-based industry.

11 Conclusion

The industrial use of xylanases could lead to new more sustainable, effective and eco-friendly manufacturing methods. Yet, their true value is not being used widely since enzymes are unstable in industrial situations, support less range of substances, respond badly to certain inhibitors and are hard to fit into complex industrial processes. Besides, worries about the environment, profitability and the image <<public perception>> make it harder for enzymes to become used on a wider scale. Although these processes are complex, recent progress in enzyme engineering, screening and modeling is allowing us to create xylanase variants that can withstand many industrial situations. In addition, improvements in immobilization techniques, process design tools and synthetic biology provide new paths for lowering expenses and raising performance in operations. In the future, xylanases will be used in industry successfully if there is progress in science, technology, interdisciplinary cooperation, appropriate rules and strong sustainable business activities. If xylanases are used to address complex challenges all around, the value they can bring will help shift industry toward greener and more reliable ecosystems.

References

1. M. Phuyal, U. Budhathoki, D. Bista, S. Shakya, R. Shrestha, A.K. Shrestha, Xylanase-Producing microbes and their real-world application. Int. J. Chem. Eng. **2023**, 3593035 (2023)
2. S.B. Ajeje, Y. Hu, G. Song, S.B. Peter, R.G. Afful, F. Sun et al., Thermostable cellulases/xylanases from thermophilic and hyperthermophilic microorganisms: current perspective. Front. Bioeng. Biotechnol. **9**, 794304 (2021)
3. F. Güler, F. Özçelik, Screening of xylanase producing Bacillus species and optimization of xylanase process parameters in submerged fermentation. Biocatal. Agric. Biotechnol. **51**, 102801 (2023)
4. J. Kaushal, A. Raina, G. Singh, M. Khatri, S.K. Arya, N. Karmegam et al., Methodical study implicating the effectiveness of microbial treatment over xylanase enzymatic treatment for pulp bio-bleaching. Environ. Technol. Innov. **28**, 102731 (2022)
5. A. Sarangi, H. Thatoi, in *Xylanase as a Promising Biocatalyst: A Review on its Production, Purification and Biotechnological Applications*. Proceedings of the National Academy of Sciences, India Section B: Biological Sciences (2024), pp. 1–16
6. R. Chaudhary, T. Kuthiala, G. Singh, S. Rarotra, A. Kaur, S.K. Arya et al., Current status of xylanase for biofuel production: a review on classification and characterization. Biomass Convers. Biorefinery **13**, 8773–8791 (2023)
7. T. Abena, A. Simachew, A review on xylanase sources, classification, mode of action, fermentation processes, and applications as a promising biocatalyst. Biotechnologia **105**, 273 (2024)
8. Z. Liu, S. Wen, G. Wu, H. Wu, Heterologous expression and characterization of Anaeromyces robustus xylanase and its use in bread making. Eur. Food Res. Technol. **248**, 2311–2324 (2022)
9. D. Kaur, A. Joshi, V. Sharma, N. Batra, A.K. Sharma, An insight into microbial sources, classification, and industrial applications of xylanases: a rapid review. Biotechnol. Appl. Biochem. **70**, 1489–1503 (2023)

10. K. Kumari, S. Nagar, S. Goyal, S. Maan, Hyper xylanase production and potential of xylooligosaccharides formation from a novel Bacillus australimaris KS2. Biocatal. Agric. Biotechnol. **54**, 102899 (2023)
11. M. Soni, C. Mathur, A. Soni, M.K. Solanki, B.K. Kashyap, D.V. Kamboj, Xylanase in waste management and its industrial applications, in *Waste to Energy: Prospects and Applications* (Springer, 2021), pp. 393–414
12. S. Verma, A. Kumar, S. Joshi, S. Gangola, A. Rani, Role of microorganisms in agricultural waste management, in *Advanced Microbial Technology for Sustainable Agriculture and Environment* (Elsevier, 2023), pp. 137–153
13. M. Çiçekler, Biopulping and biobleaching in pulp and paper industry: a review of xylanase and its effects. Forest Agric. Stud. Differ. Perspect. **2**, 21–24 (2022)
14. C. Qiu, N. Liu, X. Diao, L. He, H. Zhou, W. Zhang, Effects of cellulase and xylanase on fermentation characteristics, chemical composition and bacterial community of the mixed silage of king grass and rice straw. Microorganisms **12**, 561 (2024)
15. U. Ejaz, M. Sohail, Emerging trends in xylanase biotechnology: from lignocellulose degradation to industrial applications. Int. J. Biol. Res **12**, 80–96 (2024)

Chapter 15
Future Directions in Xylanase Research

Abdul Sattar Jatoi, Zubair Hashmi, Albert Selvakumar Jeyapaul, Shoaib Ahmed, and Abdul Karim Shah

Abstract In many applications, including making biofuels, bleaching paper and pulp, processing foods and feeds and recovering useful products from waste, xylanases are important as industrial biocatalysts. As focus worldwide turns to eco-friendly technologies, there is increasing demand for xylanases that are strong, efficient and affordable. The chapter looks at where xylanase research is headed in the future, focusing on recent progress in finding new enzymes, designing new proteins and adopting modern tools, including synthetic biology, nanotechnology and artificial intelligence. A few important trends involve the creation of blends of multiple enzymes, discoveries from hardy microbes and their environments and improvements to xylanase stability and precision for industrial needs. It also examines the idea that xylanases could promote the use of lignocellulosic biomass to make high-value products within the circular bioeconomy. Strategies for handling stability issues in enzymes, the complexity of substrates and the challenge of producing them on a large scale are discussed. Taken together, the chapter explains the impact of literature and technology on future xylanases for sustainable industry.

Keywords Xylanase engineering · Animal feed supplementation · Food processing applications · Sustainable bioprocessing

1 Introduction

Xylanases which are part of the hemicellulose family, now play an important role in numerous industrial processes such as the bioconversion of lignocellulosic biomass [1]. Because industries want both eco-friendly and reasonably priced ways to use

A. S. Jatoi (✉) · Z. Hashmi · S. Ahmed · A. K. Shah
Department of Chemical Engineering, Dawood University of Engineering and Technology, Karachi, Pakistan
e-mail: abdul.sattar@duet.edu.pk

A. S. Jeyapaul
Department of Water and Wastewater Treatment, Al Khabourah Vocational College, Al khabourah, Oman

A. Sattar Jatoi (ed.), *Xylanase from Microorganisms*,
https://doi.org/10.1007/978-981-95-0824-2_15

chemicals, the need for strong, flexible and efficient xylanases has increased greatly. Wide use of xylanases can be found in bleaching pulp and paper, baking, clarifying juices, treating textiles and fermenting hemicellulose to obtain useful sugars in biofuel production [1, 2]. Even though xylanases have been used and studied for years, there are some limitations that continue to hold back their use in industry. Problems with these enzymes are that they are unstable during harsh processing, have a narrow range of suitable materials, cannot handle inhibitors found in pretreated biomass and are hard to express at large scales. As the world shifts towards more sustainable technologies, better xylanases are more important than ever [3].

The recent achievements in metagenomics, synthetic biology, protein engineering and computational biology have created fresh opportunities for exploring xylanase research. Thanks to metagenomic methods, scientists can now investigate remote microbial habitats which has resulted in finding new xylanases with extraordinary properties [4]. Since enzymes in hot springs, salty oceans, wasted lands and vents from the deep sea are stable and effective in extreme conditions, they are often selected for industrial activities. Furthermore, synthetic biology helps produce customized xylanases with improved thermostability, pH range or what substances they can degrade [5]. Researchers can now use both rational design and driven evolution techniques to create improved chimeric and multifunctional enzymes compared to their natural forms.

At the same time, using machine learning and artificial intelligence in enzyme studies has sped up the work of categorizing and predicting enzymes [6]. These tools make it possible to identify gene candidates with useful functions from huge datasets and help structure their activity, predict it and design new enzymes from scratch [7]. When used together with prediction tools such as AlphaFold, AI-based systems have the power to change the process of discovering and optimizing xylanases. Additionally, immobilization technologies and process engineering strategies are being developed to ensure that xylanases maintain their activity and stability in continuous or large-scale operations, thereby improving the cost-efficiency of industrial biocatalytic processes [8].

Looking forward, the role of xylanases is set to expand far beyond traditional applications. Novel uses in biomedicine, diagnostics, and materials science are emerging, driven by the ability of these enzymes to generate functional oligosaccharides, aid in biofilm degradation, or participate in smart biomaterial production [9]. The integration of xylanases into the framework of biorefineries aligns with the global push toward circular bioeconomy models, where agricultural and forestry residues are transformed into value-added products. However, realizing this potential will require concerted efforts to overcome several bottlenecks [10]. These include the need for improved screening methods for enzyme activity, better expression systems for recombinant production, and clearer regulatory guidelines for the commercialization of genetically engineered enzymes [11].

This chapter explores the future directions in xylanase research, focusing on recent advancements, emerging technologies, and potential applications that will shape the next generation of xylanase-based biotechnologies. It covers the latest trends in enzyme discovery through metagenomics and functional genomics, the use of synthetic biology and enzyme engineering for tailoring xylanase properties, and the application of AI in enzyme function prediction. Part of the chapter looks at using xylanases in biorefineries, new uses in industry and medicine and what continues to be challenging. This chapter hopes to set out a plan for researchers, industry members and policymakers working in industrial biotechnology and enzyme engineering by focusing on the main areas of development and overcoming present constraints.

2 Emerging Trends in Xylanase Discovery

So far, discovering xylanase has been largely limited by the fact that 99% of microorganisms cannot be cultivated in standard lab conditions, a principle known as "the great plate count anomaly [12]. This obstacle has so far kept the range of xylanases in research and industry from growing widely. Particularly, new, quick and culture-free techniques have had a big impact on xylanase research. In particular, using approaches such as metagenomics and omics strategies has revealed a lot about the different enzymes found among the most extreme and diverse areas of nature. This part of the report discusses the most important emerging trends that will affect xylanase discovery [13].

2.1 Advances in Metagenomics and Functional Genomics

Searching for new xylanases has made metagenomics a crucial technique in recent years. Researchers can now skip cultivation and immediately extract DNA from the environment, sequence the DNA quickly, put the pieces together, annotate them and perform functional screening. If metagenomic DNA libraries are cloned and put into appropriate hosts by functional metagenomics, direct screening for activity of xylanase can take place. Alternatively, in sequence-based metagenomics, researchers use bioinformatics tools to find known parts of xylanase in the genomes they extract [14].

New bioinformatics tools and the use of metagenome-assembled genomes have greatly enhanced the accuracy in both annotating genes and finding what they do. As a consequence, metagenomic research has found highly stable, salt-loving, acid-adapted and even cold-functioning xylanases in communities of microbes from hot springs, compost heaps, seas and oceans and soils with oil waste [15].

In addition, transcriptomics and proteomics approaches from functional genomics give detailed understanding of genetic and protein activity in different microbial populations. You can find active xylanase genes using RNA-Seq; at the same time, proteomics validates that the corresponding enzymes are present and work according to plan. These approaches combined give us a broader view of how hemicellulose is broken down by microbes and help discover specific enzymes for this process [16].

2.2 Exploration of Extreme Environments

We can find many new biocatalysts in severe environments. Extremophile microorganisms (those able to tolerate high heat, unusual pH, high salt levels or great pressure) have developed useful and reliable enzymes when faced with harsh industrial environments. Pulp bleaching and breaking down biomass are possible because of thermophilic and hyperthermophilic xylanases at high temperatures [17].

Explorations at recent times in some interesting places such as geothermal springs, salt flats, vent pools of the deep sea, Antarctic permafrost and alkaline lakes have identified DNA from microbes containing special xylanases. These enzymes are naturally equipped to withstand denaturing factors, reducing the need for further protein engineering or stabilization. Psychrophilic xylanases, although less common, are gaining attention for applications in food processing and bioremediation under cold conditions [18].

As more metagenomic datasets from these ecosystems become available through global initiatives like Earth Microbiome Project and Tara Oceans, the potential to uncover rare and unexplored xylanase variants continues to grow. This expands the functional toolbox available for both fundamental research and industrial applications.

2.3 Unculturable Microbial Reservoirs

The vast majority of microbial diversity remains inaccessible through conventional cultivation techniques. However, single-cell genomics and metatranscriptomics now offer powerful methods to explore unculturable microbes. Single-cell techniques allow for the isolation and sequencing of individual microbial genomes directly from environmental samples, preserving gene context and co-regulation information [19].

These tools make it possible to mine the slowly growing or closely associated microorganisms that are missed when performing full metagenomic analyses. Characterizing the genomes of these fungi and spotting new CAZymes enables researchers to utilize enzymes that would have gone overlooked using traditional systems [20].

Additionally, many microbes that have not been cultured or described are now known to belong to previously unknown xylanase family groups. Researching these genes can result in learning about xylanases with different ways of acting and the specific substances they can break down [21] (Table 1).

Table 1 Summary of key emerging trends

Trend	Description	Potential Impact
Functional metagenomics	Cloning and screening of environmental DNA for xylanase activity	Discovery of high-performance xylanases without culturing microbes
Sequence-Based screening	In silico identification of xylanase genes using bioinformatics	High-throughput discovery from massive datasets
Extreme environment sampling	Exploration of thermophilic, halophilic, acidophilic, and psychrophilic niches	Discovery of naturally robust enzymes
Single-Cell genomics	Genomic analysis of unculturable microbes	Access to novel genes from previously hidden microbial diversity
Multi-Omics integration	Combining genomics, transcriptomics, and proteomics	Dynamic insight into microbial xylanase expression and function

3 Synthetic Biology and Enzyme Engineering

The convergence of synthetic biology and enzyme engineering is revolutionizing the field of industrial biotechnology, enabling the design and optimization of enzymes with tailor-made properties for specific applications [22]. Overcoming the limitations of heat, different molecule use and industrial exposure in xylanases is made possible through synthetic biology. In this section, we examine how the use of rational design, directed evolution and making complex enzymes is transforming the future of xylanase discovery [23].

3.1 Rational Design and Directed Evolution

In rational design, the sequence of amino acids in an enzyme is changed based on what its 3D structure and task reveals [24]. Specific parts of the enzyme involved in its stability, pH use or affinity for substrates can be switched by site-directed mutagenesis to bring about a better result. To illustrate, both changing the catalytic area and inserting new strong hydrogen-bonding amino acids may raise the protein's resistance to heat and denaturation [25].

It is very important for rational design to have good structural data which is standardly obtained through X-ray crystallography, NMR or simulations. Progress in the field, especially from AlphaFold and similar tools, has made it much easier to create new xylanase variants in the lab after successful simulations [26]. By using these models, researchers can discover regions that change easily and focus their engineering on them [27].

Unlike classical genetic engineering, directed evolution aims to reproduce what happens in nature in the laboratory. Xylanase variants are created randomly, adding many mutations along the DNA sequence and those showing increased activity or stability are then picked out through screening. Without knowing the enzyme's shape in detail, scientists can alter and improve its traits by repeating cycles of mutagenesis and selection. Several modified xylanase variants are now available, some that handle higher temperatures, different pH levels and the common inhibitors found in pretreated biomass.

The strongest results are commonly achieved by combining rational design with directed evolution—using knowledge of structure to help create the best libraries and then directly selecting better candidates [28].

3.2 Creation of Multifunctional and Chimeric Enzymes

Researchers are finding new ways in synthetic biology to create enzymes that carry proteins from various enzymes or extra parts that help them function better. When lignocellulose needs to be degraded, combining the enzyme domains of xylanases with those of cellulases, esterases or LPMOs often speeds up the process [29].

Uniting a xylanase with a CBM can result in a high affinity towards insoluble xylan and lignocellulosic substances and thus improve the catalytic capacity of the xylanase. In the same way, building enzymes that have both xylanase and β-xylosidase activities can simplify xylan breaking down into xylose and replace multi-enzyme designs [30].

Combining a steady core from a thermophilic enzyme with an active site from a mesophilic enzyme enables chimeric enzymes to improve both their stability and efficiency. This kind of hybrid enzyme is particularly important in industry, as it must survive in various and frequently extreme environments [31].

These gene circuits can be developed so that the needed enzymes are released by engineered bacteria or consortia in a way that supports the decomposition of difficult to manage substrates [32].

3.3 Customizing Enzymes for Harsh Industrial Conditions

Many commercial uses of xylanases involve having to work in high temperatures, extremely alkaline or acidic environments, salty liquids or with organic solvents and different kinds of inhibitors. They must not lose their ability to speed up reactions and must remain steady for a long time [33].

By engineering enzymes, xylanases can be modified to satisfy these problems. Strategies include:

- Gaining thermostability by adding disulfide bridges, perfecting hydrophobic attraction or shrinking flexible parts of the protein.
- Modifying the surface residues can help the structure keep its shape under both very low (acidic) and high (alkaline) pH.

By changing certain amino acids in the binding site of the enzyme or on its surface, researchers can increase the resistance of their protein to solvents or inhibitors.

Xylanases have been modified in recent times to operate well at high temperatures and pH, making them adequate for textile desizing and pulp bleaching without relying on chemicals [34].

Tying enzymes to solid structures helps increase both their stability and the number of times they can be used. Xylanases immobilized on a support can be included in continuous bioprocessing units, lasting for many cycles and being less vulnerable to denaturation or proteolytic breakdown. Polymers, silica-based products and nanoparticles comprise examples of immobilization matrices; these materials can be customized to fit the requirements of various processes [35].

4 Case Studies and Industrial Applications

Examples of successful products demonstrate that synthetic biology contributes to better xylanase results at a reduced cost to the industry and with less damage to the environment [36]. Optimase CX from DuPont is a special type of xylanase that thrives in baking and makes dough easier to handle while adding softness to the crumb, creating better end products. UltraXylan from Novozymes is developed to operate in alkaline medium which lets it control microorganisms in textile production without using harsh chemical cleaners (Table 2). Thanks to fusion technology, a bifunctional xylanase-β-xylosidase enzyme has proved very useful in the field of bioethanol by better enabling the production of fermentable sugars from hemicellulose. Such improvements prove that using synthetic biology and protein engineering allows xylanases to be customized for several important industrial applications, resulting in sustainability and cost savings [37].

5 Enzyme Engineering and Protein Engineering

These disciplines, enzyme engineering and protein engineering, aim to improve the qualities and performance of xylanases in industry [38]. Scientists change the molecular structure of enzymes and their actions by engineering them, using them in fields like biofuel, processing food or dealing with waste [38]. Directed evolution is an important technique used in enzyme engineering to help enzymes perform better by exposing them to random changes in sequence and then testing for enzymes that show the desired changes, for example, higher activity, better substrate specification

Table 2 Synthetic biology strategies in xylanase development

Strategy	Objective	Outcome
Rational design	Targeted mutagenesis using structural data	Enhanced thermal stability, substrate specificity
Directed evolution	Random mutagenesis and selection	Improved enzyme activity and robustness
Chimeric enzymes	Fusion of multiple catalytic or binding domains	Multifunctionality and synergistic effects
Immobilization	Enzyme stability and reusability	Suitable for continuous industrial processes
Synthetic gene circuits	Coordinated enzyme expression	Optimized biomass degradation pathways

or greater strength in difficult industrial conditions [39]. Using directed evolution, it has become possible to make xylanases that can tolerate temperature, pH and solvent conditions often seen in industry [40]. Rational design means changing an enzyme's structure using what is known about its activity and by using computers to help predict how altering specific amino acids can improve its function. For example, changing the site where xylanases bind can boost their capacity to split particular polysaccharides which makes them better for biofuel manufacture [41]. In addition, the use of protein engineering for xylanases has evolved to include making hybrid enzymes with several domains to allow them to handle many functions at the same time. It is most convenient when applying enzymes in biorefinery processes which work with various substances. Additionally, Progress in protein engineering now makes it possible to manufacture enzymes with better and more long-lasting properties for widespread industrial use. Many times, changing the 3D and 4D structure of an enzyme helps it better stand up to high temperatures and strong changes in pH [42]. Engineering xylanases to accomplish work in ionic liquids or other unusual solvents is a significant step that opens up many unconventional uses for these enzymes in industry. Engineered xylanases have been developed more quickly thanks to high-throughput screening systems used to examine large numbers of mutant enzymes [43]. Thanks to genetic engineering, xylanases are now being produced in yeast, plants or insect cells, making it easier and more economical to get the enzyme [44]. To make engineered xylanases useful on the market, making them more active and less costly to manufacture has become important. By combining synthetic biology and metabolic engineering, it has become possible to produce xylanases right in bioprocesses with less need for purification and smoother production steps [45]. As enzyme and protein engineering develop and advanced programs are created, more opportunities are appearing in enzyme performance, economics and flexibility, guaranteeing that xylanases and similar enzymes stay at the leading edge in many technology fields [46].

6 Emerging Trends and Future Prospects

Promising future directions and recently appearing trends are pushing xylanase research toward more creative, sustainable solutions, following the global increase in support for green solutions and circular bioeconomy ideas. There is strong interest in integrating xylanases into multi-enzyme systems so that a co-working mixture of enzymes can better break down complex lignocellulosic biomass. By using xylanases along with cellulases, ligninases and additional enzymes, researchers expect to more efficiently and affordably turn biomass into biofuels and use it in biorefineries. The use of this approach means industries require less of harsh chemicals and less energy, leading to more environmentally sound activities. Formulating and delivering xylanase with nanotechnology is now recognized as important. To increase stable and useful xylanases, we are studying nanostructures that can hold the enzymes. As an example, enzymes can be nano-encapsulated to protect them and allow the controlled release required in processing foods, producing drugs or treating wastewater.

Besides, the use of AI and ML is helping to predict how enzymes function together with their surrounding molecules. By using these tools, xylanases can be rationally developed to become more durable at higher temperatures, work over a wider pH range and show higher specificity for targeted substrates. ASSISTING technology in biochemistry speeds up the process of biocatalyst experimental development, making it far cheaper and faster than traditional trial-and-error methods. At the same time, synthetic biology is discovering different ways by producing designer microbes or even basic artificial cells that make xylanases right where they are needed. The enzymes from these bioengineered organisms can be easily produced when needed in reactors or added early to a bioconversion process, eliminating most complex downstream steps.

Both green chemistry and sustainable processing are influencing the future development of xylanase studies, mainly in the pulp and paper industry, textile manufacturing and making animal feed. Choosing to use enzymes instead of chemicals decreases harm to the environment, increases how well the process works and leads to better products. As an example, using xylanases in paper manufacture lowers chlorine use and the toxicity of wastewater. Because animal feed with thermostable xylanases lowers waste output while improving how the feed is used and digested by animals, their health improves and so does the environment.

In addition, by using metagenomics and functional genomics, it is now possible to identify new xylanases in unusual places like hot springs, compost piles, seas and the gut of herbivores and insects. These techniques allow researchers to explore the genetic potential of unculturable microbes, unlocking unique enzymes with superior traits. Advances in high-throughput sequencing, coupled with protein structure prediction tools like AlphaFold, are expediting the functional annotation and structural modeling of newly discovered enzymes. This genomics-driven approach supports the identification of "next-gen" xylanases tailored for specialized industrial roles.

Another prospective area involves **biorefineries and circular economy models**, where xylanases play a critical role in valorizing agricultural residues, forestry by-products, and food processing waste into valuable bio-based products like bioethanol, xylo-oligosaccharides, and bioplastics. This aligns with global efforts to reduce reliance on fossil fuels and move toward sustainable, low-carbon economies. Moreover, **regulatory and economic considerations** are increasingly influencing future xylanase applications, prompting the development of cost-effective, scalable, and safe production systems. Public–private partnerships, international collaborations, and open-source research platforms are expected to catalyze innovation and commercialization.

7 Conclusion

In conclusion, the emerging trends in xylanase research are being shaped by cross-disciplinary innovations and a strong emphasis on sustainability. With ongoing advancements in biotechnology, computational modeling, systems biology, and process engineering, xylanases are poised to play an even more transformative role across industrial sectors. The future of xylanase research lies not only in enhancing enzyme performance but also in integrating these biocatalysts into holistic, eco-efficient systems that contribute to a resilient and sustainable bioeconomy.

References

1. S. Liang, X. Lin, L. He, Y. Liu, W. Xiong, M. Lin, et al., Development of an efficient chimeric trifunctional xylanase/glucanase/feruloyl esterase and its application in the bioconversion of corn stover into fermentable sugars. Int. J. Biol. Macromol. **307**, 142236 (2025)
2. M. Mendonça, M. Barroca, T. Collins, Endo-1,4-β-xylanase-containing glycoside hydrolase families: characteristics, singularities and similarities. Biotechnol. Adv. **65**, 108148 (2023)
3. Y. Zhou, Z. Yang, Y. Chen, R. Zhang, C. Yang, W. Jiang, et al., Green surface modification of nylon membrane to site-specifically co-immobilize xylanase and lichenase for clean production of reducing sugar and juice clarification. J. Clean. Prod. **486**, 144438 (2025)
4. E.B. Agyekum, T. Khan, M. Tahir, S.M. Sultan, W.F. Mbasso, F.L. Rashid, et al., Evolution, trends, and future research directions of carbon capture, utilization and storage—a comprehensive bibliometric and systematic review. Energy Reports **13**, 5357–5374 (2025)
5. R.M. Maas, F. Kokou, M.C.J. Verdegem, J.W. Schrama, Enzyme supplementation (phytase and xylanase) improves low quality diets in Nile tilapia. Aquaculture Reports. **40**, 102650 (2025)
6. J. Wang, C. Bao, H. Cao, F. Huang, Y. Liu, Y. Cao, Multi-copy expression of a protease-resistant xylanase with high xylan degradation ability and its application in broilers fed wheat-based diets. Int. J. Biol. Macromol. **257**, 128633 (2024)
7. C. Li, G. Zhang, The fusions of elastin-like polypeptides and xylanase self-assembled into insoluble active xylanase particles. J. Biotechnol. **177**, 60–66 (2014)
8. Y. Huang, P.K. Busk, L. Lange, Cellulose and hemicellulose-degrading enzymes in Fusarium commune transcriptome and functional characterization of three identified xylanases. Enzyme and Microbial Technol. **73–74**, 9–19 (2015)

9. L. Thomas, A. Joseph, L.D. Gottumukkala, Xylanase and cellulase systems of Clostridium sp.: an insight on molecular approaches for strain improvement. Bioresour. Technol. **158**, 343–350 (2014)
10. L. Liu, Y. Sun, Y. Yue, J. Yang, L. Chen, J. Ashraf, et al., Composition and foam properties of whole wheat dough liquor as affected by xylanase and glucose oxidase. Food Hydrocolloids **108**, 106050 (2020)
11. S. Sahu, U. Bansal, G. Singh, S.K. Arya, Comparison of physical, chemical, physico-chemical, and enzymatic pretreatment of paddy straw for ethanol production. Sustain Chem. Environ. **10**, 100239 (2025)
12. J. Huang, D. Zhang, J.O. Omedi, Y. Lei, X. Su, M. Wu, et al., Improving the inhibitory resistance of xylanase FgXyn11C from Fusarium graminearum to SyXIP-I by site-directed mutagenesis. Int. J. Biol. Macromol. **271**, 132434 (2024)
13. P. Gao, J. Ma, R.A. Basit, X. Lou, W. Zhang, H. Song, et al., Investigation of the enzymatic characteristics of thermotolerant xylanase McXyn0243 derived from Malbranchea cinnamomea, and its application in the degradation of three agricultural residues for the production of xylooligosaccharides. Int. J. Biol. Macromol. **313**, 143973 (2025)
14. C.L.d.S. Carneiro, B. Santos, L.P.C. Monteiro, F.R. Saturnino, M.M. Olegário, A.S.G. Quadros, et al., Growth performance, phosphorus balance, gut function and bacteriome of juvenile Amazonian tambaqui (Colossoma macropomum) fed plant-based diets supplemented with phytase and/or xylanase-β-glucanase. Aquaculture. **607**, 742679 (2025)
15. G. Paës, J.-G. Berrin, J. Beaugrand, GH11 xylanases: structure/function/properties relationships and applications. Biotechnol. Adv. **30**, 564–592 (2012)
16. D. Talens-Perales, E. Jiménez-Ortega, P. Sánchez-Torres, J. Sanz-Aparicio, J. Polaina, Phylogenetic, functional and structural characterization of a GH10 xylanase active at extreme conditions of temperature and alkalinity. Comput. Struct. Biotechnol. J. **19**, 2676–2686 (2021)
17. Y. Lin, Y. Dong, X. Li, J. Cai, L. Cai, G. Zhang, Enzymatic production of xylooligosaccharide from lignocellulosic and marine biomass: a review of current progress, challenges, and its applications in food sectors. Int. J. Biol. Macromol. **277**, 134014 (2024)
18. M. Martins, W.G. Sganzerla, T. Forster-Carneiro, R. Goldbeck, Recent advances in xylo-oligosaccharides production and applications: a comprehensive review and bibliometric analysis. Biocatalysis Agric. Biotechnol. **47**, 102608 (2023)
19. A. Zerva, C. Pentari, C. Ferousi, E. Nikolaivits, A. Karnaouri, E. Topakas, Recent advances on key enzymatic activities for the utilisation of lignocellulosic biomass. Bioresour. Technol. **342**, 126058 (2021)
20. O.G. Ndochinwa, Q.-Y. Wang, O.C. Amadi, T.N. Nwagu, C.I. Nnamchi, E.S. Okeke, et al., Current status and emerging frontiers in enzyme engineering: an industrial perspective. Heliyon. **10**, e32673 (2024)
21. Z. Yi, R. Sharif, S. Gulzar, Y. Huang, T. Ning, H. Zhan, et al., Changes in hemicellulose metabolism in banana peel during fruit development and ripening. Plant Physiol. Biochem. **215**, 109025 (2024)
22. X. Qi, X. Liu, Y. Zhang, J. Hao, H. Rao, D. Zhao, et al., A novel wheat arabino-xylanase from Trichoderma gamsii: Study on biochemical characterization, substrate recognition mechanism and its application in wheat flour dough and Chinese bread. LWT **219**, 117550 (2025)
23. H. Niu, D. Zhu, J. Leng, Z. Wang, D. Liu, Y. Chen, et al., Biofilm-based immobilized fermentation of engineered Komagataella phaffii for xylanase production. Bioresour. Technol. **418**, 131918 (2025)
24. N. Taghizadeh, S. Ariaeenejad, M. Ghollasi, E. Motamedi, Development of polyvinyl alcohol-gum Arabic hydrogel nanocomposite for xylanase immobilization and its role in poultry feed enhancement. Carbohydr. Polym. Technol. Appl. **9**, 100722 (2025)
25. Z. Li, X. Li, Y. Zhou, M. Lu, M. Li, Y. Zhu, et al., Rational design of GH11 xylanase to balance the activity–stability trade-off. Int. J. Biol. Macromol. **311**, 143063 (2025)
26. Z. Li, M. Chen, W. Tian, L. Wang, X. Wu, Investigating the role of polar amino acids driven by evolution in the active site architecture of GH11 xylanase. Int. J. Biol. Macromol. 144464 (2025)

27. H. Xing, G. Zou, C. Liu, S. Chai, X. Yan, X. Li, et al., Improving the thermostability of a GH11 xylanase by directed evolution and rational design guided by B-factor analysis. Enzyme and Microbial Technol. **143**, 109720 (2021)
28. C. Fan, J. Qi, Y. Cong, C. Zhang, Enhanced L-theanine production through semi-rational design of γ-glutamylmethylamide synthetase from Methylovorus mays. Enzyme Microb. Technol. 180, 110481 (2024)
29. L.F. Ribeiro, G.P. Furtado, M.R. Lourenzoni, A.J. Costa-Filho, C.R. Santos, S.C.P. Nogueira, et al., Engineering bifunctional laccase-xylanase chimeras for improved catalytic performance*, J. Biol. Chem. **286**, 43026–43038 (2011)
30. M. Martins, T.M. Dinamarco, R. Goldbeck, Recombinant chimeric enzymes for lignocellulosic biomass hydrolysis. Enzyme Microb. Technol. **140**, 109647 (2020)
31. J.S. Jadaun, L.K. Narnoliya, A. Srivastava, S.P. Singh, Chapter 8—Chimeric enzyme designing for the synthesis of multifunctional biocatalysts, in *Biomass, Biofuels, Biochemicals*, ed by S.P. Singh, A. Pandey, R.R. Singhania, C. Larroche, Z. Li (Elsevier, 2020), pp. 119–143
32. D. Parashar, T. Satyanarayana, Engineering a chimeric acid-stable α-amylase-glucoamylase (Amy-Glu) for one step starch saccharification. Int. J. Biol. Macromol. **99**, 274–281 (2017)
33. M.P. Pinheiro, R.A.G. Reis, P. Dupree, R.J. Ward, Plant cell wall architecture guided design of CBM3-GH11 chimeras with enhanced xylanase activity using a tandem repeat left-handed β-3-prism scaffold. Comput. Struct. Biotechnol. J. **19**, 1108–1118 (2021)
34. D.T. Monterrey, I. Ayuso-Fernández, I. Oroz-Guinea, E. García-Junceda, Design and biocatalytic applications of genetically fused multifunctional enzymes. Biotechnol. Adv. **60**, 108016 (2022)
35. Q. Xu, Y. Luo, S. Y. Ding, M. E. Himmel, L. Bu, R. Lamed, et al., 3.03—multifunctional enzyme systems for plant cell wall degradation, in *Comprehensive Biotechnology (Second Edition)*, ed by M. Moo-Young (Academic Press, Burlington, 2011), pp. 15–25
36. M. Phuyal, U. Budhathoki, D. Bista, S. Shakya, R. Shrestha, A.K. Shrestha, Xylanase-Producing microbes and their real-world application. Int. J. Chem. Eng. **2023**, 3593035 (2023)
37. Y. Sürmeli, G. Şanlı-Mohamed, Engineering of xylanases for the development of biotechnologically important characteristics. Biotechnol. Bioeng. **120**, 1171–1188 (2023)
38. M. Sharma, A.K. Rai, S.P. Singh, Protein engineering for biomass hydrolyzing enzymes, in *Biomass Hydrolyzing Enzymes* (CRC Press, 2024), pp. 112–122
39. L. Du, X. Cui, H. Li, Y. Wang, L. Fan, R. He et al., Enhancing the enzymatic hydrolysis efficiency of lignocellulose assisted by artificial fusion enzyme of swollenin-xylanase. Ind. Crops Prod. **173**, 114106 (2021)
40. W. Xie, Q. Yu, R. Zhang, Y. Liu, R. Cao, S. Wang et al., Insights into the catalytic mechanism of a novel XynA and structure-based engineering for improving bifunctional activities. Biochemistry **60**, 2071–2083 (2021)
41. S. Yan, Y. Xu, X.-W. Yu, Rational engineering of xylanase hyper-producing system in Trichoderma reesei for efficient biomass degradation. Biotechnol. Biofuels **14**, 1–17 (2021)
42. J. Wang, H.M. Zabed, X. Lai, G. Zhang, X. Qi, Multifaceted engineering of xylanase guided by structural insights to enhance thermal stability and catalytic functionality. J. Agric. Food Chem. (2025)
43. Y.-Y. Lin, S. Zhao, X. Lin, T. Zhang, C.-X. Li, X.-M. Luo et al., Improvement of cellulase and xylanase production in Penicillium oxalicum under solid-state fermentation by flippase recombination enzyme/recognition target-mediated genetic engineering of transcription repressors. Biores. Technol. **337**, 125366 (2021)
44. A. Yıldırım, E. İlhan-Ayışığı, A. Düzel, S.P. Mayfield, S. Sargın, Optimization of culture conditions for the production and activity of recombinant xylanase from microalgal platform. Biochem. Eng. J. **197**, 108967 (2023)
45. A. Sharma, G. Gupta, T. Ahmad, S. Mansoor, B. Kaur, Enzyme engineering: current trends and future perspectives. Food Rev. Intl. **37**, 121–154 (2021)
46. T. Miao, A. Basit, J. Liu, F. Zheng, K. Rahim, H. Lou et al., Improved production of xylanase in Pichia pastoris and its application in xylose production from xylan. Front. Bioeng. Biotechnol. **9**, 690702 (2021)

MIX
Papier aus verantwortungsvollen Quellen
Paper from responsible sources
FSC® C105338

If you have any concerns about our products,
you can contact us on
ProductSafety@springernature.com

In case Publisher is established outside the EU,
the EU authorized representative is:
Springer Nature Customer Service Center GmbH
Europaplatz 3, 69115 Heidelberg, Germany

Printed by Libri Plureos GmbH
in Hamburg, Germany